ANTENNAS AND PROPAGATION
FOR WIRELESS COMMUNICATION SYSTEMS

ANTENNAS AND PROPAGATION FOR WIRELESS COMMUNICATION SYSTEMS

Second Edition

SIMON R. SAUNDERS,
UNIVERSITY OF SURREY, GUILDFORD, UK

ALEJANDRO ARAGÓN-ZAVALA,
TECNOLÓGICO DE MONTERREY, CAMPUS QUERÉTARO, MEXICO

1807
WILEY
2007

John Wiley & Sons, Ltd

Other Wiley Editorial Offices

John Wiley & Sons Inc., 111 River Street, Hoboken, NJ 07030, USA

Jossey-Bass, 989 Market Street, San Francisco, CA 94103-1741, USA

Wiley-VCH Verlag GmbH, Boschstr. 12, D-69469 Weinheim, Germany

John Wiley & Sons Australia Ltd, 42 McDougall Street, Milton, Queensland 4064, Australia

John Wiley & Sons (Asia) Pte Ltd, 2 Clementi Loop #02-01, Jin Xing Distripark, Singapore 129809

John Wiley & Sons Canada Ltd, 6045 Freemont Blvd, Mississauga, ONT, L5R 4J3, Canada

Wiley also publishes its books in a variety of electronic formats. Some content that appears in print may not be available
in electronic books.

Anniversary Logo Design: Richard J. Pacifico

British Library Cataloguing in Publication Data
A catalogue record for this book is available from the British Library

ISBN 978-0-470-84879-1

FSC
Mixed Sources
Product group from well-managed
forests and other controlled sources
Cert no. SGS-COC-2953
www.fsc.org
© 1996 Forest Stewardship Council

In memory of my father.
For Luke, Emily and Gráinne.

Simon Saunders

To Laura, you are my inspiration and my true love.
To Cocó, Maxi and Fimbie.

Alejandro Aragón-Zavala

Contents

Preface to the First Edition

This book has grown out of my teaching and research at the University of Surrey and out of my previous experiences in companies such as Philips, Ascom and Motorola. It is primarily intended for use by students in master's level and enhanced final-year undergraduate courses who are specialising in communication systems and wish to understand the principles and current practices of the wireless communication channel, including both antenna and propagation aspects. I have therefore included examples and problems in each chapter to reinforce the material described and to show how they are applied in specific situations. Additionally, much of the material has been used as parts of short courses run for many of the leading industrial companies in the field, so I hope that it may also be of interest to those who have a professional interest in the subject. Although there are several excellent books which cover portions of this material and which go deeper in some areas, my main motivation has been to create a book which covers the range of disciplines, from electromagnetics to statistics, which are necessary in order to understand the implications of the wireless channel on system performance. I have also attempted to bring together reference material which is useful in this field into a single, accessible volume, including a few previously unpublished research results.

For those who are intending to use this material as part of a course, a set of presentation slides, containing most of the figures from the book, is available free of charge from the World Wide Web at the following URL: ftp://ftp.wiley.co.uk/pub/books/saunders. These slides also include several of the figures in colour, which was not possible within the book in the interest of keeping the costs within reach of most students. For updated information concerning the contents of the book, related sites and software, see http://www.simonsaunders.com/apbook

I have deliberately avoided working directly with Maxwell's equations, although a verbal statement of their implications is included. This is because very few of the practical problems at the level of *systems* in this field require these equations for their solution. It is nevertheless important that the material is underpinned by basic physical principles, and this is the purpose of the first five chapters of the book. Nevertheless, I have not avoided the use of mathematics where it is actually useful in illustrating concepts, or in providing practical means of analysis or simulation.

Each chapter includes a list of references; wherever possible I have referred to journal articles and books, as these are most easily and widely available, but some more recent works only exist in conference proceedings.

The following notation is used throughout the text:

- Scalar variables are denoted by Times Roman italics, such as x and y.
- Physical vector quantities (i.e. those having magnitude and direction in three-dimensional physical space) are denoted by Times Roman boldface, such as $\mathbf{E}$ and $\mathbf{H}$.
- Unit vectors additionally have a circumflex, such as $\hat{\mathbf{x}}$ and $\hat{\mathbf{y}}$.
- Column vectors are denoted by lower case sans serif boldface, such as x and r, whereas matrix quantities are denoted by upper case sans serif boldface, such as X and R.
- The time or ensemble average of a random variable x is denoted by $E[x]$.
- The logarithm to base 10 is written log, whereas the natural logarithm is ln.
- Units are in square brackets, e.g. [metres].
- References are written in the form [firstauthor, year].
- Important *new terms* are usually introduced in italics.
- Equation numbers are given in round parentheses, e.g. (1.27)

Sincere acknowledgements are due to Mike Wilkins and Kheder Hanna of Jaybeam for providing most of the photographs of antennas and radiation patterns; to Nicholas Hollman of Cellnet for photographs of cellular masts and antenna installations; to Felipe Catedra for the GTD microcell predictions of FASPRO, to Kevin Kelly of Nortel for the scattering maps; to Heinz Mathis and Doug Pulley for providing constructive comments in the final days of production; to Mark Weller, Anthony Weller and David Pearson of Cellular Design Services for providing real-world problems, measurement data and an ideal environment in which the bulk of the work for the book was completed. I would particularly like to thank my colleagues, research assistants and students at the Centre for Communication Systems Research at the University of Surrey for providing time to complete this book and for many useful comments on the material.

I apologise in advance for any errors which may have occurred in this text, and I would be grateful to receive any comments, or suggestions about improvements for further editions.

Simon R. Saunders apbook@simonsaunders.com
Oughterard, Guildford and Ash, August 1998-June 1999

Preface to the Second Edition

Since the publication of the first edition of this book in 1999, much has changed in the wireless world. Third-generation cellular systems based on wideband CDMA have been widely deployed and are allowing high-rate applications such as video calling and music streaming to be accessed over wide areas. Wireless LAN systems, based mainly on Wi–Fi protocols and increasingly using MIMO antenna systems, have allowed access to very high data rates, particularly in indoor environments, and also increasingly in urban areas. Fixed wireless access to provide broadband services over the wide area is enjoying a resurgence of interest following the creation of the WiMax family of standards. Broadcasting is delivering increased numbers of channels, richness of content and interactivity via digitisation of both video and audio. The pace of change has increased as a result of factors such as increasing deregulation of the radio spectrum, new technologies such as software radio and greater convergence of fixed and mobile services via multimode devices for concurrent computing and communications.

Despite these changes, the fundamental importance of antennas and propagation has continued undiminished. All wireless systems are subject to the variations imposed by the wireless channel, and a good understanding of these variations is needed to answer basic questions such as "How far does it go?" "How fast can I transmit data?" and "How many users can I support?" This book aims to equip the reader with the knowledge and understanding needed to answer these questions for a very wide range of wireless systems.

The first edition of the book reached a larger audience than originally expected, including adoptions by many course tutors and by many seeking a primer in the field without being expert practitioners. At the same time many helpful comments were received, leading to the changes which have been incorporated in this revised edition. Most significantly, many people commented that the title of the book suggested that more weight should be given to antenna topics; this has been addressed via Chapters 4 and 14, devoted to the fundamentals of antennas and to their applications in mobile systems. Chapter 19 has also been added, giving practical details of channel measurement techniques for mobile systems. Throughout the book, enhancements and corrections have been made to reflect the current practice and to address specific comments from readers.

In addition to the acknowledgements of the first edition, I am particularly grateful to my co-author, Dr. Alejandro Aragón -Zavala of Tecnológico de Monterrey, Campus Querétaro in Mexico, who did most of the hard work on the updates to allow this second edition to be produced in a reasonably timely fashion despite my efforts to the contrary. Thanks are also due to many friends, colleagues, customers and suppliers for continued insights into the real world of wireless systems. Particular thanks for contributions and comments in this edition to Tim Brown, Abdus Owadally, Dave Draffin, Steve Leach, Stavros Stavrou, Rodney Vaughan, Jørgen Bach Andersen and Constantine Balanis. Lastly to Sarah Hinton at Wiley for patience above and beyond the call of duty.

Updates and further information regarding this book, including presentation slides, are available from the following web site:

http://www.simonsaunders.com/apbook

In addition, a solutions manual is available to lecturers at http://www.wiley.com/go/saunders.

Comments and suggestions are gratefully received via email to:
apbook@simonsaunders.com

Simon R. Saunders

1 Introduction: The Wireless Communication Channel

'I think the primary function of radio is that people want company.'
Elise Nordling

1.1 INTRODUCTION

Figure 1.1 shows a few of the many interactions between electromagnetic waves, the antennas which launch and receive them and the environment through which they propagate. All of these effects must be accounted for, in order to understand and analyse the performance of wireless communication systems. This chapter sets these effects in context by first introducing the concept of the wireless communication *channel*, which includes all of the antenna and propagation effects within it. Some systems which utilise this channel are then described, in order to give an appreciation of how they are affected by, and take advantage of, the effects within the channel.

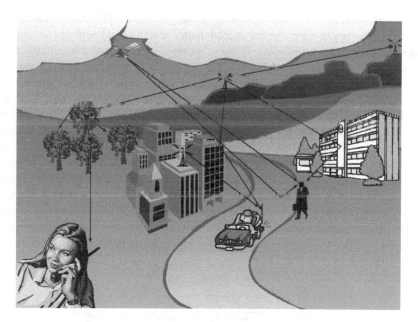

Figure 1.1: The wireless propagation landscape

Antennas and Propagation for Wireless Communication Systems Second Edition   Simon R. Saunders and Alejandro Aragón-Zavala
© 2007 John Wiley & Sons, Ltd

1.2 CONCEPT OF A WIRELESS CHANNEL

An understanding of the wireless channel is an essential part of the understanding of the operation, design and analysis of any wireless system, whether it be for cellular mobile phones, for radio paging or for mobile satellite systems. But what exactly is meant by a *channel*?

The architecture of a generic communication system is illustrated in Figure 1.2. This was originally described by Claude Shannon of Bell Laboratories in his classic 1948 paper

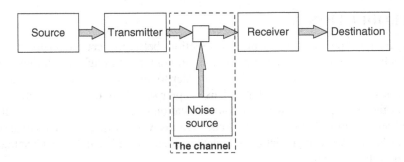

Figure 1.2: Architecture of a generic communication system

'*A Mathematical Theory of Communication*' [Shannon, 48]. An information source (e.g. a person speaking, a video camera or a computer sending data) attempts to send information to a destination (a person listening, a video monitor or a computer receiving data). The data is converted into a signal suitable for sending by the transmitter and is then sent through the channel. The channel itself modifies the signal in ways which may be more or less unpredictable to the receiver, so the receiver must be designed to overcome these modifications and hence to deliver the information to its final destination with as few errors or distortions as possible.

This representation applies to all types of communication system, whether wireless or otherwise. In the wireless channel specifically, the noise sources can be subdivided into multiplicative and additive effects, as shown in Figure 1.3. The additive noise arises from the noise generated within the receiver itself, such as thermal and shot noise in passive and active components and also from external sources such as atmospheric effects, cosmic radiation and interference from other transmitters and electrical appliances. Some of these interferences may be intentionally introduced, but must be carefully controlled, such as when channels are reused in order to maximise the capacity of a cellular radio system.

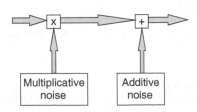

Figure 1.3: Two types of noise in the wireless communication channel

The multiplicative noise arises from the various processes encountered by transmitted waves on their way from the transmitter antenna to the receiver antenna. Here are some of them:

- The directional characteristics of both the transmitter and receiver antennas;
- reflection (from the smooth surfaces of walls and hills);
- absorption (by walls, trees and by the atmosphere);
- scattering (from rough surfaces such as the sea, rough ground and the leaves and branches of trees);
- diffraction (from edges, such as building rooftops and hilltops);
- refraction (due to atmospheric layers and layered or graded materials).

It is conventional to further subdivide the multiplicative processes in the channel into three types of fading: *path loss*, *shadowing* (or *slow fading*) and *fast fading* (or *multipath fading*), which appear as time-varying processes between the antennas, as shown in Figure 1.4. All of these processes vary as the relative positions of the transmitter and receiver change and as any contributing objects or materials between the antennas are moved.

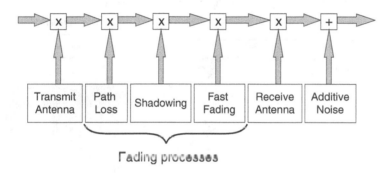

Figure 1.4: Contributions to noise in the wireless channel

An example of the three fading processes is illustrated in Figure 1.5, which shows a simulated, but nevertheless realistic, signal received by a mobile receiver moving away from a transmitting base station. The path loss leads to an overall decrease in signal strength as the distance between the transmitter and the receiver increases. The physical processes which cause it are the outward spreading of waves from the transmit antenna and the obstructing effects of trees, buildings and hills. A typical system may involve variations in path loss of around 150 dB over its designed coverage area. Superimposed on the path loss is the shadowing, which changes more rapidly, with significant variations over distances of hundreds of metres and generally involving variations up to around 20 dB. Shadowing arises due to the varying nature of the particular obstructions between the base and the mobile, such as particular tall buildings or dense woods. Fast fading involves variations on the scale of a half-wavelength (50 cm at 300 MHz, 17 cm at 900 MHz) and frequently introduces variations as large as 35–40 dB. It results from the constructive and destructive interference between multiple waves reaching the mobile from the base station.

Each of these variations will be examined in depth in the chapters to come, within the context of both fixed and mobile systems. The path loss will be described in basic concept in

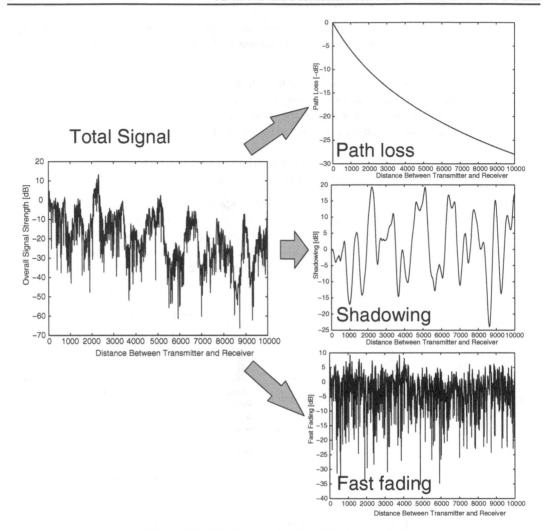

Figure 1.5: The three scales of mobile signal variation

Chapter 5 and examined in detail in Chapters 6, 7 and 8 in the context of fixed terrestrial links, fixed satellite links and terrestrial macrocell mobile links, respectively. Shadowing will be examined in Chapter 9, while fast fading comes in two varieties, narrowband and wideband, investigated in Chapters 10 and 11, respectively.

1.3 THE ELECTROMAGNETIC SPECTRUM

The basic resource exploited in wireless communication systems is the electromagnetic spectrum, illustrated in Figure 1.6. Practical radio communication takes place at frequencies from around 3 kHz [kilohertz] to 300 GHz [gigahertz], which corresponds to wavelengths in free space from 100 km to 1 mm.

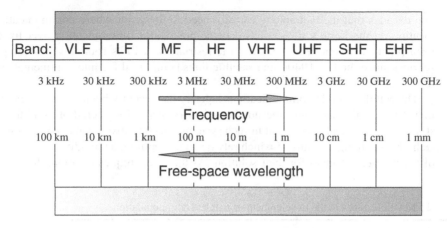

Figure 1.6: The electromagnetic spectrum

Table 1.1 defines two conventional ways of dividing the spectrum into frequency bands . The frequencies chosen for new systems have tended to increase over the years as the demand for wireless communication has increased; this is because enormous bandwidths are available at the higher frequencies. This shift has created challenges in the technology needed to support reliable communications, but it does have the advantage that antenna structures can be smaller in absolute size to support a given level of performance. This book will be concerned only with communication at VHF frequencies and above, where the wavelength is typically small compared with the size of macroscopic obstructions such as hills, buildings and trees. As the size of obstructions relative to a wavelength increases, their obstructing effects also tend to increase, reducing the range for systems operated at higher frequencies.

1.4 HISTORY

Some of the key milestones in the development of wireless communications are listed in Table 1.2. Mobile communication has existed for over a hundred years, but it is only in the last

Table 1.1: Naming conventions for frequency bands

Band name	Frequency range	Band name	Frequency range [GHz]
Very low frequency	3–30 kHz	L band	1–2
Low frequency (long wave)	30–300 kHz	S band	2–4
Medium frequency (medium wave)	0.3–3.0 MHz	C band	4–8
High frequency (short wave)	3–30 MHz	X band	8–12
Very high frequency	30–300 MHz	Ku band	12–18
Ultra high frequency	0.3–3.0 GHz	K band	18–26
Super high frequency (centimetre wave)	3–30 GHz	Ka band	26–40
Extra high frequency (millimetre wave)	30–300 GHz	V band	40–75
		W band	75–111

two decades that the technology has advanced to the point where communication to every location on the Earth's surface has become practical. Communication over fixed links has been practical for rather longer, with terrestrial fixed links routinely providing telephone services since the late 1940s, and satellite links being used for intercontinental communication since the 1960s.

The cellular mobile communications industry has recently been one of the fastest growing industries of all time, with the number of users increasing incredibly rapidly. As well as stimulating financial investment in such systems, this has also given rise to a large number of technical challenges, many of which rely on an in-depth understanding of the characteristics of the wireless channel for their solution. As these techniques develop, different questions

Table 1.2: Key milestones in the development of wireless communication

1873	Maxwell predicts the existence of electromagnetic waves
1888	Hertz demonstrates radio waves
1895	Marconi sends first wireless signals a distance of over a mile
1897	Marconi demonstrates mobile wireless communication to ships
1898	Marconi experiments with a land 'mobile' system – the apparatus is the size of a bus with a 7 m antenna
1916	The British Navy uses Marconi's wireless apparatus in the Battle of Jutland to track and engage the enemy fleet
1924	US police first use mobile communications
1927	First commercial phone service between London and New York is established using long wave radio
1945	Arthur C. Clarke proposes geostationary communication satellites
1957	Soviet Union launches Sputnik 1 communication satellite
1962	The world's first active communications satellite 'Telstar' is launched
1969	Bell Laboratories in the US invent the cellular concept
1978	The world's first cellular phone system is installed in Chicago
1979	NTT cellular system (Japan)
1988	JTACS cellular system (Japan)
1981	NMT (Scandinavia)
1983	AMPS cellular frequencies allocated (US)
1985	TACS (Europe)
1991	USDC (US)
1991	GSM cellular system deployed (Europe)
1993	DECT & DCS launched (Europe)
1993	Nokia engineering student Riku Pihkonen sends the world's first SMS text message
1993	PHS cordless system (Japan)
1995	IS95 CDMA (US)
1998	Iridium global satellite system launched
1999	Bluetooth short-range wireless data standard agreed
1999	GPRS launched to provide fast data communication capabilities (Europe)
2000	UK government runs the world's most lucrative spectrum auction as bandwidth for 3G networks is licensed for £22.5 billion
2001	First third-generation cellular mobile network is deployed (Japan)
2002	Private WLAN networks are becoming more popular (US)
2003	WCDMA third-generation cellular mobile systems deployed (Europe)
2004	First mobile phone viruses found
2006	GSM subscriptions reach two billion worldwide. The second billion took just 30 months.

concerning the channel behaviour are asked, ensuring continuous research and development in this field.

Chapter 20 contains some predictions relating to the future of antennas and propagation. For a broader insight into the future development of wireless communications in general, see [Webb, 07].

1.5 SYSTEM TYPES

Figure 1.7 shows the six types of wireless communication system which are specifically treated in this book. The principles covered will also apply to many other types of system.

- *Satellite fixed links (chapter 7)*: These are typically created between fixed earth stations with large dish antennas and geostationary earth-orbiting satellites. The propagation effects are largely due to the Earth's atmosphere, including meteorological effects such as rain. Usually operated in the SHF and EHF bands.
- *Terrestrial fixed links (chapter 6)*: Used for creating high data rate links between points on the Earth, for services such as telephone and data networks, plus interconnections between base stations in cellular systems. Also used for covering wide areas in urban and suburban environments for telephone and data services to residential and commercial buildings. Meteorological effects are again significant, together with the obstructing effects of hills, trees and buildings. Frequencies from VHF through to EHF are common.
- *Megacells (chapter 14)*: These are provided by satellite systems (or by high-altitude platforms such as stratospheric balloons) to mobile users, allowing coverage of very wide areas with reasonably low user densities. A single satellite in a low earth orbit would typically cover a region of 1000 km in diameter. The propagation effects are dominated by objects close to the user, but atmospheric effects also play a role at higher frequencies. Most systems operate at L and S bands to provide voice and low-rate data services, but

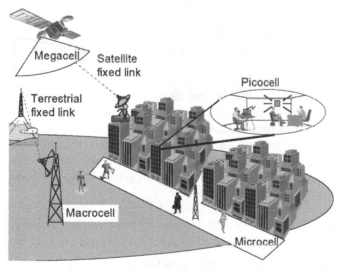

Figure 1.7: Wireless communication system types

systems operating as high as Ka band can be deployed to provide Internet access at high data rates over limited areas.

- *Macrocells (chapter 8)*: Designed to provide mobile and broadcast services (including both voice and data), particularly outdoors, to rural, suburban and urban environments with medium traffic densities. Base station antenna heights are greater than the surrounding buildings, providing a cell radius from around 1 km to many tens of kilometres. Mostly operated at VHF and UHF. May also be used to provide fixed broadband access to buildings at high data rates, typically at UHF and low SHF frequencies.
- *Microcells (chapter 12)*: Designed for high traffic densities in urban and suburban areas to users both outdoors and within buildings. Base station antennas are lower than nearby building rooftops, so coverage area is defined by street layout. Cell length up to around 500 m. Again mostly operated at VHF and UHF, but services as high as 60 GHz have been studied.
- *Picocells (chapter 13)*: Very high traffic density or high data rate applications in indoor environments. Users may be both mobile and fixed; fixed users are exemplified by wireless local area networks between computers. Coverage is defined by the shape and characteristics of rooms, and service quality is dictated by the presence of furniture and people.

Used together, these six system types provide networks capable of delivering an enormous range of service to locations anywhere on the Earth.

1.6 AIMS OF CELLULAR SYSTEMS

The complexity of systems that allow wide area coverage, particularly cellular systems, influences the parameters of the channel which have the most significance. These systems have three key aims:

- *Coverage and mobility*: The system must be available at all locations where users wish to use it. In the early development of a new system, this implies outdoor coverage over a wide area. As the system is developed and users become more demanding, the *depth of coverage* must be extended to include increasing numbers of indoor locations. In order to operate with a single device between different systems, the systems must provide *mobility* with respect to the allocation of resources and support of interworking between different standards.
- *Capacity*: As the number of users in a mobile system grows, the demands placed on the resources available from the allocated spectrum grow proportionately. These demands are exacerbated by increasing use of high data rate services. This necessitates the assignment of increasing numbers of channels and thus dense reuse of channels between cells in order to minimise problems with *blocked* or *dropped* calls. If a call is blocked, users are refused access to the network because there are no available channels. If a call is dropped, it may be interrupted because the user moves into a cell with no free channels. Dropped calls can also arise from inadequate coverage.
- *Quality*: In a mature network, the emphasis is on ensuring that the services provided to the users are of high quality – this includes the perceived speech quality in a voice system and the bit error rate (BER), throughput, latency and jitter in a data system.

Subsequent chapters will show that path loss and shadowing dominate in establishing good coverage and capacity, while quality is particularly determined by the fast-fading effects.

1.7 CELLULAR NETWORKS

Figure 1.8 shows the key elements of a standard cellular network. The terminology used is taken from GSM, the digital cellular standard originating in Europe, but a similar set of elements exists in many systems. The central hub of the network is the mobile switching centre (MSC), often simply called the *switch*. This provides connection between the cellular network and the public switched telephone network (PSTN) and also between cellular subscribers. Details of the subscribers for whom this network is the home network are held on a database called the home location register (HLR), whereas the details of subscribers who

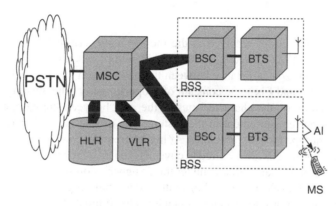

Figure 1.8: Elements of a standard cellular system, using GSM terminology

have entered the network from elsewhere are on the visitor location register (VLR). These details include authentication and billing details, plus the current location and status of the subscriber. The coverage area of the network is handled by a large number of base stations. The base station subsystem (BSS) is composed of a base station controller (BSC) which handles the logical functionality, plus one or several base transceiver stations (BTS) which contain the actual RF and baseband parts of the BSS. The BTSs communicate over the air interface (AI) with the mobile stations (MS). The AI includes all of the channel effects as well as the modulation, demodulation and channel allocation procedures within the MS and BTS. A single BSS may handle 50 calls, and an MSC may handle some 100 BSSs.

1.8 THE CELLULAR CONCEPT

Each BTS, generically known as a base station (BS), must be designed to cover, as completely as possible, a designated area or *cell* (Figure 1.9). The power loss involved in transmission between the base and the mobile is the *path loss* and depends particularly on antenna height, carrier frequency and distance. A very approximate model of the path loss is given by

$$\frac{P_R}{P_T} = \frac{1}{L} = k\frac{h_m h_b^2}{r^4 \ f^2} \tag{1.1}$$

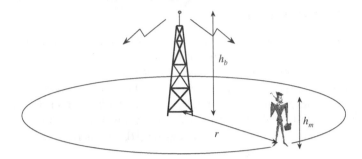

Figure 1.9: Basic geometry of cell coverage

where P_R is the power received at the mobile input terminals [W]; P_T is the base station transmit power [W]; h_m and h_b are the mobile and base station antenna heights, respectively [m]; r is the horizontal distance between the base station and the mobile [m]; f is the carrier frequency [Hz] and k is some constant of proportionality. The quantity L is the path loss and depends mainly on the characteristics of the path between the base station and the mobile rather than on the equipment in the system. The precise dependencies are functions of the environment type (urban, rural, etc.) At higher frequencies the range for a given path loss is reduced, so more cells are required to cover a given area. To increase the cell radius for a given transmit power, the key variable under the designer's control is the antenna height: this must be large enough to clear surrounding clutter (trees, buildings, etc.), but not so high as to cause excessive interference to distant co-channel cells. It must also be chosen with due regard for the environment and local planning regulations. Natural terrain features and buildings can be used to increase the effective antenna height to increase coverage or to control the limits of coverage by acting as shielding obstructions.

When multiple cells and multiple users are served by a system, the system designer must allocate the available channels (which may be frequencies, time slots or other separable resources) to the cells in such a way as to minimise the interaction between the cells. One approach would be to allocate completely distinct channels to every cell, but this would limit the total number of cells possible in a system according to the spectrum which the designer has available. Instead, the key idea of cellular systems is that it is possible to serve an *unlimited* number of subscribers, distributed over an *unlimited* area, using only a *limited* number of channels, by efficient *channel reuse*. A set of cells, each of which operates on a different channel (or group of channels), is grouped together to form a *cluster*. The cluster is then repeated as many times as necessary to cover a very wide area. Figure 1.10 illustrates the use of a seven-cell cluster. The use of hexagonal areas to represent the cells is highly idealised, but it helps in establishing basic concepts. It also correctly represents the situation when path loss is treated as a function of distance only, within a uniform environment. In this case, the hexagons represent the areas within which a given base station transmitter produces the highest power at a mobile receiver.

The smaller the cluster size, therefore, the more efficiently the available channels are used. The allowable cluster size, and hence the *spectral efficiency* of the system, is limited by the level of interference the system can stand for acceptable quality. This level is determined by the smallest ratio between the wanted and interfering signals which can be tolerated for reasonable quality communication in the system. These levels depend on the types of

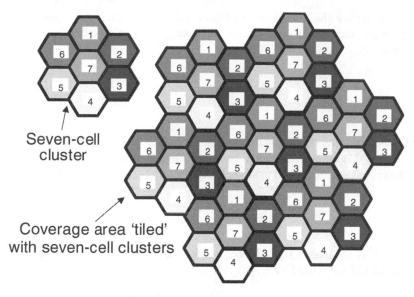

Figure 1.10: Cellular reuse concept

modulation, coding and synchronisation schemes employed within the base station and the mobile. The ratio is called the *threshold carrier-to-interference power ratio* (C/I or CIR). Figure 1.11 illustrates a group of co-channel cells, in this case the set labelled 3 in Figure 1.10. There will be other co-channel cells spread over a wider area than illustrated, but those shown here represent the first *tier*, which are the nearest and hence most significant interferers. Each cell has a radius R and the centres of adjacent cells are separated by a distance D, the *reuse distance*.

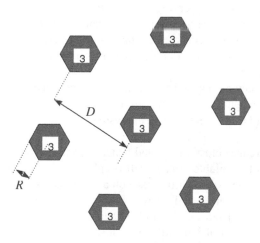

Figure 1.11: A group of co-channel cells

Considering the central cell in Figure 1.11 as the wanted cell and the other six as the interferers, the path loss model from (1.1) suggests that a mobile located at the edge of the wanted cell experiences a C/I of

$$\frac{C}{I} \approx \frac{1}{R^4} \Bigg/ \sum_{k=1}^{6} \frac{1}{D^4} = \frac{1}{6}\left(\frac{D}{R}\right)^4 \tag{1.2}$$

This assumes that the distances between the interferers and the mobile are all approximately equal and that all the base stations have the same heights and transmit powers. The geometry of hexagons sets the relationship between the cluster size and the reuse distance as:

$$\frac{D}{R} = \sqrt{3N} \tag{1.3}$$

where N is the cluster size. Hence, taking (1.2) and (1.3) together, the cluster size and the required C/I are related by

$$\frac{C}{I} = \frac{1}{6}(3N)^2 \tag{1.4}$$

For example, if the system can achieve acceptable quality provided the C/I is at least 18 dB, then the required cluster size is

$$N = \sqrt{\frac{2}{3} \times \frac{C}{I}} = \sqrt{\frac{2}{3} \times 10^{18/10}} = 6.5 \tag{1.5}$$

Hence a cluster size of $N = 7$ would suffice. Not all cluster sizes are possible, due to the restrictions of the hexagonal geometry. Possible values include 3, 4, 7, 12, 13, 19 and 27. The smaller the value of C/I, the smaller the allowed cluster size. Hence the available channels can be reused on a denser basis, serving more users and producing an increased capacity. Had the dependence on r in (1.1) been slower (i.e. the path loss exponent was less than 4), the required cluster size would have been greater than 7, so the path loss characteristics have a direct impact on the system capacity. Practical path loss models in various cell types are examined in depth in Chapters 8, 12–14.

Note this is only an approximate analysis. In practice, other considerations such as the effect of terrain height variations require that the cluster size is different from the theoretical value, or is varied in different parts of the system in order to suit the characteristics of the local environment.

One way to reduce cluster size, and hence increase capacity, is to use *sectorisation*. The group of channels available at each cell is split into say three subgroups, each of which is confined in coverage to one-third of the cell area by the use of directional antennas, as shown in Figure 1.12. Chapters 4 and 15 will describe how this directionality is achieved. Interference now comes from only 2, rather than 6, of the first-tier interfering sites, reducing interference by a factor of 3 and allowing cluster size to be increased by a factor of $3^{0.5} = 1.7$ in theory. Sectorisation has three disadvantages:

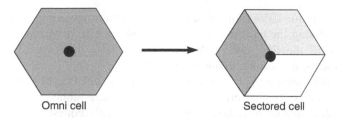

Figure 1.12: Sectorisation of an omnidirectional cell into three sectors

- More base station equipment is needed, especially in the radio frequency (RF) domain.
- Mobiles have to change channels more often, resulting in an increased signalling load on the system.
- The available pool of channels has to be reduced by a factor 3 for a mobile at any particular location; this reduces the trunking efficiency (Section 1.9).

Despite these issues, sectorisation is used very widely in modern cellular systems, particularly in areas needing high traffic density. More than three sectors can be used to further improve the interference reduction; the ultimate is to have very narrow-beam antennas which track the position of the mobile, and these are examined in Chapter 18.

As the mobile moves through the system coverage area, it crosses cell boundaries and thus has to change channels. This process is *handover* or *handoff* and it must be performed quickly and accurately. Modern fast-switching frequency synthesisers and digital signal processing have allowed this process to be performed with no significant impact on call quality.

The handover process needs to be carefully controlled: if handover occurs as soon as a new base station becomes stronger than the previous one, then 'chatter' or very rapid switching between the two BSs will occur, especially when the mobile moves along a cell boundary. An element of hysteresis is therefore introduced into the handover algorithm: the handover occurs only when the new BS is stronger than the old one by at least some *handover margin*. If this margin is too large, however, the mobile may move far into the coverage area of a new cell, causing interference to other users and itself suffering from poor signal quality. The optimum handover margin is set crucially by the level of shadowing in the system, as this determines the variation of signal level along the cell boundaries (Chapter 9). Handover accuracy is usually improved by *mobile-assisted handover*, also known as MAHO, in which the mobile monitors both the current cell and several neighbouring cells, and sends signal strength and quality reports back to the current serving BS.

1.9 TRAFFIC

The number of channels which would be required to guarantee service to every user in the system is impractically large. It can, however, be reduced by observing that, in most cases, the number of users needing channels simultaneously is considerably smaller than the total number of users. The concept of *trunking* can then be applied: a common pool of channels is created and is shared among all the users in a cell. Channels are allocated to particular users when they request one at the start of a call. At the end of the call, the channel is returned to the pool. This means there will be times when a user requests a channel and none is left in the pool: the call is then *blocked*. The probability of blocking for which a system is designed is the *grade of service*. Traffic is measured in *erlangs*: one erlang (E) is equivalent to one user making a call for 100% of the time. A typical cellular voice user generates around 2–30 mE of traffic during the busiest hour of the system, that is, a typical user is active for around 0.2–3.0% of the time during the busy-hour. These figures tend to increase for indoor environments, fluctuating around 50–60 mE.

The *traffic per user* A_u is required if the traffic per cell is to be computed. Therefore, a *user traffic profile* is often described, in which an average mobile phone user makes λ calls of duration H during the busy-hour. λ is known as the call request rate and H is the holding time. Hence, the average traffic per user is

$$A_u = \lambda H \tag{1.6}$$

For U users in the cell, the total *carried traffic A* is given by:

$$A = U A_u \tag{1.7}$$

To predict the number of channels needed to support a given number of users to a certain grade of service, it is usual to apply the *Erlang-B formula*:

$$\Pr(\text{blocking occurs}) = \frac{A^C/C!}{\displaystyle\sum_{k=0}^{C} \frac{A^k}{k!}} \tag{1.8}$$

where A is the total offered traffic in erlangs and C is the total number of channels available. This formula is plotted in Figure 1.13. The Erlang-B formula is idealised in that it makes particular assumptions about the call request rate and holding time, but it provides a useful starting point for estimating the number of channels required.

In summary, *capacity* for a cellular system can be dimensioned if blocking, number of channels and offered traffic can be estimated. A network is often dimensioned for busy-hour operation, as network congestion limits the number of available resources. Example 1.1 illustrates how this capacity dimensioning is performed in practice.

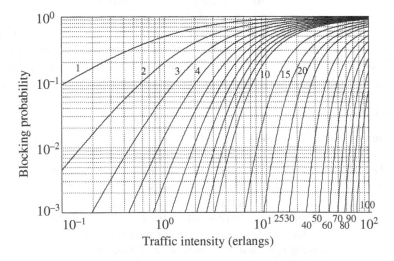

Figure 1.13: Erlang-B graph: each curve is marked with the number of channels

Example 1.1

A cellular operator is interested in providing GSM coverage at 900 MHz in an international airport. Surveys show that approximately 650 000 passengers make use of the airport every year, of which it is believed that around 80% are mobile phone users. The airport layout is shown in Figure 1.14.

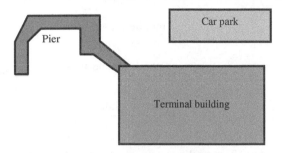

Figure 1.14: Airport layout for example 1.1

The following assumptions apply for this airport:

(a) Busy-hour traffic takes about 25% of the total daily traffic.
(b) The traffic in the airport is distributed in the following proportions: 70% is carried in the terminal building, 20% in the pier and 10% in the car park.
(c) Each airport user makes an average of three phone calls of 2 min of duration during the busy hour.
(d) Three cells are required for this system: one in the terminal, one in the car park and the other one in the pier.
(e) This operator has a market penetration in the airport of 28%.

Determine the required number of channels per cell, if only 2% of the users attempting to make a phone call are to be blocked when no free channels are available.

Solution

The busy-hour number of mobile phone users needs to be calculated. If 650 000 passengers per year use the airport, then:

$$\frac{650\,000 \text{ passengers}}{\text{Year}} \times \frac{1 \text{ year}}{365 \text{ days}} \approx 1781 \text{ passengers/day}$$

As only 80% of the passengers use a mobile phone, the cellular operator with 28% of market penetration carries

$$1781 \times 0.8 \times 0.28 \approx 399 \text{ mobile phone users/day}$$

This number represents only the number of mobile phone users per day for this cellular operator. As capacity needs to be dimensioned on a per busy-hour basis:

$$\frac{399 \text{ mobile phone users}}{\text{Day}} \times 0.25 \approx 100 \frac{\text{Mobile phone users}}{\text{Busy-hour}}$$

Next, it is necessary to split this number of users per cell. As the terminal takes 70% of the traffic, the car park only 10% and the pier, 20%, then:

$$U_{\text{terminal}} = 100 \frac{\text{mobile phone users}}{\text{busy-hour}} \times 0.7 = 70 \text{ mobile users}$$

$$U_{\text{pier}} = 100 \frac{\text{mobile phone users}}{\text{busy-hour}} \times 0.2 = 20 \text{ mobile users}$$

$$U_{\text{car park}} = 100 \frac{\text{mobile phone users}}{\text{busy-hour}} \times 0.1 = 10 \text{ mobile users}$$

The offered traffic per cell must now be estimated. For this, an average traffic profile per user has been given, from which an average traffic per user can be computed. Given that a mobile user makes three phone calls of 2 min of duration during the busy-hour, the traffic per user is

$$A_u = \lambda H = \frac{3 \text{ calls}}{\text{hour}} \times 2 \min \times \frac{1 \text{ hour}}{60 \min} = 100 \text{ mE/user}$$

Given this number of users per cell, and the traffic per user, it is now possible to compute the total traffic per cell:

$$A_{\text{terminal}} = 70 \frac{\text{users}}{\text{cell}} \times 100 \frac{\text{mE}}{\text{user}} = 7 \text{ Erlangs}$$

$$A_{\text{pier}} = 20 \frac{\text{users}}{\text{cell}} \times 100 \frac{\text{mE}}{\text{user}} = 2 \text{ Erlangs}$$

$$A_{\text{car park}} = 10 \frac{\text{users}}{\text{cell}} \times 100 \frac{\text{mE}}{\text{user}} = 1 \text{ Erlang}$$

With a 2% blocking probability, an estimate of the required number of channels per cell is possible, as the traffic per cell is available. Referring to the Erlang-B graph in Figure 1.13:

$$C_{\text{terminal}} \approx 13 \text{ channels}$$

$$C_{\text{pier}} \approx 6 \text{ channels}$$

$$C_{\text{car park}} \approx 4 \text{ channels}$$

Note that it is necessary to round to the next highest integer in all cases, to guarantee a minimum number of channels to provide the required traffic A. It is also worth noting that the Erlang-B formula can be applied only on a *per-cell* basis, as Eq. (1.8) is not linear.

1.10 MULTIPLE ACCESS SCHEMES AND DUPLEXING

Given a portion of the frequency spectrum, multiple users may be assigned channels within that portion according to various techniques, known as *multiple access schemes*. The three most common schemes are examined here. The *duplexing scheme* is also examined, whereby simultaneous two-way communication is enabled from the user's point of view. The multiple access schemes are

- frequency division multiple access (FDMA)
- time division multiple access (TDMA)
- code division multiple access (CDMA).

Chapters 17 and 18 will introduce two further multiple access schemes, orthogonal frequency division multiple access (OFDMA) and space division multiple access (SDMA). The duplexing schemes described here are

- frequency division duplex (FDD)
- time division duplex (TDD).

1.10.1 Frequency Division Multiple Access

Figure 1.15 illustrates a system using FDMA and FDD. The total bandwidth available to the system operator is divided into two sub-bands and each of these is further divided into a number of frequency channels. Each mobile user is allocated a pair of channels, separated by the *duplex spacing*, one in the uplink sub-band, for transmitting to the base station, the other in the downlink sub-band, for reception from the base station.

This scheme has the following features:

- Transmission and reception are simultaneous and continuous, so RF duplexers are needed at the mobile to isolate the two signal paths, which increase cost.
- The carrier bandwidth is relatively narrow, so equalisers are not usually needed (Chapter 17).
- The baseband signal processing has low complexity.
- Little signalling overhead is required.
- Tight RF filtering is needed to avoid adjacent channel interference.
- Guard bands are needed between adjacent carriers and especially between the sub-bands.

FDMA is the most common access scheme for systems based on analogue modulation techniques such as frequency modulation (FM), but it is less commonly used in modern digital systems.

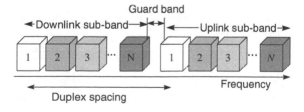

Figure 1.15: Frequency division multiple access used with FDD

1.10.2 Time Division Multiple Access

The scheme illustrated in Figure 1.16 combines TDMA and FDD, requiring two frequencies to provide duplex operation, just as in FDMA. Time is divided into *frames* and each frame is further divided into a number of slots (four in this case). Mobiles are allocated a pair of *time slots*, one at the uplink frequency and the other at the downlink frequency, chosen so that they do not coincide in time. The mobile transmits and receives *bursts*, whose duration is slightly less than the time slot to avoid overlap and hence interference between users.

This scheme has several features:

- Transmission and reception are never simultaneous at the mobile, so duplexers are not required.
- Some bits are wasted due to burst start and stop bits and due to the guard time needed between bursts.
- The wide channel bandwidth needed to accommodate several users usually leads to a need for equalisation (Chapter 17).
- The time between slots is available for handover monitoring and channel changing.
- The receiver must resynchronise on each burst individually.
- Different bit rates can be flexibly allocated to users by allocating multiple time slots together.

TDMA can also be used with TDD, by allocating half the slots to the uplink and half to the downlink, avoiding the need for frequency switching between transmission and reception and permitting the uplink channel to be estimated from the downlink, even in the presence of frequency-dependent fading (Chapter 11).

1.10.3 Code Division Multiple Access

In CDMA or *spread spectrum* systems, each user occupies a bandwidth much wider than is needed to accommodate their data rate. In the form usually used for cellular mobile systems, this is achieved by multiplying the source of data by a *spreading code* at a much higher rate, the *chip rate*, thereby increasing the transmitted signal bandwidth (Figure 1.17). At the receiver, the reverse process is performed to extract the original signal, known as *despreading*. This is *direct sequence* spread spectrum, described in more detail in Chapter 17. When this is applied to multiple access, the users are each given different codes, and the codes are specially

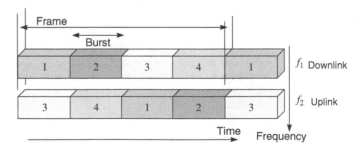

Figure 1.16: Time division multiple access used with FDD

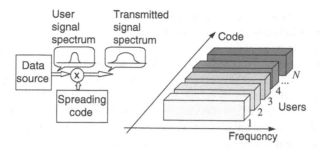

Figure 1.17: Code division multiple access

chosen to produce low *multiple access interference* following the despreading process. The duplex method is usually FDD. Here are some of its features:

- Increasing the number of users increases interference gradually, so there is no specific limit to the number of users, provided that codes with low mutual interference properties are chosen. Instead, system performance degrades progressively for all users.
- The high bandwidth leads to a requirement for an equaliser-like structure, a *Rake receiver* (Chapter 17). This allows multipath diversity gain to be obtained (Chapter 16).
- The power of all users must be equal at the base station to allow the despreading process to work effectively, so some complex power control is required.
- The baseband processing may be complex compared with FDMA and TDMA, but this is less important with modern silicon integration densities.

CDMA is increasingly being applied in modern cellular mobile systems, placing increasing emphasis on the need for characterisation of wideband channel effects (Chapter 11).

1.11 AVAILABLE DATA RATES

The access schemes introduced in the previous section allow multiple users to access portions of the available system bandwidth. Another relevant consideration, particularly for multi-media and data services, is the data rate available to each user. This is sometimes loosely referred to as the 'bandwidth', but in fact the user data rate is not a simple function of the bandwidth occupied, being influenced by three important elements:

- the spectrum efficiency of the modulation scheme employed;
- the error correction and detection schemes in use;
- the channel quality.

The modulation scheme efficiency for digital signals is measured as a ratio of the data rate transmitted over the air to the bandwidth occupied and is thus typically measured in bits per second per Hertz [bit s^{-1} Hz^{-1} or bps Hz^{-1}]. For example, the simple binary phase shift keying (BPSK) scheme analysed in Chapter 10 achieves a spectral efficiency of 1 bit s^{-1} Hz^{-1}, whereas a quaternary phase shift keying (QPSK) scheme signals two binary bit in the same period as a single BPSK bit and thus achieves twice the channel data rate in the same bandwidth, i.e. 2 bit s^{-1} Hz^{-1} [Proakis, 89]. This process can be continued almost indefinitely, with sixty-four level quadrature amplitude modulation (64-QAM), for example, achieving 6 bit s^{-1} Hz^{-1}. The price for this increased bit rate, however, is an increased sensitivity to noise and interference, so that more errors are caused and the useful error-free data rate may actually be reduced by signalling faster.

In order to increase the robustness of the modulation scheme, error correction and detection schemes are applied. At a basic level, this is achieved by adding extra checksum bits at the transmitter, which allow the receiver to detect whether some bits have been received in error and to correct at least some of those errors. This process of 'forward error correction' (FEC) adds redundancy according to some coding rate, so every user data bit is represented in the transmission by multiple coded data bits. This ratio is the coding rate of the coder. For example, if there are two coded bits for every one data bit, the coding rate = 1/2. Thus the user data rate is the channel data rate multiplied by the coding rate. Selection of the most appropriate FEC scheme involves a trade-off between acceptable error rates and the transmission rate.

When errors are detected by the coding scheme and cannot be corrected by FEC, it is common for the receiver to request retransmission of the suspect part of the transmitted data. Such *automatic repeat request* or ARQ schemes can produce very reliable data transmission, at the expense of further reduced user data rates. The actual user data rate achieved depends directly on the channel quality.

The upper limit for the useful data rate, or *channel capacity* C [bit s^{-1}] achieved in a channel of bandwidth B [Hz] at a signal power to noise power ratio S/N was predicted by [Shannon, 48] as:

$$C = \log_2\left(1 + \frac{S}{N}\right) \qquad (1.9)$$

This implies that, for an ideal system, the bit error rate can be reduced to zero by the application of appropriate coding schemes, provided the user data rate is less than the channel capacity. Shannon did not, however, provide any constructive techniques for creating such codes and the five decades following his original paper saw researchers expending very significant effort on constructing codes which approached ever closer to this limit. The practical impact of this effort is illustrated by the increasing user data rate available from real-world systems in Figure 1.18, which appears to double approximately every 18 months, the same as the rate predicted by Moore's law for the increase in integrated circuit complexity [Cherry, 04].

One of the developments which has allowed such high data rates was the development of *turbo codes* in 1993, a form of FEC which allows the attainment of performance virtually at the Shannon capacity, albeit at the expense of high decoder complexity [Berrou, 93]. The fact that user data rates have actually continued to increase even beyond this limit may seem mysterious, but arises directly from a detailed understanding of the characteristics of both antennas and the propagation channel. See Chapter 18 for further details of the MIMO techniques which permit this – but the reader is advised to study the intervening chapters to gain the best possible appreciation of these!

1.12 STRUCTURE OF THIS BOOK

The preceding sections have given an indication of the significance and effect of antennas and propagation in wireless communication systems. Following on from this introduction, the book is loosely structured into five major sections.

The first section is concerned with the key features of radio wave propagation and antennas which are common to all wireless communication systems. Chapter 2 examines the

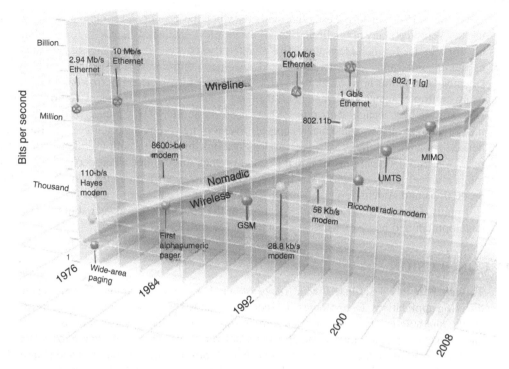

Figure 1.18: Edholm's law of bandwidth (Reproduced by permission of IEEE, © 2004 IEEE)

fundamental properties of electromagnetic waves travelling in uniform media. These are developed in Chapter 3 in order to describe the basic mechanisms of propagation which occur when waves encounter boundaries between different media. Chapter 4 then shows how such waves can be launched and received using antennas, which have certain key requirements and properties and for which there are a number of generic types.

Starting a new section, Chapter 5 introduces the concept of a propagation model, together with the basic techniques for analysing communication systems; these must be understood in order to predict system performance. The theory in Chapter 5 and the first section is then used to analyse two practical types of fixed communication system – terrestrial fixed links in Chapter 6 and satellite fixed links in Chapter 7.

The next section (Chapters 8–11) examines propagation and channel effects in macrocell mobile systems. Chapter 8 describes the path loss models which establish the basic range and interference levels from such systems. Chapter 9 describes the shadowing effects which cause statistical variations relative to these levels and which determine the percentage of locations in a given area at which the system is available. Chapter 10 explains the mechanisms and statistical characterisation of the narrowband fast-fading effects in macrocells, which affect the quality of individual links between mobiles and macrocell base stations. These are generalised to wideband systems in Chapter 11.

The macrocell concepts are then broadened in the next section to microcells, picocells and megacells in Chapters 12–14, respectively, including all the major differences in propagation and antennas for such cells.

Chapters 15–19 inclusive establish methods which the system and equipment designer can use to overcome many of the limitations of the wireless channel. Chapter 15 first explains the antennas in use for mobile systems. This chapter also covers other specifics of antennas such as health and safety considerations, and specific requirements for mobile applications. Chapter 16 explains how *diversity* provides a means of overcoming narrowband fast fading using multiple antennas. Chapter 17 then provides a similar treatment for overcoming wideband fading, using *equalisation, Rake receivers* and *OFDM*. Chapter 18 then indicates the potential of adaptive antenna systems for overcoming co-channel interference effects and hence for increasing the capacity of wireless systems, in terms of both user density and data rates. Chapter 19 highlights essential practical issues to consider when performing mobile radio channel measurements, which affect the accuracy of both radio wave propagation predictions and system designs.

Finally, Chapter 20 gives some indications of the developments expected in the research, development and usage of wireless channels in the future.

1.13 CONCLUSION

This chapter has described a wide range of wireless communication systems, each with differing applications, technologies and requirements. All of them are unified by their reliance on the characteristics of the wireless communication channel, incorporating both antennas and propagation, to accurately and efficiently deliver information from source to destination. For further details of these systems, texts such as [Molisch, 05], [Parsons, 00] and [Maral, 02] are recommended. The rest of this book will detail the principles and applied practical techniques needed to understand, predict and evaluate the channel effects which have impact on these systems.

REFERENCES

[Berrou, 93] C. Berrou, A. Glavieux and P. Thitimajshima, Near Shannon Limit error-correcting coding and decoding: Turbo-codes, *IEEE International Communications Conference*, 1064–1070, 1993.

[Cherry, 04] S. Cherry, Edholm's Law of Bandwidth, *IEEE Spectrum*, July 2004, 58–60.

[Maral, 02] G. Maral and M. Bousquet, *Satellite communications systems: systems, techniques and technology*, John Wiley & Sons, Ltd, Chichester, ISBN 0-471-49654-5, 2002.

[Molisch, 05] A. Molisch, *Wireless communications*, John Wiley & Sons, Ltd, Chichester, ISBN 0-470-84888-X, 2005.

[Parsons, 00] J. D. Parsons, *The mobile radio propagation channel*, John Wiley & Sons, Ltd, Chichester, ISBN 0-471-98857-X, 2000.

[Proakis, 89] J. G. Proakis, *Digital communications*, 2nd edn, McGraw-Hill, New York, ISBN 0-07-100269-3, 1989.

[Shannon, 48] C. Shannon, A mathematical theory of communication, *Bell system technical journal*, **27**, 379–423 and 623–56, 1948. Reprinted in *Claude Elwood Shannon: collected papers*, edited by N.J.A. Sloane and A.D. Wyner, IEEE Press, New York, 1993, ISBN 0-78-0304349.

[Webb, 07] W. Webb (editor), *Wireless communications: The Future*, John Wiley & Sons, Ltd, Chichester, ISBN 0-470-03312-6, 2007.

PROBLEMS

1.1 A cellular operator has created a system for 900 MHz which covers the desired service area with 500 cells and a base station antenna height of 15 m. The operator is given a new frequency allocation at 1800 MHz. How many cells would be needed at the new frequency using the same antenna heights and transmit powers? How much higher would the antennas need to be to produce the same range as the 900 MHz system?

1.2 What is the ratio of the cluster sizes needed for two systems with a 9 dB C/I requirement and a 19 dB C/I requirement?

1.3 A certain cell has 4000 users, who generate an average of 2 mE of busy-hour traffic each. How many channels are needed to serve these users with no more than 2% blocking probability? How many users can be served, with the same number of channels and grade of service, if the cell is divided into three sectors?

1.4 A leading cellular operator wants to provide cellular coverage in one of the busiest international airports in the world. The cellular operator's research indicates that around 31 650 potential mobile users per month during the high season pass through the airport for domestic and international flights. Radio measurements have shown that sufficient coverage is provided with three cells: one in the main terminal building, one in the international flights building and one in the domestic flights building, as shown in Figure 1.19.

Clipboard research shows that mobile users at this airport make two phone calls of 3 min of duration on average.

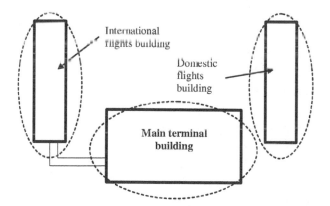

Figure 1.19: International airport scenario for Problem 1.4

(a) Determine the number of required channels to deploy at each base station per cell, if the system is to be dimensioned for a maximum blocking of 2% for an Erlang-B system. Assume that the main terminal carries the 60% of the total traffic and each other building carries only 20%. Approximately 25% of the daily traffic is carried during the busy-hour.

(b) What is the total traffic that is carried at this airport?

(c) Suggest a method for making a more efficient use of the radio spectrum if the traffic is variable in the three cells, as this traffic tends to be bursty.

2 Properties of Electromagnetic Waves

'I have no picture of this electromagnetic field that is in any sense accurate...
I see some kind of vague, shadowy, wiggling lines...
So if you have some difficulty in making such a picture, you should not be worried that your
difficulty is unusual'.
Richard P. Feynman, Nobel Prize Laureate in Physics

2.1 INTRODUCTION

Some of the key properties of electromagnetic waves travelling in free space and in other uniform media are introduced in this chapter. They establish the basic parameters and relationships which are used as standard background when considering problems in antennas and propagation in later chapters. This chapter does not aim to provide rigorous or complete derivations of these relationships; for a fuller treatment see books such as, [Kraus, 98] or [Hayt, 01]. It does, however, aim to show that any uniform medium can be specified by a small set of descriptive parameters and that the behaviour of waves in such media may easily be calculated.

2.2 MAXWELL'S EQUATIONS

The existence of propagating electromagnetic waves can be predicted as a direct consequence of Maxwell's equations [Maxwell, 1865]. These equations specify the relationships between the variations of the vector electric field **E** and the vector magnetic field **H** in time and space within a medium. The **E** field strength is measured in volts per metre and is generated by either a time-varying magnetic field or a free charge. The **H** field is measured in amperes per metre and is generated by either a time-varying electric field or a current. Maxwell's four equations can be summarised in words as

An electric field is produced by a time-varying magnetic field

A magnetic field is produced by a time-varying electric field or by a current

Electric field lines may either start and end on charges, or are continuous $\qquad$ (2.1)

Magnetic field lines are continuous

The first two equations, Maxwell's *curl* equations, contain constants of proportionality which dictate the strengths of the fields. These are the permeability of the medium μ in henrys per

Antennas and Propagation for Wireless Communication Systems Second Edition   Simon R. Saunders and Alejandro Aragón-Zavala
© 2007 John Wiley & Sons, Ltd

metre and the permittivity of the medium ε in farads per metre. They are normally expressed relative to the values in free space:

$$\mu = \mu_0 \mu_r \tag{2.2}$$

$$\varepsilon = \varepsilon_0 \varepsilon_r \tag{2.3}$$

where μ_0 and ε_0 are the values in free space, given by

$$\mu_0 = 4\pi \times 10^{-7} \ \mathrm{H\,m^{-1}} \tag{2.4}$$

$$\varepsilon_0 = 8.854 \times 10^{-12} \approx \frac{10^{-9}}{36\pi} \mathrm{F\,m^{-1}} \tag{2.5}$$

and μ_r, ε_r are the relative values (i.e. $\mu_r = \varepsilon_r = 1$ in free space). Free space strictly indicates a vacuum, but the same values can be used as good approximations in dry air at typical temperatures and pressures.

2.3 PLANE WAVE PROPERTIES

Many solutions to Maxwell's equations exist and all of these solutions represent fields which could actually be produced in practice. However, they can all be represented as a sum of *plane waves*, which represent the simplest possible time varying solution.

Figure 2.1 shows a plane wave, propagating parallel to the z-axis at time $t = 0$.

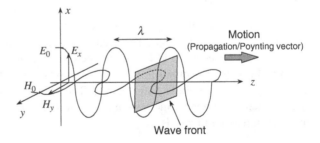

Figure 2.1: A plane wave propagating through space at a single moment in time

The electric and magnetic fields are perpendicular to each other and to the direction of propagation of the wave; the direction of propagation is along the z axis; the vector in this direction is the *propagation vector* or *Poynting vector.* The two fields are in phase at any point in time or in space. Their magnitude is constant in the xy plane, and a surface of constant phase (a *wavefront*) forms a plane parallel to the xy plane, hence the term *plane wave*. The oscillating electric field produces a magnetic field, which itself oscillates to recreate an electric field and so on, in accordance with Maxwell's curl equations. This interplay between the two fields stores energy and hence carries power along the Poynting vector. Variation, or *modulation*, of the properties of the wave (amplitude, frequency or phase) then allows information to be carried in the wave between its source and destination, which is the central aim of a wireless communication system.

2.3.1 Field Relationships

The electric field can be written as

$$\mathbf{E} = E_0 \cos(\omega t - kz)\hat{\mathbf{x}} \tag{2.6}$$

where E_0 is the field amplitude [V m^{-1}], $\omega = 2\pi f$ is the angular frequency in radians for a frequency f [Hz], t is the elapsed time [s], k is the wavenumber [m^{-1}], z is distance along the z-axis (m) and $\hat{\mathbf{x}}$ is a unit vector in the positive x direction. The wavenumber represents the rate of change of the phase of the field with distance; that is, the phase of the wave changes by kr radians over a distance of r metres. The distance over which the phase of the wave changes by 2π radians is the wavelength λ. Thus

$$k = \frac{2\pi}{\lambda} \tag{2.7}$$

Similarly, the magnetic field vector $\mathbf{H}$ can be written as

$$\mathbf{H} = H_0 \cos(\omega t - kz)\hat{\mathbf{y}} \tag{2.8}$$

where H_0 is the magnetic field amplitude and $\hat{\mathbf{y}}$ is a unit vector in the positive y direction.

In both Eqs. (2.6) and (2.8), it has been assumed that the medium in which the wave travels is lossless, so the wave amplitude stays constant with distance. Notice that the wave varies sinusoidally in both time and distance.

It is often convenient to represent the phase and amplitude of the wave using complex quantities, so Eqs. (2.6) and (2.8) become

$$\mathbf{E} = E_0 \exp[j(\omega t - kz)]\hat{\mathbf{x}} \tag{2.9}$$

and

$$\mathbf{H} = H_0 \exp[j(\omega t - kz)]\hat{\mathbf{y}} \tag{2.10}$$

The real quantities may then be retrieved by taking the real parts of Eqs. (2.9) and (2.10). Complex notation will be applied throughout this book.

2.3.2 Wave Impedance

Equations (2.6) and (2.8) satisfy Maxwell's equations, provided the ratio of the field amplitudes is a constant for a given medium,

$$\frac{|\mathbf{E}|}{|\mathbf{H}|} = \frac{E_x}{H_y} = \frac{E_0}{H_0} = \sqrt{\frac{\mu}{\varepsilon}} = Z \tag{2.11}$$

where Z is called the *wave impedance* and has units of ohms. In free space, $\mu_r = \varepsilon_r = 1$ and the wave impedance becomes

$$Z = Z_0 = \sqrt{\frac{\mu_0}{\varepsilon_0}} \approx \sqrt{4\pi \times 10^{-7} \times \frac{36\pi}{10^{-9}}} = 120\pi \approx 377\,\Omega \tag{2.12}$$

Thus, in free space or any uniform medium, it is sufficient to specify a single field quantity together with Z in order to specify the total field for a plane wave.

2.3.3 Poynting Vector

The Poynting vector $\mathbf{S}$, measured in watts per square metre, describes the magnitude and direction of the power flow carried by the wave per square metre of area parallel to the xy plane, i.e. the *power density* of the wave. Its instantaneous value is given by

$$\mathbf{S} = \mathbf{E} \times \mathbf{H}^* \tag{2.13}$$

Usually, only the time average of the power flow over one period is of concern,

$$\mathbf{S}_{av} = \frac{1}{2} E_0 H_0 \hat{\mathbf{z}} \tag{2.14}$$

The direction vector in Eq. (2.14) emphasises that $\mathbf{E}$, $\mathbf{H}$ and $\mathbf{S}_{av}$ form a right-hand set, i.e. $\mathbf{S}_{av}$ is in the direction of movement of a right-handed corkscrew, turned from the $\mathbf{E}$ direction to the $\mathbf{H}$ direction.

2.3.4 Phase Velocity

The velocity of a point of constant phase on the wave, the *phase velocity v* at which wave fronts advance in the $\mathbf{S}$ direction, is given by

$$v = \frac{\omega}{k} = \frac{1}{\sqrt{\mu\varepsilon}} \tag{2.15}$$

Hence the wavelength λ is given by

$$\lambda = \frac{v}{f} \tag{2.16}$$

This book is concerned entirely with frequencies from around 30 MHz to 300 GHz, i.e. free space wavelengths from 10 m to 1 mm.

In free space the phase velocity becomes

$$v = c = \frac{1}{\sqrt{\mu_0 \varepsilon_0}} \approx 3 \times 10^8 \ \mathrm{m\,s^{-1}} \tag{2.17}$$

Note that light is an example of an electromagnetic wave, so c is the speed of light in free space.

2.3.5 Lossy Media

So far only lossless media have been considered. When the medium has significant conductivity, the amplitude of the wave diminishes with distance travelled through the medium as energy is removed from the wave and converted to heat, so Eqs. (2.9) and (2.10) are then replaced by

$$\mathbf{E} = E_0 \exp[j(\omega t - kz) - \alpha z]\hat{\mathbf{x}} \tag{2.18}$$

and

$$\mathbf{H} = H_0 \exp[j(\omega t - kz) - \alpha z]\hat{\mathbf{y}} \tag{2.19}$$

The constant α is known as the *attenuation constant*, with units of per metre [m^{-1}], which depends on the permeability and permittivity of the medium, the frequency of the wave and the *conductivity* of the medium, σ, measured in siemens per metre or per-ohm-metre [Ωm]$^{-1}$. Together σ, μ and ε are known as the *constitutive parameters* of the medium.

In consequence, the field strength (both electric and magnetic) diminishes exponentially as the wave travels through the medium as shown in Figure 2.2. The distance through which the wave travels before its field strength reduces to $e^{-1} = 0.368 = 36.8\%$ of its initial value is its *skin depth δ*, which is given by

$$\delta = \frac{1}{\alpha} \tag{2.20}$$

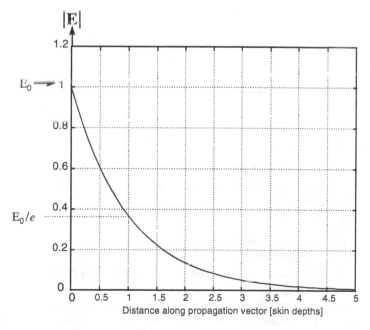

Figure 2.2: Field attenuation in lossy medium

Thus the amplitude of the electric field strength at a point z compared with its value at $z = 0$ is given by

$$E(z) = E(0)e^{-z/\delta} \qquad (2.21)$$

Table 2.1 gives expressions for α and k which apply in both lossless and lossy media. Note that the expressions may be simplified depending on the relative values of σ and $\omega\varepsilon$. If σ dominates, the material is a *good conductor*; if σ is very small, the material is a *good insulator* or *good dielectric*.

Table 2.1: Attenuation constant, wave number, wave impedance, wavelength and phase velocity for plane waves in lossy media (after [Balanis, 89])

$n = ck/\omega$ in all cases	Exact expression	Good dielectric (insulator) $(\sigma/\omega\varepsilon)^2 \ll 1$	Good conductor $(\sigma/\omega\varepsilon)^2 \gg 1$
Attenuation constant α [m^{-1}]	$\omega\sqrt{\dfrac{\mu\varepsilon}{2}\left[\sqrt{1+\left(\dfrac{\sigma}{\omega\varepsilon}\right)^2}-1\right]}$	$\approx \dfrac{\sigma}{2}\sqrt{\dfrac{\mu}{\varepsilon}}$	$\approx \sqrt{\dfrac{\omega\mu\sigma}{2}}$
Wave number k [m^{-1}]	$\omega\sqrt{\dfrac{\mu\varepsilon}{2}\left[\sqrt{1+\left(\dfrac{\sigma}{\omega\varepsilon}\right)^2}+1\right]}$	$\approx \omega\sqrt{\mu\varepsilon}$	$\approx \sqrt{\dfrac{\omega\mu\sigma}{2}}$
Wave impedance Z [Ω]	$\sqrt{\dfrac{j\omega\mu}{\sigma+j\omega\varepsilon}}$	$\approx \sqrt{\dfrac{\mu}{\varepsilon}}$	$\approx \sqrt{\dfrac{\omega\mu}{2\sigma}}(1+j)$
Wavelength λ [m]	$\dfrac{2\pi}{k}$	$\approx \dfrac{2\pi}{\omega\sqrt{\mu\varepsilon}}$	$\approx 2\pi\sqrt{\dfrac{2}{\omega\mu\sigma}}$
Phase velocity v [m s^{-1}]	$\dfrac{\omega}{k}$	$\approx \dfrac{1}{\sqrt{\mu\varepsilon}}$	$\approx \sqrt{\dfrac{2\omega}{\mu\sigma}}$

Example 2.1

A linearly polarised plane wave at 900 MHz travels in the positive z direction in a medium with constitutive parameters $\mu_r = 1, \varepsilon_r = 3$ and $\sigma = 0.01$ S m^{-1}. The electric field magnitude at $z = 0$ is 1 V m^{-1}.
Calculate:

(a) the wave impedance;
(b) the magnitude of the magnetic field at $z = 0$;
(c) the average power available in a 0.5 m^2 area perpendicular to the direction of propagation at $z = 0$;
(d) the time taken for the wave to travel through 10 cm;
(e) the distance travelled by the wave before its field strength drops to one tenth of its value at $z = 0$.

Solution

Referring to Table 2.1,

$$\frac{\sigma}{\omega\varepsilon} = \frac{0.01}{2\pi \times 900 \times 10^6 \times 3 \times \frac{10^{-9}}{36\pi}} \approx 0.07 \ll 1$$

so the material can clearly be regarded as a good insulator.

(a) For an insulator, the wave impedance is given by

$$Z \approx \sqrt{\frac{\mu}{\varepsilon}} = \sqrt{\frac{\mu_0}{\varepsilon_r \varepsilon_0}} = \frac{Z_0}{\sqrt{\varepsilon_r}} \approx \frac{377}{\sqrt{3}} \approx 218\,\Omega$$

(b) From Eq. (2.11) the magnetic field amplitude is given by

$$H = \frac{E}{Z} = \frac{1}{218} \approx 4.6\,\mathrm{mA\,m}^{-1}$$

(c) The available average power is the magnitude of the time-average Poynting vector multiplied by the collection area, i.e.

$$P = SA = \frac{EH}{2}A = \frac{1 \times 0.005}{2} \times 0.5 = 1.25\,\mathrm{mW}$$

(d) The time taken to travel a given distance is simply the distance divided by the phase velocity

$$t = \frac{d}{v} = d\sqrt{\mu\varepsilon} = d\sqrt{\mu_0\varepsilon_r\varepsilon_0} = \frac{d}{c}\sqrt{\varepsilon_r} \approx \frac{0.1}{3 \times 10^8}\sqrt{3} \approx 0.6\,\mathrm{ns}$$

(e) Rearranging Eq. (2.21) yields

$$z = -\delta\ln\frac{E(z)}{E(0)}$$

where

$$\delta = \frac{1}{\alpha} \approx \frac{2}{\sigma}\sqrt{\frac{\varepsilon}{\mu}} = \frac{2}{\sigma Z}$$

and from Table 2.1 the approximation holds for good insulators. Thus

$$z = -\frac{2}{\sigma Z}\ln\frac{E(z)}{E(0)} = -\frac{2}{0.01 \times 218}\ln\frac{1}{10} = 2.11\,\mathrm{m}$$

2.4 POLARISATION

2.4.1 Polarisation States

The alignment of the electric field vector of a plane wave relative to the direction of propagation defines the *polarisation* of the wave. In Figure 2.1 the electric field is parallel to the *x* axis, so this wave is *x* polarised. This wave could be generated by a straight wire antenna parallel to the *x* axis. An entirely distinct *y*-polarised plane wave could be generated with the same direction of propagation and recovered independently of the other wave using pairs of transmit and receive antennas with perpendicular polarisation. This principle is sometimes used in satellite communications to provide two independent communication channels on the same earth satellite link. If the wave is generated by a vertical wire antenna (**H** field horizontal), then the wave is said to be *vertically polarised*; a wire antenna parallel to the ground (**E** field horizontal) primarily generates waves that are *horizontally polarised*.

The waves described so far have been *linearly polarised*, since the electric field vector has a single direction along the whole of the propagation axis. If two plane waves of equal amplitude and orthogonal polarisation are combined with a 90° phase difference, the resulting wave will be *circularly polarised* (CP), in that the motion of the electric field vector will describe a circle centred on the propagation vector. The field vector will rotate by 360° for every wavelength travelled. Circularly polarised waves are most commonly used in satellite communications, since they can be generated and received using antennas which are oriented in any direction around their axis without loss of power. They may be generated as either right-hand circularly polarised (RHCP) or left-hand circularly polarised (LHCP); RHCP describes a wave with the electric field vector rotating clockwise when looking in the direction of propagation. In the most general case, the component waves could be of unequal amplitudes or at a phase angle other than 90°. The result is an *elliptically polarised* wave, where the electric field vector still rotates at the same rate but varies in amplitude with time, thereby describing an ellipse. In this case, the wave is characterised by the ratio between the maximum and minimum values of the instantaneous electric field, known as the axial ratio, *AR*,

$$AR = \frac{E_{\text{maj}}}{E_{\text{min}}} \tag{2.22}$$

AR is defined to be positive for left-hand polarisation and negative for right-hand polarisation. These various polarisation *states* are illustrated in Figure 2.3.

2.4.2 Mathematical Representation of Polarisation

All of the polarisation states illustrated in Figure 2.3 can be represented by a compound electric field vector **E** composed of *x* and *y* linearly polarised plane waves with amplitudes E_x and E_y,

$$\mathbf{E} = E_x \hat{\mathbf{x}} + E_y \hat{\mathbf{y}} \tag{2.23}$$

The relative values of E_x and E_y for the six polarisation states in Figure 2.3 are as shown in Table 2.2, assuming that the peak amplitude of the wave is E_0 in all cases and where the

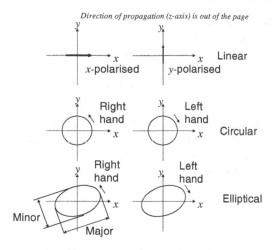

Figure 2.3 Possible polarisation states for a z-directed plane wave

complex constant a depends upon the axial ratio. The axial ratio is given in terms of E_x and E_y as follows [Siwiak, 1998]:

$$AR = \left[\frac{\left| 1 + \left| \frac{E_y}{E_x} \right| \cos[\arg(E_y) - \arg(E_x)] \right|^2}{\left| \frac{E_y}{E_x} \sin[\arg(E_y) - \arg(E_x)] \right|} \right]^{\pm 1} \tag{2.24}$$

The exponent in Eq. (2.24) is chosen such that $AR \geq 1$.

2.4.3 Random Polarisation

The polarisation states considered in the previous section involved the sum of two linearly polarised waves whose amplitudes were constant. These waves are said to be *completely polarised*, in that they have a definite polarisation state which is fixed for all time. In some cases, however, the values of E_x and E_y may vary with time in a random fashion. This could

Table 2.2: Relative electric field values for the polarisation states illustrated in Figure 2.3

Polarisation State	E_x	E_y
Linear x	$E_0/\sqrt{2}$	0
Linear y	0	$E_0/\sqrt{2}$
Right-hand circular	$-E_0/\sqrt{2}$	$jE_0/\sqrt{2}$
Left-hand circular	$E_0/\sqrt{2}$	$jE_0/\sqrt{2}$
Right-hand elliptical	$-aE_0/\sqrt{2}$	$jE_0/\sqrt{2}$
Left-hand elliptical	$aE_0/\sqrt{2}$	$jE_0/\sqrt{2}$

happen if the fields were created by modulating random noise onto a carrier wave of a given frequency. If the resultant fields are completely uncorrelated, then the wave is said to be *completely unpolarised*, and the following condition holds:

$$E[E_x E_y^*] = 0 \qquad (2.25)$$

where $E[.]$ indicates the time-averaged value of the quantity in brackets, or its *expectation*; see Appendix A for a definition of *correlation*. In the most general case, when E_x and E_y are partially correlated, the wave can be expressed as the sum of an unpolarised wave and a completely polarised wave. It is then said to be *partially polarised*.

2.5 CONCLUSION

Propagation of waves in uniform media can conveniently be described by considering the properties of plane waves, whose interactions with the medium are entirely specified by their frequency and polarisation and by the constitutive parameters of the medium. Not all waves are plane, but all waves can be described by a sum of plane waves with appropriate amplitude, phase, polarisation and Poynting vector. Later chapters will show how the characteristics of propagation and antennas in a wireless communication system can be described in terms of the behaviour of plane waves in random media.

REFERENCES

[Balanis, 89] C. A. Balanis, *Advanced engineering electromagnetics*, John Wiley & Sons, Inc., New York, ISBN 0-471-62194-3, 1989.

[Hayt, 01] W. H. Hayt Jr and J. Buck, *Engineering electromagnetics*, 6th edn, McGraw-Hill, New York, ISBN 0-07-230424-3, 2001.

[Kraus, 98] J. D. Kraus and K. Carver, *Electromagnetics*, McGraw-Hill, New York, ISBN 0-07-289969-7, 1998.

[Maxwell, 1865] J. Clerk Maxwell, *A dynamical theory of the electromagnetic field*, Scientific Papers, 1865, reprinted by Dover, New York, 1952.

[Siwiak, 98] K. Siwiak, *Radiowave propagation and antennas for personal communications*, 2nd edn, Artech House, Norwood MA, ISBN 0–89006-975-1, 1998.

PROBLEMS

2.1 Prove Eq. (2.14).

2.2 How far must a plane wave of frequency 60 GHz travel in order for the phase of the wave to be retarded by $180°$ in a lossless medium with $\mu_r = 1$ (a *non-magnetic* medium) and $\varepsilon_r = 3.5$?

2.3 What is the average power density carried in a plane wave with electric field amplitude 10 V m^{-1}? What electric and magnetic field strengths are produced when the same power density is carried by a plane wave in a lossless non-magnetic medium with $\varepsilon_r = 4.0$?

2.4 By what proportion is a 400 MHz plane wave reduced after travelling 1.5 m through a non-magnetic material with constants $\sigma = 1000$, $\varepsilon_r = 10$?

2.5 A plane wave travels through free space and has an average Poynting vector of magnitude $10\,\mathrm{W\,m^{-2}}$. What is the peak electric field strength?

2.6 Calculate the distance required for the electric field of a 5 GHz propagating plane wave to diminish to 13.5% (e^{-2}) given $\varepsilon_r = 3$, $\mu_r = 2$ and $\sigma = 100$?

2.7 Repeat Example 2.1(e) for the case when $\sigma = 10$. Compare your answer with the 2.11 m found in Example 2.1(e), and use this to explain why the surfaces of copper conductors in high-frequency circuits are often gold-plated.

2.8 Compare the attenuation of a plane wave travelling 1 m through a non-magnetic medium with $\sigma = 10^{-4}\,\mathrm{S\,m^{-1}}$ and $\varepsilon_r = 3$ at 100 MHz, 1 GHz and 10 GHz.

2.9 Describe, in your own words, the physical meaning of Maxwell's equations and why they are important for wireless communications.

2.10 A vertically polarised plane wave at 1900 MHz travels in the positive z direction in a medium with constitutive parameters $\mu_r = 1$, $\varepsilon_r = 3$ and $\sigma = 10\,\mathrm{S\,m^{-1}}$. The electric field magnitude at $z = 0$ is $1.5\,\mathrm{V\,m^{-1}}$. Calculate: (a) the wave impedance; (b) the magnitude of the magnetic field at $z = 0$; (c) the average power available in a $1.3\,\mathrm{m^2}$ area perpendicular to the direction of propagation at $z = 0$; (d) the time taken for the wave to travel through 15 cm; (e) the distance travelled by the wave before its field strength drops to one fifth of its value at $z = 0$.

3 Propagation Mechanisms

'Wireless is all very well but I'd rather send a message by a boy on a pony'!
Lord Kelvin

3.1 INTRODUCTION

In Chapter 2, the media in which waves were propagating consisted of regions within which the constitutive parameters did not vary in space, being infinite in extent in all directions. In practice, however, we must consider the boundaries between media (between air and the ground, between buildings and the air, from Earth to space, etc.). These boundary effects give rise to changes in the amplitude, phase and direction of propagating waves. Almost all these effects can be understood in terms of combinations of simple mechanisms operating on plane waves. These propagation mechanisms are now described and will later be used to analyse wave propagation in the real world.

3.2 REFLECTION, REFRACTION AND TRANSMISSION

3.2.1 Lossless Media

Figure 3.1 shows a plane wave incident onto a plane boundary between two media with different permeabilities and permittivities. Both media are assumed lossless for the moment. The electric field vector may be in any direction perpendicular to the propagation vector. The propagation vector is at an angle θ_i to the surface normal at the point of incidence.

If Maxwell's equations are solved for this situation, the result is that two new waves are produced, each with the same frequency as the incident wave. Both the waves have their Poynting vectors in the plane which contains both the incident propagation vector and the normal to the surface (i.e. normal to the plane of the paper in Figure 3.1). This is called the *scattering plane*. The first wave propagates within medium 1 but moves away from the boundary. It makes an angle θ_r to the normal and is called the *reflected* wave. The second wave travels into medium 2, making an angle θ_t to the surface normal. This is the *transmitted* wave, which results from the mechanism of *refraction*. When analysing reflection and refraction, it is convenient to work in terms of *rays*; in a homogeneous medium rays are drawn parallel to the Poynting vector of the wave at the point of incidence. They are always perpendicular to the wave fronts.

The angle of the reflected ray is related to the incidence angle as follows:

$$\theta_i = \theta_r \tag{3.1}$$

Equation (3.1) is *Snell's law of reflection*, which may be used to find the point of reflection given by any pair of source (transmitter) and field (receiver) points as shown in Figure 3.2.

Antennas and Propagation for Wireless Communication Systems Second Edition   Simon R. Saunders and Alejandro Aragón-Zavala
© 2007 John Wiley & Sons, Ltd

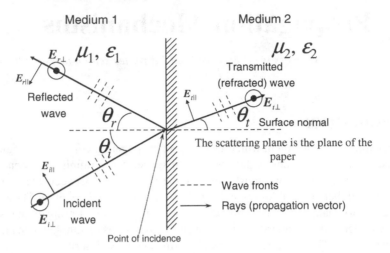

Figure 3.1: Plane wave incident onto a plane boundary

This law is one consequence of a deeper truth, *Fermat's principle*, which states that every ray path represents an extremum (maximum or minimum) of the total electrical length kd of the ray, usually a minimum. In Figure 3.2, the actual ray path is simply the path which minimises the distance $(d_1 + d_2)$, because the wave number is the same for the whole ray.

Fermat's principle can also be used to find the path of the refracted ray. In this case, the wave number in the two media is different, so the quantity which is minimised is $(k_1 d_1 + k_2 d_t)$, where d_t is the distance from the point of reflection to the field point in medium 2. The result is Eq. (3.2) *Snell's law of refraction*,

$$\frac{\sin \theta_i}{\sin \theta_t} = \frac{k_2}{k_1} \tag{3.2}$$

Equation (3.2) is consistent with the observation that the phase velocity of the wave in the medium with higher permittivity and permeability (the *denser* medium) is reduced, causing the

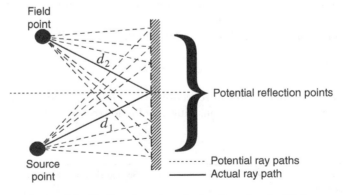

Figure 3.2: Finding the point of reflection using Snell's law

transmitted wave to bend towards the surface normal.[1] This change in velocity can be expressed in terms of the *refractive index*, n, which is the ratio of the free space phase velocity, c, to the phase velocity in the medium,

$$n = \frac{c}{v} = \frac{ck}{\omega}$$

(3.3)

Thus Snell's law of refraction can be expressed as

$$\frac{\sin \theta_i}{\sin \theta_t} = \frac{n_2}{n_1}$$

(3.4)

Note that the frequency of the wave is unchanged following reflection and transmission; instead, the ratio $v/\lambda = f$ is maintained everywhere. For example, a wave within a dense medium will have a smaller phase velocity and longer wavelength than that in free space.

In addition to the change of direction in accordance with Eqs. (3.2) and (3.4), the interaction between the wave and the boundary also causes the energy in the incident wave to be split between the reflected and transmitted waves. The amplitudes of the reflected and transmitted waves are given relative to the incident wave amplitude by the *Fresnel* [frā - nel] *reflection and transmission coefficients*, which arise from the solution of Maxwell's equations at the boundary. These express the ratio of the transmitted and reflected electric fields to the incident electric field. The coefficients are different for cases when the electric field is parallel and normal to the scattering plane which are denoted by subscripts $\parallel$ and $\perp$, respectively. The reflection coefficients are denoted by R and the transmission coefficients by T. The coefficients depend on the impedances of the media and on the angles,

$$R_{\parallel} = \frac{E_{r\parallel}}{E_{i\parallel}} = \frac{Z_1 \cos \theta_i - Z_2 \cos \theta_t}{Z_2 \cos \theta_t + Z_1 \cos \theta_i} \quad R_{\perp} = \frac{E_{r\perp}}{E_{i\perp}} = \frac{Z_2 \cos \theta_i - Z_1 \cos \theta_t}{Z_2 \cos \theta_i + Z_1 \cos \theta_t}$$

(3.5)

$$T_{\parallel} = \frac{E_{t\parallel}}{E_{i\parallel}} = \frac{2Z_2 \cos \theta_i}{Z_2 \cos \theta_t + Z_1 \cos \theta_i} \quad T_{\perp} = \frac{E_{t\perp}}{E_{i\perp}} = \frac{2Z_2 \cos \theta_i}{Z_2 \cos \theta_i + Z_1 \cos \theta_t}$$

(3.6)

where Z_1 and Z_2 are the wave impedances of medium 1 and medium 2, respectively and the E fields are defined in the directions shown in Figure 3.1.

The total reflected electric field is therefore given by

$$\mathbf{E}_r = E_{r\parallel}\mathbf{a}_{\parallel} + E_{r\perp}\mathbf{a}_{\perp} = E_{i\parallel}R_{\parallel}\mathbf{a}_{\parallel} + E_{i\perp}R_{\perp}\mathbf{a}_{\perp}$$

(3.7)

where $\mathbf{a}_{\parallel}$ and $\mathbf{a}_{\perp}$ are unit vectors parallel and normal to the scattering plane, respectively, and the incident electric field is permitted to take any polarisation state expressed as

$$\mathbf{E}_i = \mathbf{E}_{i\parallel}\mathbf{a}_{\parallel} + \mathbf{E}_{i\perp}\mathbf{a}_{\perp}$$

(3.8)

---

[1]Imagine a person walking across a beach into the sea at an angle, with the sea to their left. Their left leg gets wet first and slows down, causing their direction to shift more directly into the sea, closer to the normal to the sea edge. In this analogy the sea represents the denser medium and the beach is the less dense.

Equation (3.7) is sometimes represented in matrix form to make calculations involving mixed polarisation easier,

$$\mathbf{E}_r = \mathbf{R}\mathbf{E}_i \tag{3.9}$$

where

$$\mathbf{E}_r = \begin{bmatrix} E_{r\parallel} \\ E_{r\perp} \end{bmatrix}, \mathbf{R} = \begin{bmatrix} R_{\parallel} & 0 \\ 0 & R_{\perp} \end{bmatrix} \quad \text{and} \quad \mathbf{E}_i = \begin{bmatrix} E_{i\parallel} \\ E_{i\perp} \end{bmatrix} \tag{3.10}$$

Similarly, the total transmitted field is given by

$$\mathbf{E}_t = \mathbf{T}\mathbf{E}_i \tag{3.11}$$

where

$$\mathbf{E}_t = \begin{bmatrix} E_{t\parallel} \\ E_{t\perp} \end{bmatrix}, \mathbf{T} = \begin{bmatrix} T_{\parallel} & 0 \\ 0 & T_{\perp} \end{bmatrix} \quad \text{and} \quad \mathbf{E}_i = \begin{bmatrix} E_{i\parallel} \\ E_{i\perp} \end{bmatrix} \tag{3.12}$$

It is often useful to express the Fresnel coefficients in terms of θ_i only, avoiding the need to calculate θ_t: for dielectric materials with $\sigma_1 = \sigma_2 = 0$ and $\mu_1 = \mu_2$, Snell's law of refraction implies that

$$R_{\parallel} = \frac{Z_1 \cos\theta_i - Z_2\sqrt{1 - \left(\dfrac{Z_2}{Z_1}\right)^2 \sin^2\theta_i}}{Z_2\sqrt{1 - \left(\dfrac{Z_2}{Z_1}\right)^2 \sin^2\theta_i} + Z_1 \cos\theta_i} \tag{3.13}$$

$$R_{\perp} = \frac{Z_2 \cos\theta_i - Z_1\sqrt{1 - \left(\dfrac{Z_2}{Z_1}\right)^2 \sin^2\theta_i}}{Z_2 \cos\theta_i + Z_1\sqrt{1 - \left(\dfrac{Z_2}{Z_1}\right)^2 \sin^2\theta_i}} \tag{3.14}$$

$$T_{\parallel} = \frac{2Z_2 \cos\theta_i}{Z_2\sqrt{1 - \left(\dfrac{Z_2}{Z_1}\right)^2 \sin^2\theta_i} + Z_1 \cos\theta_i} \tag{3.15}$$

$$T_{\perp} = \frac{2Z_2 \cos\theta_i}{Z_2 \cos\theta_i + Z_1\sqrt{1 - \left(\dfrac{Z_2}{Z_1}\right)^2 \sin^2\theta_i}} \tag{3.16}$$

Other useful relationships can be deduced by considering the conservation of energy at the point of incidence: since the power in the incident wave is divided between the reflected and transmitted waves, the following relationships hold:

$$|T_{\parallel}|^2 = 1 - |R_{\parallel}|^2 \text{ and } |T_{\perp}|^2 = 1 - |R_{\perp}|^2 \tag{3.17}$$

3.2.2 Lossy Media

In lossy media, Snell's law of refraction no longer holds in its standard form (3.4), because the phase velocity of the transmitted wave (and the attenuation constant) depends on the incidence angle as well as on the constitutive parameters. If a wave is incident from a dielectric onto a conductor, increasing the conductivity causes the refraction angle θ_t to decrease towards zero whereas the attenuation constant increases, so the penetration of the wave into the conductor decreases. See [Balanis, 89] for more details.

Snell's law of reflection still holds in lossy media, however and the Fresnel coefficients can still be applied as in Eqs. (3.5) and (3.6) using the correct values of the wave impedance from Table 2.1, i.e.

$$Z = \sqrt{\frac{j\omega\mu}{\sigma + j\omega\varepsilon}} \tag{3.18}$$

Example 3.1

A linearly polarised plane wave in free space with an electric field amplitude 30 V m^{-1} is incident onto the plane boundary of a lossless non-magnetic medium with $\varepsilon_r = 2$ from an incidence angle of $30°$. The electric field is parallel to the plane of incidence. Calculate:

(a) the angle of reflection;
(b) the angle of refraction;
(c) the amplitude of the transmitted electric and magnetic fields;
(d) the amplitude of the reflected electric field.

Solution

(a) Since both the wave and the boundary are plane, Snell's law of reflection (3.1) applies, and the angle of reflection is therefore $30°$.
(b) Rearranging Eq. (3.4) yields

$$\theta_t = \sin^{-1}\left(\frac{n_1}{n_2}\sin\theta_i\right)$$

$$= \sin^{-1}\left(\frac{\sqrt{\mu_{r1}\varepsilon_{r1}}}{\sqrt{\mu_{r2}\varepsilon_{r2}}}\sin\theta_i\right)$$

$$= \sin^{-1}\left(\frac{1}{\sqrt{2}}\sin 30°\right)$$

$$= 20.7°$$

(c)   The incident wave is parallel polarised with respect to the plane of polarisation, so the expression for $T_{\parallel}$ in (3.6) is applied, with

$$Z_1 = Z_0 = 377\Omega \quad \text{and} \quad Z_2 = \frac{Z_0}{\sqrt{\varepsilon_{r2}}} = \frac{377}{\sqrt{2}}\Omega$$

Hence $\dfrac{Z_1}{Z_2} = \sqrt{2}$, and (3.6) yields

$$T_{\parallel} = \frac{2\cos\theta_i}{\cos\theta_t + \frac{Z_1}{Z_2}\cos\theta_i} = \frac{2\cos 30°}{\cos 20.7° + \sqrt{2}\cos 30°} \approx 0.80$$

So $E_t = E_i T_{\parallel} = 0.80 \times 30 = 24.0\,\text{V m}^{-1}$
From Eq. (2.11) the magnetic field amplitude is

$$H_t = \frac{E_t}{Z_2} = \frac{24\sqrt{2}}{377} \approx 0.09\,\text{A m}^{-1}$$

(d)   The amplitude of the electric field is

$$E_r = R_{\parallel} E_i = \frac{\frac{Z_1}{Z_2}\cos\theta_i - \cos\theta_t}{\cos\theta_t + \frac{Z_1}{Z_2}\cos\theta_i} E_i \approx -8.54 \times 30$$

$$= -256.2\,\text{V m}^{-1}$$

The negative sign indicates that the field undergoes a 180° phase change on reflection.

3.2.3 Typical Reflection and Transmission Coefficients

Representative values for the constitutive parameters of various non-magnetic materials are given in Table 3.1.

The magnitudes of the Fresnel coefficients for two of these materials are shown in Figures 3.3 and 3.4 for a wave propagating in free space and impinging onto dry and wet ground, respectively. In the case of reflection from the ground, the ∥ parts represent vertical

Table 3.1: Typical constitutive parameters, from [ITU, 527]

Surface	Conductivity, σ [S m^{-1}]	Relative dielectric constant, ε_r
Dry ground	0.001	4–7
Average ground	0.005	15
Wet ground	0.02	25–30
Sea water	5	81
Fresh water	0.01	81

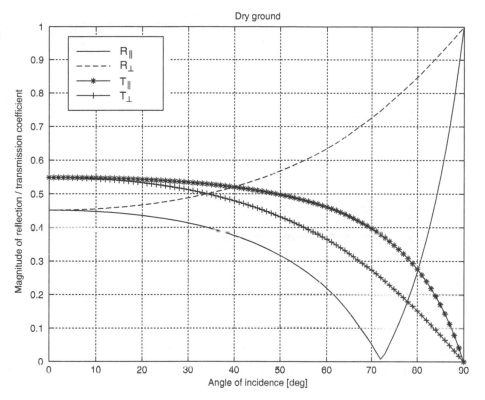

Figure 3.3: Reflection and transmission coefficients for dry ground at 100 MHz with ε_r 7, $\sigma = 0.0013 \, \text{m}^{-1}$

polarisation and the $\perp$ parts are horizontally polarised. Notice that in both cases the vertically polarised reflection goes to zero at one angle. Waves of random polarisation incident at this angle will be reflected to have a purely horizontal component, illustrating that the polarisation state is not preserved following reflection or transmission. The angle is called the *Brewster angle*, $\theta_B = \theta_i$, and is given by

$$\theta_B = \tan^{-1}\frac{n_2}{n_1} \tag{3.19}$$

Figures 3.3 and 3.4 also show that, as the angle of incidence goes closer to 90°, the reflection coefficient approaches -1 in all cases, independent of polarization, while the transmission coefficient drops to zero. This situation is known as *grazing incidence* and is closely approximated in practice when a transmitter antenna is low in height compared with the distance to the receiver. This is most apparent in Figure 3.5, where Figure 3.3 has been redrawn assuming that the conductivity of the ground is zero, so the coefficients always take real values.

For angles of incidence greater than the Brewster angle, both $R_\parallel$ and $R_\perp$ are negative, so the reflected wave undergoes a 180° phase change. If the incident wave has components with both

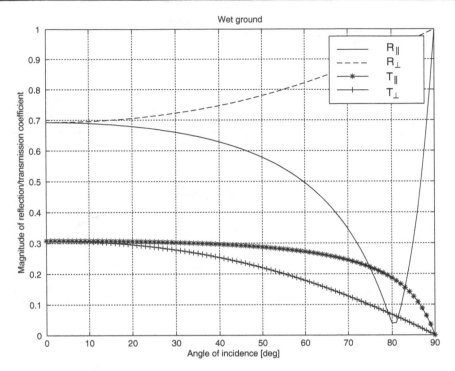

Figure 3.4: Reflection and transmission coefficients for wet ground at 100 MHz with $\varepsilon_r = 30$, $\sigma = 0.02\,\text{S m}^{-1}$

polarisations, as in circular or elliptical polarisation, this phase change causes a reversal of the sense of rotation. Reversal occurs together with a differential change in the amplitudes of the two component linear polarisations, which also changes the axial ratio. These changes are summarised in Table 3.2. The impact of circular polarisation can also be computed via co-polar and cross-polar reflection coefficients as follows:

$$R_{co} = \frac{1}{2}(R_{\|} + R_{\perp}) \quad \text{and} \quad R_{cx} = \frac{1}{2}(R_{\|} - R_{\perp}) \tag{3.20}$$

Table 3.2: Change of polarisation state on reflection

Incident polarisation	Reflected polarisation	
	$\theta_i < \theta_B$	$\theta_i > \theta_B$
Right-hand circular	Left-hand elliptical	Right-hand elliptical
Left-hand circular	Right-hand elliptical	Left-hand elliptical
Right-hand elliptical	Left-hand elliptical (*AR* changed)	Right-hand elliptical (*AR* changed)
Left-hand elliptical	Right-hand elliptical (*AR* changed)	Left-hand elliptical (*AR* changed)

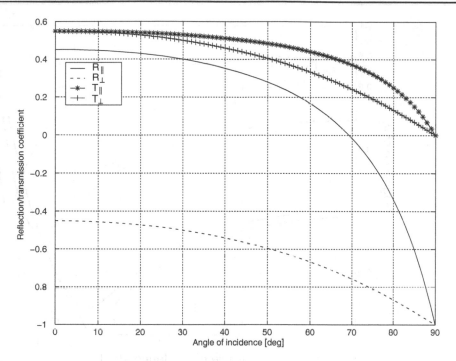

Figure 3.5. Reflection and transmission coefficients for dry ground of zero conductivity

When the conductivity of the medium is large, the $R_{\parallel}$ component never quite goes to zero but there is still a clearly defined minimum known as the pseudo-Brewster angle (Figure 3.6).

3.3 ROUGH SURFACE SCATTERING

The reflection processes discussed so far have been applicable to smooth surfaces only; this is termed *specular reflection*. When the surface is made progressively rougher, the reflected wave becomes *scattered* from a large number of positions on the surface, broadening the scattered energy (Figure 3.7). This reduces the energy in the specular direction and increases the energy radiated in other directions. The degree of scattering depends on the angle of incidence and on the roughness of the surface in comparison to the wavelength. The apparent roughness of the surface is reduced as the incidence angle comes closer to grazing incidence $(\theta_i = 90°)$ and as the wavelength is made larger[2].

If a surface is to be considered smooth, then waves reflected from the surface must be only very slightly shifted in phase with respect to each other. If there is a height difference Δh between two points on the surface, then waves reflected from those points will have a relative

---

[2]Try looking along the surface of a sheet of blank white paper towards a light: you will observe a bright reflection from the paper. This reflection is not apparent at normal viewing angles because the surface roughness of the paper is large compared with the wavelength of light.

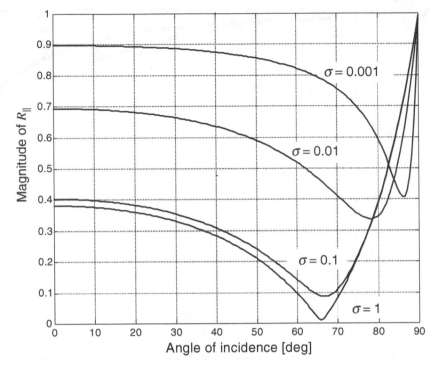

Figure 3.6: Pseudo-Brewster angle: $R_{\parallel}$ never goes to zero $(\varepsilon_{r1} = 1, \varepsilon_{r2} = 5)$

phase difference of

$$\Delta\phi = \frac{4\pi\Delta h \cos\theta_i}{\lambda} \qquad (3.21)$$

A reasonable criterion for considering a surface smooth is if this phase shift is less than 90°, which leads to the *Rayleigh criterion*,

$$\Delta h < \frac{\lambda}{8\cos\theta_i} \qquad (3.22)$$

This is illustrated in Figure 3.8. For accurate work, it is suggested that surfaces should only be considered smooth if the roughness is less than one-quarter of the value indicated by the

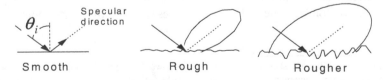

Figure 3.7: The effect of surface roughness on reflection

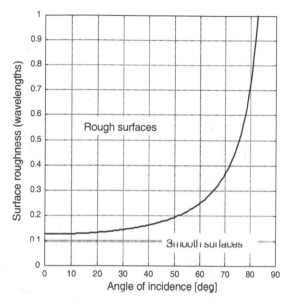

Figure 3.8: Rayleigh criterion for surface roughness: surfaces above the curve cannot be accurately modelled using the Fresnel reflection coefficients alone

Rayleigh criterion (i.e. phase difference less than $\pi/8$). Note that surfaces of any roughness may be considered smooth for $\theta_i = 90°$, since all the reflected rays arrive with the same phase shift. When the surface is rough, the reduction in the amplitude of the specular component may be accounted for by multiplying the corresponding value of R by a roughness factor f, which depends on the angle of incidence and on the standard deviation of the surface height σ_s. One formulation for this factor is [Beckmann, 63],

$$f(\sigma_s) = \exp\left[-\frac{1}{2}\left(\frac{4\pi\sigma_s\cos\theta}{\lambda}\right)^2\right] \qquad (3.23)$$

This is plotted in Figure 3.9. The effective reflection coefficient is then $R_{rough} = Rf(\sigma_s)$

3.4 GEOMETRICAL OPTICS

3.4.1 Principles

This chapter has described ways in which the interactions between plane waves and infinite plane surfaces can be calculated with high accuracy, provided that the constitutive parameters are known. In practice, however, these conditions are rarely fulfilled and approximations must be made. For example, Figure 3.10 represents a real propagation situation in which it is desired to calculate the field at a field point (i.e. receiver) within a building, illuminated from a source point (transmitter) on top of another building. The transmitter radiates *spherical waves*, in which the wavefronts are spheres centred on the source point and radiating outwards in all directions. The surfaces are finite in extent and

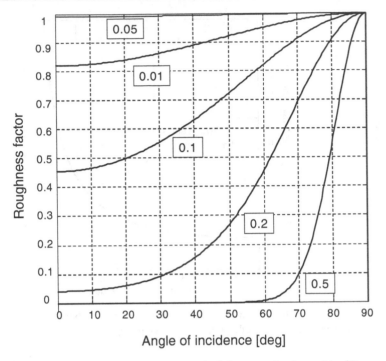

Figure 3.9: Roughness factor $f(\sigma_s)$ for several values of (σ_s/λ)

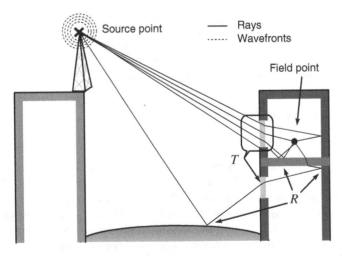

Figure 3.10: Combination of propagation mechanisms in the real environment: different shading indicates materials with potentially different constitutive parameters

involve boundaries between media with differing constitutive parameters. Finally, the surface of the ground is curved.

Despite these complications, *geometrical optics* represents a simple way to find an *approximate* value for the field at the receiver using the plane wave propagation mechanisms described in this chapter. The following steps are involved:

1. Calculate all possible ray paths between the source and field points which are consistent with Snell's laws of reflection and refraction (*ray tracing*).
2. Calculate the Fresnel reflection and transmission coefficients at each of the points of reflection and transmission *as if* the incident waves were plane and the boundaries were plane and infinite (see for example, the reflection points labelled R and the transmission points labelled T in Figure 3.10).
3. For each ray path, correct the amplitude to account for wavefront curvature from the source and due to the curvature of any boundaries.
4. Sum all ray paths, paying due regard to both amplitude and phase.

Step 1 is, in principle, a straightforward procedure, but requires a large amount of computational effort in order to identify all possible rays. Much effort has been devoted to creating efficient techniques for performing ray tracing and they typically rely on tracing only the strongest rays and on techniques taken from the study of computer graphics [Cátedra, 99]. The strongest rays are usually those involving the fewest interactions on their journey from source to field points.

Step 2 relies on the waves being *locally plane* at the points of interaction and on the surfaces with given constitutive parameters being reasonably large. This is the central assumption of geometrical optics and requires that the wavelength is short in comparison with all of the following lengths:

- distance between the source and the first interactions along each ray path;
- distance between individual interactions;
- dimensions of any of the individual materials.

Step 3 similarly requires that the curvature of any of the boundaries is not too large compared with the wavelength.

Despite these limitations, geometrical optics is useful for solving a large number of high-frequency problems in both antennas and propagation with reasonable accuracy.

3.4.2 Formulation

Following the procedure outlined in Section 3.4.1, the total geometrical optics field is given by

$$\mathbf{E} = \mathbf{E}_0 A_0 e^{-jk_0 r_0} + \sum_{i=1}^{N_r} \mathbf{R} \mathbf{E}_i A_i e^{-jk_i r_i} + \sum_{j=1}^{N_t} \mathbf{T} \mathbf{E}_j A_j e^{-jk_j r_j} \qquad (3.24)$$

where

N_r, N_t are the total numbers of reflected and transmitted rays which have been traced from the source point to the field point, respectively.

r_n	is the distance along the nth ray.
k_i	is the wave number associated with the medium in which the ith ray propagates.
A_i	is the *spreading factor* for the ith ray.
$\mathbf{E}_{i,j}$	is the incident field immediately adjacent to the corresponding transmission or reflection point.

The spreading factor depends on the distances to the source and field points and on the curvature of the reflection or transmission boundary. For spherical waves and plane boundaries, $A_i \propto 1/r_i$, whereas for plane waves and plane boundaries, A_i does not vary with distance. Parameters with a subscript 0 account for the direct ray from source to field point, if unobstructed. Equation (3.24) only shows contributions from rays which have been reflected or transmitted once, but it can easily be extended to multiple interactions along a ray by simply multiplying the appropriate coefficients together. See [Balanis, 89] and [James 80] for more details.

3.5 DIFFRACTION

3.5.1 Principle

The geometrical optics field described in Section 3.4 is a very useful description, accurate for many problems where the path from transmitter to receiver is not blocked. However, such a description leads to entirely incorrect predictions when considering fields in the *shadow region* behind an obstruction, since it predicts that no field whatsoever exists in the shadow region as shown in Figure 3.11. This suggests that there is an infinitely sharp transition from the shadow region to the illuminated region outside. In practice, however, shadows are never completely sharp, and some energy does propagate into the shadow region. This effect is *diffraction* and can most easily be understood by using *Huygen's principle.*

1. Each element of a wavefront at a point in time may be regarded as the centre of a secondary disturbance, which gives rise to spherical wavelets.
2. The position of the wavefront at any later time is the envelope of all such wavelets.

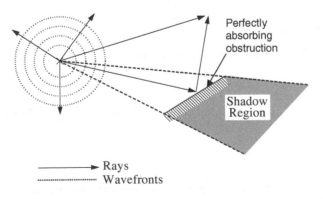

Figure 3.11: Geometrical optics incorrectly predicts zero field in the shadow region

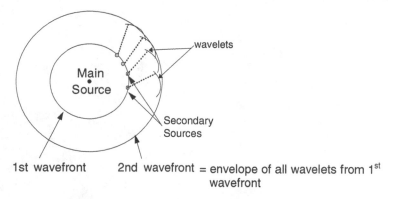

Figure 3.12: Huygen's principle for a spherical source

Figure 3.12 shows how a number of secondary sources on a spherical wavefront give rise to wavelets whose envelope is another spherical wavefront of larger radius.

3.5.2 Single Knife-Edge Diffraction

We now use Huygen's principle to predict diffraction of a plane wave over an absorbing plane or *knife-edge*. This is a wide screen which allows no energy to pass through it. Figure 3.13 shows how plane wavefronts impinging on the edge from the left become curved by the edge so that, deep inside the geometrical shadow region, rays appear to emerge from a point close to the edge, filling in the shadow region with diffracted rays. The same effect can be viewed in nature following diffraction of water waves around a coastline (Figure 3.14).

Huygen's principle may be applied in a mathematical form to predict the actual field strength which is diffracted by the knife-edge. The contributions from an infinite number of secondary sources in the region above the edge are summed, paying due regard to their relative amplitudes and phases.

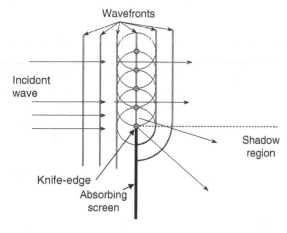

Figure 3.13: Huygen's principle for knife-edge diffraction

Figure 3.14: Knife-edge diffraction of waves on the sea surface

The final result can be expressed as a *propagation loss*, which expresses the reduction in the field strength due to the knife-edge diffraction process in decibels, in terms of a *diffraction parameter, v,*

$$L_{ke}(v) = -20 \log \left(\left| \frac{E_d}{E_i} \right| \right) = -20 \log |F(v)| \tag{3.25}$$

where E_d is the diffracted field, E_i is the incident field and

$$F(v) = \frac{1+j}{2} \int_v^\infty \exp \left(-\frac{j\pi t^2}{2} \right) dt \tag{3.26}$$

An alternative form is

$$|F(v)| = \frac{1}{2} \left(\frac{1}{2} + C^2(v) - C(v) + S^2(v) - S(v) \right) \tag{3.27}$$

where $C(v)$ and $S(v)$ are the Fresnel cosine and sine integrals, which can be evaluated using the methods given in Appendix B.

Notice how the integration limits in Eq. (3.26) indicate the nature of the summation of secondary sources from the top of the knife-edge, with parameter v, up to infinity. The eventual result L_{ke} is illustrated in Figure 3.15. It can be numerically evaluated using standard routines for calculating Fresnel integrals (Appendix B), or approximated for $v > 1$ (i.e. well within the shadow region) with accuracy better than 1 dB,

$$L_{ke}(v) \approx -20 \log \frac{1}{\pi v \sqrt{2}} \approx -20 \log \frac{0.225}{v} \tag{3.28}$$

Another significant value is $L_{ke}(0) = 6$ dB, i.e. the received power is reduced by a factor of 4 when the knife-edge is situated exactly on the direct path between the transmitter and the

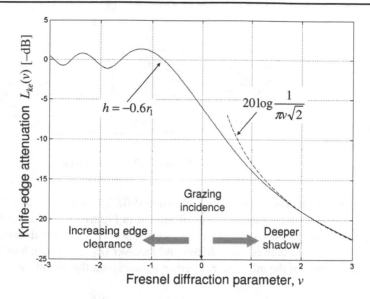

Figure 3.15: Knife-edge diffraction attenuation: (——) exact (– –) large v approximation

receiver. The parameter v can be expressed in terms of the geometrical parameters defined in Figure 3.16 as

$$v = h' \sqrt{\frac{2(d_1' + d_2')}{\lambda d_1' d_2'}} = \alpha \sqrt{\frac{2 d_1' d_2'}{\lambda (d_1' + d_2')}} \qquad (3.29)$$

where h' is the *excess height* of the edge above the straight line from source to field points. For many practical cases, $d_1, d_2 \gg h$, so the diffraction parameter v can be approximated in terms of the distances measured along the ground rather than along the direct wave,

$$v \approx h \sqrt{\frac{2(d_1 + d_2)}{\lambda d_1 d_2}} = \alpha \sqrt{\frac{2 d_1 d_2}{\lambda (d_1 + d_2)}} \qquad (3.30)$$

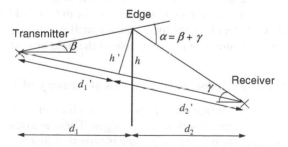

Figure 3.16: Knife-edge diffraction parameters

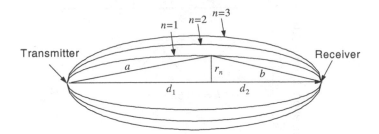

Figure 3.17: Fresnel zones

Another useful way to consider knife-edge diffraction is in terms of the obstruction of *Fresnel zones* around the direct ray as illustrated in Figure 3.17. The nth Fresnel zone is the region inside an ellipsoid defined by the locus of points where the distance $(a + b)$ is larger than the direct path between transmitter and receiver $(d_1 + d_2)$ by n half-wavelengths. Hence the radius of the nth zone r_n is given by applying the condition

$$a + b = d_1 + d_2 + \frac{n\lambda}{2} \tag{3.31}$$

If we assume that $r_n \ll d_1$ and $r_n \ll d_2$, then to a good approximation

$$r_n \approx \sqrt{\frac{n\lambda d_1 d_2}{d_1 + d_2}} \tag{3.32}$$

The Fresnel zones can be thought of as containing the main propagating energy in the wave. Contributions within the first zone are all in phase, so any absorbing obstructions which do not enter this zone will have little effect on the received signal. The *Fresnel zone clearance* (h/r_n) can be expressed in terms of the diffraction parameter v as follows:

$$v \approx h\sqrt{\frac{2(d_1 + d_2)}{\lambda d_1 d_2}} = \frac{h}{r_n}\sqrt{2n} \tag{3.33}$$

When the obstruction occupies 0.6 times the first Fresnel zone, the v parameter is then approximately -0.8. Referring to Figure 3.15, the obstruction loss is then 0 dB. This clearance is often used as a criterion to decide whether an object is to be treated as a significant obstruction. Thus, the shaded region in Figure 3.18 can be considered as a 'forbidden' region; if this region is kept clear then the total path attenuation will be practically the same as in the unobstructed case. Figure 3.19 is an example of the field around an absorbing knife-edge.

3.5.3 Other Diffracting Obstacles: Geometrical Theory of Diffraction

In many situations, diffraction over obstructions, such as hills and buildings, may be treated as if those obstructions were absorbing knife-edges. There are other cases, however, when it is necessary to account for the structure of the obstruction in a more specific way, including its shape and the materials of which it is built. One approach is to extend geometrical optics to

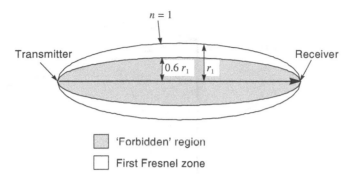

Figure 3.18: 0.6 times first Fresnel zone clearance defines significant obstructions

include diffraction, producing a formulation similar to Section 3.4. This results in the *geometrical theory of diffraction* (GTD). GTD was first devised by Joseph Keller in the 1950s [Keller, 62]. In the past, GTD has been applied mainly to the analysis of small shapes, such as antennas, or for calculating the radar cross-sections of complex objects. More recently, it has been successfully applied to the modelling of terrain [Luebbers, 84] and of buildings.

The central idea of GTD is that an extended version of Fermat's principle may be used to predict the existence of diffracted rays, which may then be treated with the ease of any

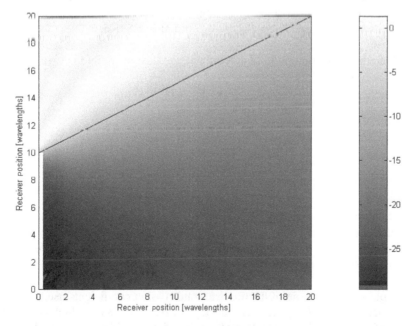

Figure 3.19: Single knife-edge attenuation: the transmitter is at $(-10,5)$, the top of the edge is at $(0,10)$, the colour indicates the attenuation in decibels and the black line represents the shadow boundary

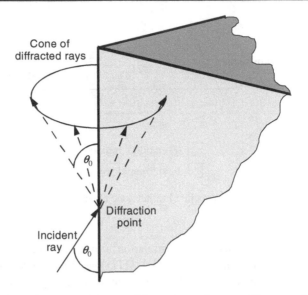

Figure 3.20: Generation of edge-diffracted rays from a wedge according to GTD

other ray in geometrical optics. Figure 3.20 shows a ray obliquely incident upon the edge of an obstacle at an angle θ_0 to the edge. Fermat's principle for edge diffraction predicts that a cone of diffracted rays will be produced, where the cone has semi-angle θ_0. This contrasts with reflected and transmitted rays, where only a single ray is produced at each interaction. In the simpler case of normal incidence, $\theta_0 = \pi/2$ and the cone reduces to a disc.

Once the diffracting point is determined, the diffracted field is given by

$$\mathbf{E}_d = \mathbf{D}\mathbf{E}_i A_d \tag{3.34}$$

where

$$\mathbf{E}_d = \begin{bmatrix} E_{d\parallel} \\ E_{d\perp} \end{bmatrix}, \mathbf{D} = \begin{bmatrix} D_\parallel & 0 \\ 0 & D_\perp \end{bmatrix} \quad \text{and} \quad \mathbf{E}_i = \begin{bmatrix} E_{i\parallel} \\ E_{i\perp} \end{bmatrix} \tag{3.35}$$

Here $E_{i\parallel}$ and $E_{i\perp}$ are measured parallel and perpendicular to the *plane of incidence*, which contains the incident ray and the diffracting edge, while $E_{d\parallel}$ and $E_{d\perp}$ are measured parallel and perpendicular to the *plane of diffraction*, which contains the diffracting edge and the diffracted ray. By analogy with Eq. (3.24), A_d is a spreading factor which depends on the distances to the source and field points and on the curvature of the diffracting edge.

In place of the Fresnel reflection and transmission coefficients used earlier, we now have coefficients $D_\parallel$ and $D_\perp$. These are the *diffraction coefficients*, which describe the characteristics of the diffracting obstacle at the point of diffraction, assuming the frequency is high enough that the diffraction characteristics can be determined with regard to local characteristics only, rather than by those of the entire obstacle. Terms such as Eq. (3.34) can now be added to the geometrical optics field of Eq. (3.24), yielding a field which has the proper behaviour within the shadow region. GTD also provides a simple explanation for the fluctuations in the single knife-edge diffraction (Figure 3.15) for negative v: it arises from constructive and destructive interferences between the direct ray and the diffracted ray.

The diffraction coefficients are determined from one of a number of *canonical problems*. These are diffraction problems for simple scattering objects, such as a half-plane [Volakis, 86], a wedge or a cone, which have been solved using exact methods for solving Maxwell's equations for plane waves incident on these objects. The resulting solutions are reduced, via asymptotic assumptions, to terms which correspond to a ray description of the field. As with geometrical optics, these assumptions are only valid if the obstacle dimensions are large compared to a wavelength and if the spatial variation of the scattered field is not too rapid. Keller's original formulation of GTD, though simple, had the disadvantage that it did not predict the field correctly for field points in a Fresnel-zone-like region close to the shadow boundary – the *transition region*. GTD was therefore extended to the uniform GTD (UTD) which applies at all points in space [Kouyoumjian, 74].

Consider the wedge diffraction situation as shown in Figure 3.21. The case shown is perpendicular polarisation ($\perp$). For parallel polarisation (||), replace E_i and E_d in the figure with H_i and H_d, respectively, and H_i and H_d in the figure with $-E_i$ and $-E_d$, respectively. Assuming the wedge to be a perfect conductor, the UTD diffraction coefficient for the case when the incident wave is normal to the diffracting edge is then as follows [Kouyoumjian, 74]. Note that the UTD diffraction coefficient is polarisation dependent, in contrast to the simpler theory presented in Section 3.5.2.

$$D_{\perp,||} = \frac{1}{n\sqrt{8\pi jk}} \left[\begin{array}{c} \left\{ \cot\left(\frac{\pi + \Phi^i}{2n}\right) F[kra^+(\Phi^i)] + \cot\left(\frac{\pi - \Phi^i}{2n}\right) F[kra^-(\Phi^i)] \right\} \mp \\ \left\{ \cot\left(\frac{\pi + \Phi^r}{2n}\right) F[kra^+(\Phi^r)] + \cot\left(\frac{\pi - \Phi^r}{2n}\right) F[kra^-(\Phi^r)] \right\} \end{array} \right] \quad (3.36)$$

where

$$\Phi^{i,r} = \phi \pm \phi_0; \quad 0 \leq \phi, \phi_0 \leq 2\pi; \quad n = \frac{2\pi - \beta}{\pi} \quad (3.37)$$

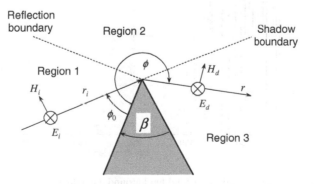

Figure 3.21: Geometry for wedge diffraction coefficient

The function $F(x)$ is an alternative definition of the Fresnel integral,

$$F(x) = 2j\sqrt{x}e^{jx} \int_{t=\sqrt{x}}^{\infty} e^{jt^2} dt \tag{3.38}$$

where $a^{\pm}$ are the integers which most nearly satisfy

$$2\pi na^{\pm} - \Phi = \pm\pi \tag{3.39}$$

The spreading factor A_d in Eq. (3.34) is given by

$$A_d = \begin{cases} 1/\sqrt{r} & \text{for plane cylindrical wave incidence} \\ \sqrt{r_i/r(r+r_i)} & \text{for spherical wave incidence} \end{cases} \tag{3.40}$$

Keller's GTD diffraction coefficient takes exactly the same form, but with the Fresnel integrals replaced by 1, since this is a reasonable approximation outside of the transition regions. The total field at any point in space is thus given by

$$\mathbf{E}_t = \mathbf{E}_i(A_0 + U_r\mathbf{R}A_r + \mathbf{D}A_d) \tag{3.41}$$

where A_0, A_r and A_d are appropriate spreading factors and $U_r = 1$ when a reflection exists and $U_r = 0$ otherwise. The total field around a $0°$ wedge (i.e. a conducting half-plane) is illustrated in Figure 3.22. The influence of the three regions marked in Figure 3.21 is clearly

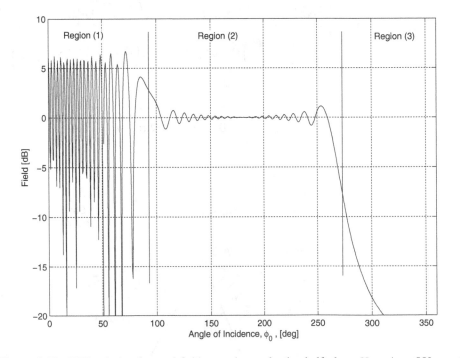

Figure 3.22: UTD solution for total field around a conducting half-plane. Here $\phi_0 = 90°, r = 10\lambda$ and the polarisation is parallel

visible. In region 1, the *visible region*, the field is the sum of the direct ray and the diffracted ray, plus a ray reflected from the surface. Since the wedge is perfectly conducting, the reflected wave has the same amplitude as the incident ray and complete cancellation of the two fields occurs at intervals, since the diffracted ray has negligibly small amplitude by comparison. In region 2, there is no reflection point on the wedge which can obey Snell's law, so no reflection exists, and the diffraction is still very small, so the field is nearly at its free space value. In region 3, the *shadow region*, only the diffracted ray is present, and it diminishes in amplitude in a similar way to the simple knife-edge approximation (compared with Figure 3.15).

UTD will be applied to practical propagation problems in Chapters 12 and 13.

3.6 CONCLUSION

All of the major propagation mechanisms have been introduced in this chapter, namely, reflection, refraction or transmission, scattering and diffraction. These complement the basic effects of attenuation and phase velocity which were introduced in Chapter 2 and provide a basis for practical prediction of wireless communication system performance in the rest of this book. Combinations of these mechanisms will be seen later to account for all of the observed effects in the wireless channel.

REFERENCES

[Balanis, 89] C. A. Balanis, *Advanced engineering electromagnetics*, John Wiley & Sons, Inc., New York, ISBN 0-471-62194-3, 1989.

[Beckmann, 63] P. Beckmann and A. Spizzichino, *The scattering of electromagnetic waves from rough surfaces*, Macmillan, New York, ISBN 0-63010-108, 1963.

[Cátedra, 99] M. F. Cátedra and J. Pérez-Arriaga, *Cell planning for wireless communications*, Artech House, ISBN 0-89006-601-9, 1999.

[ITU, 527] International Telecommunication Union, *ITU-R Recommendation 527-3: Electrical characteristics of the surface of the earth*, Geneva, 1992.

[James, 80] G. L. James, *geometrical theory of diffraction for electromagnetic waves*, Peter Peregrinus, London, ISBN 0-906048-34-6, 1980.

[Keller, 62] J. B. Keller, Geometrical theory of diffraction, *Journal of the Optical Society of America*, 52 (2), 116–130, 1962.

[Kouyoumjian, 74] R. G. Kouyoumjian and P. H. Pathak, A uniform geometrical theory of diffraction for an edge in a perfectly conducting surface, *Proceedings of IEEE*, 62, 1448–1461, 1974.

[Luebbers, 84] R. J. Luebbers, Finite conductivity uniform GTD versus knife edge diffraction in prediction of propagation path loss, *IEEE Transactions on Antennas and Propagation*, 32 (1), 70–76, 1984.

[Volakis, 86] J. L. Volakis, A uniform geometrical theory of diffraction for an imperfectly conducting half-plane, *IEEE Transactions on Antennas and Propagation*, 34 (2), 172–180, 1986.

PROBLEMS

3.1 Prove Eqs. (3.13) and (3.15).

3.2 Repeat Example 3.1 for the case of perpendicular polarisation and for the case where the polarisation is at 45° to the scattering plane.

3.3 Prove the expression for the Brewster angle (3.19).

3.4 When the angle of incidence of a plane wave onto a plane boundary between media is equal to the Brewster angle, all of the energy in the incident wave becomes transmitted into the second medium. At what incidence angle (the *critical* angle) does all of the energy in the incident wave become reflected?

3.5 Prove expressions (3.21) and (3.22) for the Rayleigh criterion.

3.6 Polaroid sunglasses reduce glare from road surfaces by permitting only one polarisation to be transmitted. Using the Fresnel reflection coefficients, explain whether vertical or horizontal polarisation should be transmitted.

3.7 A plane wave at 900 MHz is incident from free space onto a material with relative dielectric constant of 4. What are the phase velocity and wavelength of the refracted wave?

3.8 A transmitter and a receiver separated by 10 km operate at 400 MHz and are at the same height above the Earth. Relative to the transmitter, how much lower must an absorbing diffracting obstacle situated at the centre of the path be for negligible diffraction loss? Calculate the diffraction loss produced when the obstacle is increased to 10 m above the transmitter height.

3.9 Use Keller's GTD to repeat Problem 3.7, for both horizontal and vertical polarisations, assuming a conducting knife-edge in place of the absorbing knife-edge.

3.10 A microwave link is to be deployed in an urban area at 17 GHz. The transmitter antenna is to be located on a rooftop at 15 m. The receiver antenna is to be installed at 5 m above ground level. Determine the maximum height of a building at the centre of the path if transmitter and receiver antennas are separated by 5 km.

3.11 Determine the surface height variation required to consider a surface rough if a 900 MHz wave is incident on the surface at 30° to the normal.

3.12 Repeat problem 3.10 for the case when the radiowave is at a frequency of 60 GHz. Explain your answer.

4 Antenna Fundamentals

'I do not think that the wireless waves I have discovered will have any practical application'.
Heinrich R. Hertz

4.1 INTRODUCTION

Chapters 2 and 3 described how waves interact with materials in various ways. This chapter introduces the antennas responsible for generating and receiving these waves. It looks at the fundamental behaviour and parameters which characterise an antenna as a component within an overall wireless communication system, together with a description of the key types of antenna. More details of practical antennas based on these structures will be given in later chapters, particularly Chapter 15.

4.2 PRINCIPLES

4.2.1 What is an Antenna?

Most fundamentally, an antenna is a way of converting the guided waves present in a waveguide, feeder cable or transmission line into radiating waves travelling in free space, or vice versa. Figure 4.1 shows how the fields trapped in the transmission line travel in one dimension towards the antenna, which converts them into radiating waves, carrying power away from the transmitter in three dimensions into free space.

The art of antenna design is to ensure this process takes place as efficiently as possible, with the antenna radiating as much power from the transmitter into useful directions, particularly the direction of the intended receiver, as can practically be achieved.

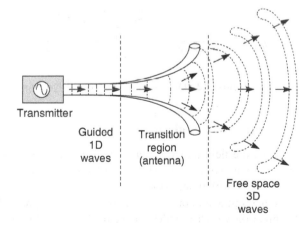

Figure 4.1: The antenna as a transition region between guided and propagating waves

---

Antennas and Propagation for Wireless Communication Systems Second Edition   Simon R. Saunders and Alejandro Aragón-Zavala

4.2.2 Necessary Conditions for Radiation

A question then arises as to what distinguishes the current in an antenna from the current in a guided wave structure. As Figure 4.2(a) shows, and as a direct consequence of Maxwell's equations, a group of charges in uniform motion (or stationary charges) do not produce radiation. In Figure 4.2(b)–(d), however, radiation does occur, because the velocity of the charges is changing in time. In Figure 4.2(b) the charges are reaching the end of the wire and reversing direction, producing radiation. In Figure 4.2(c) the speed of the charges remains constant, but their direction is changing, thereby creating radiation. Finally, in Figure 4.2(d), the charges are oscillating in periodic motion, causing a continuous stream of radiation. This is the usual practical case, where the periodic motion is excited by a sinusoidal transmitter. Antennas can therefore be seen as devices which cause charges to be accelerated in ways which produce radiation with desired characteristics. Similarly, rapid changes of direction in structures which are designed to guide waves may produce undesired radiation, as is the case when a printed circuit track carrying high-frequency currents changes direction over a short distance.

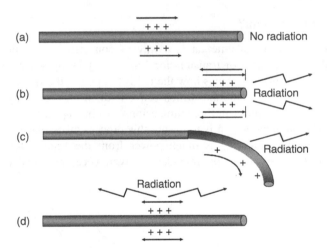

Figure 4.2: Only accelerating charges produce radiation

4.2.3 Near-Field and Far-Field Regions

Close to an antenna, the field patterns change very rapidly with distance and include both radiating energy and reactive energy, which oscillates towards and away from the antenna, appearing as a reactance which only stores, but does not dissipate, energy. Further away, the reactive fields are negligible and only the radiating energy is present, resulting in a variation of power with direction which is independent of distance (Figure 4.3). These regions are conventionally divided at a radius R given by

$$R = \frac{2L^2}{\lambda} \tag{4.1}$$

where L is the diameter of the antenna or of the smallest sphere which completely encloses the antenna [m] and λ is the wavelength [m]. Within that radius is the *near-field* or *Fresnel region*,

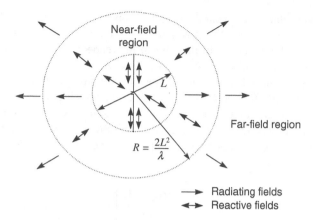

Figure 4.3: Definition of field regions

while beyond it lies the *far-field* or *Fraunhofer region*. Within the far-field region the wave fronts appear very closely as spherical waves, so that only the power radiated in a particular direction is of importance, rather than the particular shape of the antenna. Measurements of the power radiated from an antenna have either to be made well within the far field, or else special account has to be taken of the reactive fields. In siting an antenna, it is particularly important to keep other objects out of the near field, as they will couple with the currents in the antenna and change them, which in turn may greatly alter the designed radiation and impedance characteristics.

4.2.4 Far-Field Radiation from Wires

Many antenna types are composed only of wires with currents flowing on them. In this section, and in more detail in Section 4.4, we illustrate how the radiation from an antenna in the far field may be calculated from a knowledge of the current distribution on the wires.

In Figure 4.4, an appropriate coordinate system is defined. It is usually most convenient to work in spherical coordinates (r, θ, ϕ) rather than Cartesian coordinates, with the antenna

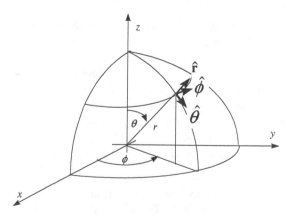

Figure 4.4: Coordinate system for antenna calculations

under analysis placed at or near the origin. Often the z-axis is taken to be the vertical direction and the x–y plane is horizontal, in which case ϕ denotes the *azimuth* angle. Here $\hat{r}$, $\hat{\theta}$ and $\hat{\phi}$ are unit vectors in the directions of increase of the respective coordinates.

The simplest wire antenna is a *Hertzian dipole* or *infinitesimal dipole*, which is a piece of straight wire whose length L and diameter are both very much less than one wavelength, carrying a current $I(0)$ which is uniform along its length, surrounded by free space. If this dipole is placed along the z-axis at the origin, then, in accordance with Maxwell's equations, it radiates fields which are given as follows. Note that a phase term $e^{j\omega t}$ has been dropped from these equations for simplicity, and all of the fields are actually varying sinusoidally in time.

$$E_\theta = jZ_0 \frac{kI(0)Le^{-jkr}}{4\pi r} \left[1 + \frac{1}{jkr} \frac{1}{(kr)^2} \right] \sin\theta$$

$$E_r = Z_0 \frac{I(0)Le^{-jkr}}{2\pi r^2} \left[1 + \frac{1}{jkr} \right] \cos\theta \tag{4.2}$$

$$H_\phi = j\frac{kI(0)Le^{-jkr}}{4\pi r} \left[1 + \frac{1}{jkr} \right] \sin\theta$$

$$H_r = 0, H_\theta = 0, E_\phi = 0$$

In the far field, the terms in r^2 and higher and can be neglected, so the fields are given by

$$E_\theta = jZ_0 \frac{kI(0)Le^{-jkr}}{4\pi r} \sin\theta$$

$$H_\phi = j\frac{kI(0)Le^{-jkr}}{4\pi r} \sin\theta \tag{4.3}$$

$$E_r = 0, E_\phi = 0, H_r = 0, H_\theta = 0$$

The following points are important to note from Eq. (4.3):

- The only non-zero fields are E_θ and H_ϕ. Hence the total electric and magnetic fields are transverse to each other everywhere and the antenna produces pure linear polarisation in the θ direction.
- The ratio $E_\theta/H_\phi = Z_0$, so the fields are in phase and the wave impedance is 120π ohms, just as for the plane waves in free space of Chapter 2.
- The field is inversely proportional to r.
- The directions of E, H and r form a right-handed set, so the Poynting vector is in the r-direction and carries power away from the origin in all directions.
- The fields are all zero at $\theta = 0$ and π, but reach a maximum at $\theta = \pi/2$, i.e. in the x–y plane.

These points justify the assumptions of geometrical optics which were introduced in Chapter 3: the waves are *locally plane* in the far field, and differ from plane waves only in that their amplitude is inversely proportional to the distance from the antenna. Another way of saying the same thing is that the wave fronts are actually spherical, but that a small section of the wave fronts is indistinguishable from a plane wave if observed sufficiently far from the antenna.

4.3 ANTENNA PARAMETERS

4.3.1 Radiation Pattern

The *radiation pattern* of an antenna is a plot of the far-field radiation from the antenna. More specifically, it is a plot of the power radiated from an antenna per unit solid angle, or its *radiation intensity U* [watts per unit solid angle]. This is arrived at by simply multiplying the power density at a given distance by the square of the distance r, where the power density S [watts per square metre] is given by the magnitude of the time-averaged Poynting vector:

$$U = r^2 S \qquad (4.4)$$

This has the effect of removing the effect of distance and of ensuring that the radiation pattern is the same at all distances from the antenna, provided that r is within the far field. The simplest example is an idealised antenna which radiates equally in all directions, an *isotropic* antenna. If the total power radiated by the antenna is P, then the power is spread over a sphere of radius r, so the power density at this distance and in any direction is

$$S = \frac{P}{\text{area}} = \frac{P}{4\pi r^2} \qquad (4.5)$$

The radiation intensity is then

$$U = r^2 S = \frac{P}{4\pi} \qquad (4.6)$$

which is clearly independent of r.

In the case of the infinitesimal dipole, the time-averaged Poynting vector is given by

$$\mathbf{S}_{av} = \frac{1}{2} E_\theta H_\phi^* \hat{\mathbf{r}} \qquad (4.7)$$

Hence, using Eq. (4.3), the radiation pattern is

$$U = r^2 \times \frac{1}{2} \frac{|E_\theta|^2}{Z_0} = \frac{Z_0}{2} \left(\frac{KI(0)L}{4\pi} \right)^2 \sin^2 \theta \qquad (4.8)$$

Radiation patterns are usually plotted by normalising the radiation intensity by its maximum value and plotting the result. This maximum value is at $\theta = \pi/2$. Equation (4.8) then yields

$$\frac{U}{U_{\text{max}}} = \frac{\dfrac{Z_0}{2} \left(\dfrac{KI(0)L}{4\pi} \right)^2 \sin^2 \theta}{\dfrac{Z_0}{2} \left(\dfrac{KI(0)L}{4\pi} \right)^2} = \sin^2 \theta \qquad (4.9)$$

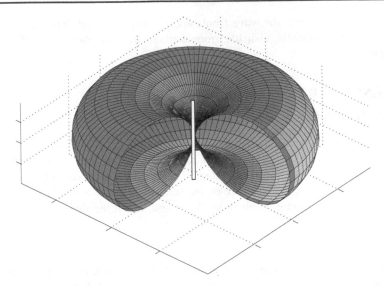

Figure 4.5: Radiation pattern of a Hertzian dipole. The section from $\phi = 0$ to $\phi = \pi/2$ has been cut away to reveal detail. The units are linear radiation intensity

This pattern is plotted as a surface in Figure 4.5 with a cutaway portion to reveal the detail. A radiation pattern plot for a generic directional antenna is shown in Figure 4.6, illustrating the *main lobe*, which includes the direction of maximum radiation (sometimes called the *bore-sight* direction), a *back lobe* of radiation diametrically opposite the main lobe and several *side lobes* separated by *nulls* where no radiation occurs. The Hertzian dipole has nulls along the *z*-axis.

Some common parameters used to compare radiation patterns are defined as follows:

- The *half-power beamwidth* (HPBW), or commonly the *beamwidth*, is the angle subtended by the half-power points of the main lobe. The pattern of the Hertzian dipole falls by one-half at $\theta = \pi/4$ and $\theta = 3\pi/4$, so its half-power beamwidth is $\pi/2 = 90°$.
- The *front-back ratio* is the ratio between the peak amplitudes of the main and back lobes, usually expressed in decibels.
- The *sidelobe level* is the amplitude of the biggest sidelobe, usually expressed in decibels relative to the peak of the main lobe.

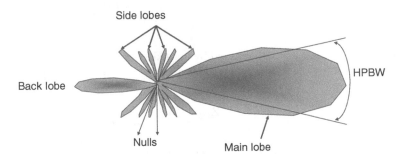

Figure 4.6: Radiation pattern of a generic antenna

Two special cases of radiation patterns are often referred to. The first is the *isotropic antenna*, a hypothetical antenna which radiates power equally in all directions. This cannot be achieved in practice, but acts as a useful point of comparison. More practical is the *omnidirectional antenna*, whose radiation pattern is constant in, say, the horizontal plane but may vary vertically. The Hertzian dipole is thus clearly omnidirectional in the *x–y* plane as illustrated in Figure 4.5.

4.3.2 Directivity

The *directivity D* of an antenna, a function of direction, is defined by

$$D(\theta, \phi) = \frac{\text{Radiation intensity of antenna in direction } (\theta, \phi)}{\text{Mean radiation intensity in all directions}}$$

$$= \frac{\text{Radiation intensity of antenna in direction } (\theta, \phi)}{\text{Radiation intensity of isotropic antenna radiating the same total power}}$$

(4.10)

Sometimes directivity is specified without referring to a direction. In this case the term 'directivity' implies the maximum value of $D(\theta, \phi) = D_{max}$. It is also common to express the directivity in decibels. The use of the isotropic antenna as a reference in the second line of Eq. (4.10) is then emphasised by giving the directivity units of dBi:

$$D[\text{dBi}] = 10 \log D \tag{4.11}$$

In the case of the Hertzian dipole, the directivity can be shown to be $D - 3/2$, or approximately 1.8 dBi [Kraus, 01].

4.3.3 Radiation Resistance and Efficiency

The equivalent circuit of a transmitter and its associated antenna is shown in Figure 4.7. The resistive part of the antenna impedance is split into two parts, a *radiation resistance R_r* and a *loss resistance R_l*. The power dissipated in the radiation resistance is the power actually radiated by the antenna, and the loss resistance is power lost within the antenna itself. This may be due to losses in either the conducting or the dielectric parts of the antenna.

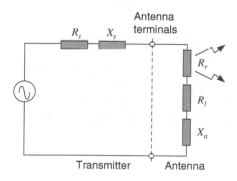

Figure 4.7: Equivalent circuit of transmitting antenna

Although only the radiated power normally serves a useful purpose, it is useful to define the radiation efficiency e of the antenna as

$$e = \frac{\text{Power radiated}}{\text{Power accepted by antenna}} = \frac{R_r}{R_r + R_l} \tag{4.12}$$

An antenna with high radiation efficiency therefore has high associated radiation resistance compared with the losses. The antenna is said to be *resonant* if its input reactance $X_a = 0$. If the source impedance, $Z_s = R_s + jX_s$, and the total antenna impedance, $Z_a = R_r + R_l + jX_a$, are complex conjugates, i.e. $Z_s = Z_a^*$ then the source is *matched* to the antenna and a maximum of the source power is delivered to the antenna. If the match is not ideal, then the degree of mismatch can be measured using the *reflection coefficient* ρ, defined by

$$\rho = \frac{V_r}{V_i} = \frac{Z_a - Z_s}{Z_a + Z_s} \tag{4.13}$$

where V_r and V_i are the amplitudes of the waves reflected from the antenna to the transmitter and incident from the transmitter onto the antenna terminals, respectively. Note that this definition of ρ is exactly analogous to the reflection coefficient R defined in Chapter 3, but is here applied to the guided waves in the transmission line of the antenna rather than to waves propagating in other media. It is also common to measure the mismatch via the *voltage standing wave ratio* (VSWR), with an optimum value of 1:

$$\text{VSWR} = \frac{1 + |\rho|}{1 - |\rho|} \tag{4.14}$$

It is common to design antennas to a standard input impedance of either 50 Ω or 75 Ω.

4.3.4 Power Gain

The *power gain* G, or simply the *gain*, of an antenna is the ratio of its radiation intensity to that of an isotropic antenna radiating the same total power as accepted by the real antenna. When antenna manufacturers specify simply the gain of an antenna they are usually referring to the maximum value of G. From the definition of efficiency (4.12), the directivity and the power gain are then related by

$$G(\theta, \phi) = eD(\theta, \phi) \tag{4.15}$$

Gain may be expressed in decibels to emphasise the use of the isotropic antenna as reference, but see Eq. (4.31) for alternative units.

Although the gain is, in principle, a function of both θ and ϕ together, it is common for manufacturers to specify patterns in terms of the gain in only two orthogonal planes, usually called *cuts*. In such cases the gain in any other direction may be estimated by assuming that the pattern is separable into the product of functions G_θ and G_ϕ which are functions of only θ and ϕ, respectively. Thus

$$G(\theta, \phi) \approx G_\theta(\theta) G_\phi(\phi) \tag{4.16}$$

Figure 4.8 shows a practical example.

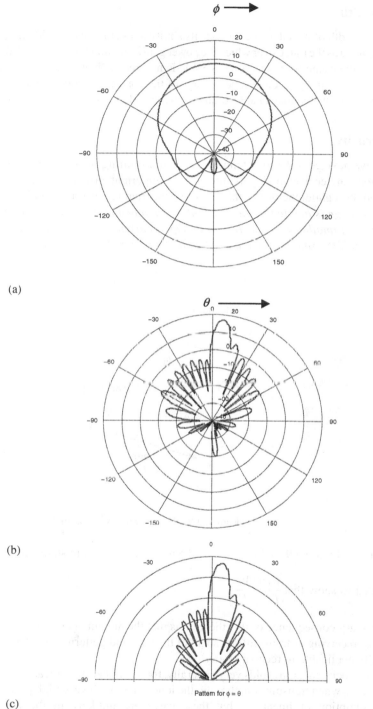

(a)

(b)

(c)

Figure 4.8: Composite pattern calculation for a typical cellular base station antenna: (a) azimuthal pattern, $0 < \phi < 2\pi, \theta = \pi/2$; (b) elevation pattern $\phi = 0, 0 < \theta < 2\pi$; and (c) composite pattern in the plane $\phi = \theta$ from Eq. (4.16) (reproduced by permission of Jaybeam Wireless)

4.3.5 Bandwidth

The bandwidth of an antenna expresses its ability to operate over a wide frequency range. It is often defined as the range over which the power gain is maintained to within 3 dB of its maximum value, or the range over which the VSWR is no greater than 2:1, whichever is smaller. The bandwidth is usually given as a percentage of the nominal operating frequency. The radiation pattern of an antenna may change dramatically outside its specified operating bandwidth.

4.3.6 Reciprocity

So far we have considered antennas only as transmitting devices. In order to consider their behaviour in receive mode, we use a very important principle: the *reciprocity theorem*. Applied throughout the rest of this book, the theorem is illustrated in Figure 4.9 and it states:

If a voltage is applied to the terminals of an antenna A and the current measured at the terminals of another antenna B then an equal current will be obtained at the terminals of antenna A if the same voltage is applied to the terminals of antenna B.

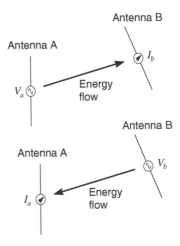

Figure 4.9: The reciprocity theorem

Thus, in Figure 4.9, if $V_a = V_b$ then the reciprocity theorem states that $I_a = I_b$ and can be extended to show that $\dfrac{V_a}{I_a} = \dfrac{V_b}{I_b}$.

A necessary consequence of this theorem is that the antenna gain must be the same whether used for receiving or transmitting, so all of the gain and pattern characteristics derived so far are fully applicable in receive mode.

The reciprocity theorem holds very generally, for any linear time-invariant medium. There are some cases when transmission through the ionosphere with very high powers can contravene the assumption of linearity,[1] but these cases are unlikely in the context of practical

---

[1]This occurs because the wave energy causes ionisation of atoms in the ionosphere, so that the constitutive parameters of the medium change with the power in the wave. Chapter 7 gives more details of ionospheric propagation effects.

transmission systems operated at VHF or above. Note how the reciprocity theorem does not state that the current distribution on the two antennas will be the same when receiving or transmitting, or that the way in which the field changes with respect to time or space at the two antennas will be the same.

4.3.7 Receiving Antenna Aperture

If an antenna is used to receive a wave with a power density S [W m^{-2}], it will produce a power in its terminating impedance (usually a receiver input impedance) of P_r watts. The constant of proportionality between P_r and S is A_e, the *effective aperture* of the antenna in square metres:

$$P_r = A_e S \tag{4.17}$$

For some antennas, such as horn or dish antennas, the aperture has an obvious physical interpretation, being almost the same as the physical area of the antenna, but the concept is just as valid for all antennas. The effective aperture may often be very much larger than the physical area, especially in the case of wire antennas. Note, however, that the effective aperture will reduce as the efficiency of an antenna decreases.

The antenna gain G is related to the effective aperture as follows [Balanis, 97]:

$$G = \frac{4\pi}{\lambda^2} A_e \tag{4.18}$$

4.3.8 Beamwidth and Directivity

The directivity of an antenna increases as its beamwidth is made smaller, as the energy radiated is concentrated into a smaller solid angle. For large antennas, with a single major lobe, the half-power beamwidths of the antenna in the $\hat{\theta}$ and $\hat{\phi}$ directions may be related to its directivity by the following approximate formula:

$$D \approx \frac{41,000}{\theta^{\circ}_{HP} \phi^{\circ}_{HP}} \tag{4.19}$$

where θ°_{HP} and ϕ°_{HP} are the half-power beamwidths in degrees.

4.3.9 The Friis Formula: Antennas in Free Space

Assume that antenna A and antenna B in Figure 4.9 are arranged such that their directions of maximum gain are aligned, their polarisations are matched and they are separated by a distance r, great enough that the antennas are in each other's far-field regions. If the power input to antenna A is P_t, then the power density incident on antenna B in accordance with (4.5) is

$$S = \frac{P_t G_a}{4\pi r^2} \tag{4.20}$$

where G_a is the maximum gain of antenna A. Applying Eq. (4.17), the received power at the terminals of antenna B is then

$$P_r = \frac{P_t G_a A_{eb}}{4\pi r^2} \tag{4.21}$$

where A_{eb} is the effective aperture of antenna B.

If Eq. (4.18) is substituted for A_{eb}, then the received and transmitted powers are related as follows:

$$\frac{P_r}{P_t} = G_a G_b \left(\frac{\lambda}{4\pi r}\right)^2 \tag{4.22}$$

This is the *Friis transmission formula* and its consequences for the range of wireless communication systems will be thoroughly explored in later chapters. If this same derivation is applied, but with *B* transmitting and *A* receiving, then exactly the same result is obtained as in (4.22), in accordance with the reciprocity theorem. The Friis formula exhibits an *inverse square law*, where the received power diminishes with the square of the distance between the antennas.

In practical free space conditions, the received power may be less than that predicted by (4.22) if the polarisation states of the antennas are not matched or if the source and load impedances do not match the antenna impedances.

4.3.10 Polarisation Matching

The Friis formula (4.22) must be modified in the general case to account for the mismatch between the polarisation state of the incoming wave and that of the receiving antennas. In Chapter 2 it was shown that the polarisation state of a plane wave could be expressed in terms of the relative amplitudes of the orthogonal components of the electric field:

$$\mathbf{E} = E_0(E_\theta \hat{\theta} + E_\phi \hat{\phi}) = E_0 \mathbf{p}_w \tag{4.23}$$

where E_0 is the electric field amplitude and $\mathbf{p}_w$ is the *polarisation vector* of the wave. For example, circular polarisation can be described by

$$\mathbf{p}_w = \begin{cases} \dfrac{\hat{\theta} - j\hat{\phi}}{\sqrt{2}} & \text{for right-hand circular polarisation} \\[2mm] \dfrac{\hat{\theta} + j\hat{\phi}}{\sqrt{2}} & \text{for left-hand circular polarisation} \end{cases} \tag{4.24}$$

Similarly, the polarisation state of an antenna can be written in the same way, with $\mathbf{p}_a$ denoting the polarisation vector of the far field produced by an antenna in a given direction. By applying reciprocity, it is clear that the polarisation response of the antenna is the same when used in receive mode. The *polarisation mismatch loss* is the ratio between the power received by the antenna and the power which would be received by an antenna perfectly matched to the incident wave [Stutzman, 81]:

$$L = \frac{P_{received}}{P_{matched}} = \left| \frac{\mathbf{p}_w \bullet \mathbf{p}_a^*}{|\mathbf{p}_w| \times |\mathbf{p}_a|} \right|^2 \qquad (4.25)$$

where $\bullet$ denotes the vector dot product. The Friis formula (4.22) must be multiplied by this formula whenever $\mathbf{p}_a \neq \mathbf{p}_w$.

In free space, the polarisation state of the received wave is the same as that of the transmitter antenna. In more complicated media, which may involve polarisation-sensitive phenomena such as reflection, refraction and diffraction, the wave polarisation is modified during propagation in accordance with the principles in Chapter 3.

4.4 PRACTICAL DIPOLES

4.4.1 Dipole Structure

In Section 4.2.4, the Hertzian dipole was assumed to have a uniform current distribution. In this section the current distribution of dipoles of varying lengths is examined and shown to modify the radiation pattern.

If a length of two-wire transmission line is fed from a source at one end and left open-circuit at the other, then a wave is reflected from the far end of the line. This returns along the line, interfering with the forward wave. The resulting interference produces a standing wave pattern on the line, with peaks and troughs at fixed points on the line (upper half of Figure 4.10). The current is zero at the open-circuit end and varies sinusoidally, with zeros of current spaced half a wavelength apart. The current flows in opposite directions in the two wires, so the radiation from the two elements is almost exactly cancelled, yielding no far-field radiation.

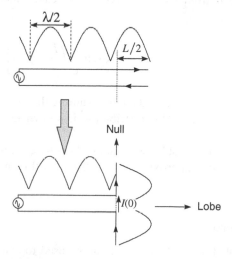

Figure 4.10: From transmission line to dipole

If a short section of length $L/2$ at the end of the transmission line is bent outwards, it forms a dipole perpendicular to the original line and of length L (lower half of Figure 4.10). The currents on the bent section are now in the same direction, and radiation occurs. Although this radiation does change the current distribution slightly, the general shape of the current distribution remains the same and a sinusoidal approximation may be used to analyse the resulting radiation pattern.

Some qualitative results may be deduced before the full analysis:

- As the dipole is rotationally symmetric around its axis, it must be omnidirectional, whatever the current distribution.
- In a plane through the transmission line and perpendicular to the plane of the page in Figure 4.10, the distance from the arms of the dipole to all points is equal. Hence the radiation contributions from all parts of the dipole will add in phase and a lobe will always be produced.
- The current always points directly towards or away from all points on the axis of the dipole, so no radiation is produced and a null appears at all such points.

For the Hertzian dipole, the radiated field was proportional to the dipole length in Eq. (4.8). It is desirable to increase the overall field of the antenna by increasing its directivity. Two steps are required in order to calculate the radiated field of a longer dipole; first an expression for the current distribution is determined (Section 4.4.2), then the effects of short sections ($\ll \lambda$) of the dipole are summed in the far field as if they were individual Hertzian dipoles to determine the total field (Section 4.4.3).

4.4.2 Current Distribution

The uniform transmission line illustrated in Figure 4.10 has a sinusoidal current distribution before it is bent. An exact calculation of the current distribution on a dipole must account for the variation in capacitance and inductance along the line as well as the effect of the radiation away from the dipole. However, the current distribution on the dipole may initially be assumed unchanged from the transmission line case. Standard transmission line theory then gives

$$I(z) = I(0) \sin\left[k\left(\frac{L}{2} - |z|\right)\right] \tag{4.26}$$

where $I(0)$ is the current at the feed point, assuming that the dipole is aligned with the z-axis and centred on the origin. This distribution is shown in Figure 4.11 for various dipole lengths.

It turns out that these results are exact if the wire forming the dipole is infinitesimally thin, and they are good approximations if the wire thickness is small compared with its length.

4.4.3 Radiation Pattern

The current distributions of Figure 4.11 can be used to calculate the corresponding dipole radiation patterns by summing the small field contributions dE_θ, dE_r and dH_ϕ from a series of

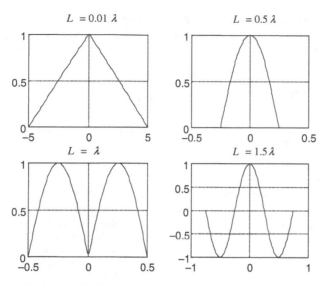

Figure 4.11: Dipole current distribution for various lengths; y-axis is the current normalised to the feed-point current, $I(0)$

Hertzian dipoles of length dL, with current contributions set by their position on the dipole in accordance with Eq. (4.26) and their radiation by Eq. (4.3). Thus the far-field contribution from a Hertzian dipole located on the z-axis at position z is

$$dE_\theta = jZ_0 \frac{kI(z)e^{-jkr}}{4\pi r} e^{jkz\cos\theta} \sin\theta dL$$

$$dE_r = 0; dH_r = 0; dH_\theta = 0 \tag{4.27}$$

$$dH_\phi = j \frac{kI(z)e^{-jkr}}{4\pi r} e^{jkz\cos\theta} \sin\theta dL$$

where the $e^{jkz\cos\theta}$ phase term accounts for the extra path length associated with an element at z compared with one at the origin.

The total E_θ field is therefore

$$E_\theta = \int_{-L/2}^{L/2} dE_\theta = jZ_0 \frac{ke^{-jkr}}{4\pi r} \sin\theta \int_{-L/2}^{L/2} I(z)e^{jkz\cos\theta} dz \tag{4.28}$$

The result of these calculations is

$$E_\theta = \frac{jZ_0 I(0)e^{-jkr}}{2\pi r} \left[\frac{\cos\left(\frac{kL}{2}\cos\theta\right) - \cos\left(\frac{kL}{\lambda}\right)}{\sin\theta} \right] \tag{4.29}$$

This pattern is omnidirectional with nulls along the z-axis, just as for the Hertzian dipole. A similar calculation can be performed for H_ϕ, but it is simpler to use the relation established in Section 4.2.4.

$$\frac{E_\theta}{H_\phi} = Z_0 \qquad (4.30)$$

Equation (4.29) is plotted in Figure 4.12 for the same dipole lengths as Figure 4.11. In the case of $L = 0.01\lambda$ the radiation pattern is essentially identical to the Hertzian dipole: despite the current distribution being non-uniform, the radiation contributions are so closely located that they sum as if the current was indeed constant along the dipole length. As the length increases through 0.5λ to λ, the HPBW increases through 78° to 47°. Above around $L = 1.2\lambda$, however, the pattern becomes multilobed, which is rarely useful in practice. Note also that the magnitude of the field is itself strongly dependent upon the

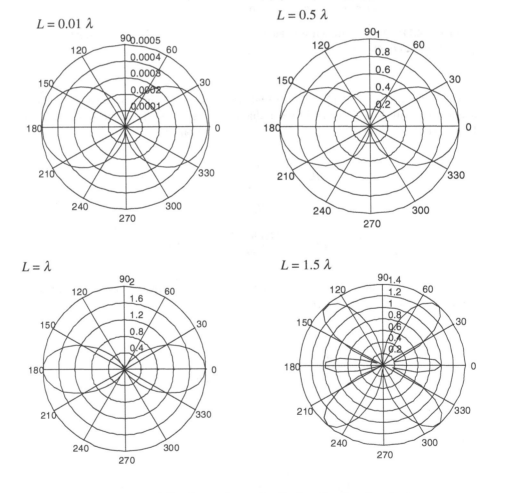

Figure 4.12: Pattern factors for dipoles of various lengths

antenna length. The most common length is the half-wave dipole ($L = 0.5\lambda$), which has significant directivity, high efficiency and relatively compact size. The directivity of the half-wave dipole is 1.64 or 2.15 dBi. As the half-wave dipole is an easily controlled, realisable antenna, whereas the isotropic antenna is not, it is common for antenna manufacturers to measure and quote antenna gains and directivities referenced to the half-wave dipole, in which case a suffix 'd' is used. Thus

$$0 \text{ dBd} = 2.15 \text{ dBi} \tag{4.31}$$

The above analysis of the dipole illustrates one general means of analysing antennas: first the current distribution is determined, and then contributions from infinitesimal elements are summed to find the radiation pattern. The first step is usually the most complicated, as simple approximations like the transmission line analogy above cannot always be used. Specific methods are beyond the scope of this book; see [Balanis, 97] for more details.

4.4.4 Input Impedance

A thin, lossless dipole, exactly half a wavelength long, has an input impedance $Z_a = 73 + j42.5\Omega$. It is desirable to make it exactly resonant, which is usually achieved in practice by reducing its length to around 0.48λ, depending slightly on the exact conductor radius and on the size of the feed gap. This also reduces the radiation resistance.

4.5 ANTENNA ARRAYS

4.5.1 Introduction

The maximum directivity available from a single dipole was shown in Figure 4.12 to be limited to that which can be achieved with a dipole a little over one wavelength. In some applications this may not be sufficient. One approach to improving on this is to combine *arrays* of dipoles, or of other antenna elements, where the amplitude and phase with which each element is fed may be different. The fields produced by the elements then combine with different phases in the far field, and the radiation pattern is changed. This also allows the radiation pattern to be tailored according to the particular application, or varied to allow beam scanning without any physical antenna motion. If the amplitude and phase weights are controlled electronically, then the beam can be scanned very rapidly to track changes in the communication channel. This is the topic of Chapter 18.

4.5.2 Linear and Planar Arrays

Arrays may be linear or planar, as shown in Figure 4.13. A linear array allows beam steering in one dimension, permitting directivity to be obtained in a single plane, hence an omnidirectional pattern can be synthesised. A planar array has two dimensions of control, permitting a narrow pencil beam to be produced.

4.5.3 The Uniform Linear Array

The simplest array type is the *uniform linear array*, which is a linear array with equal interelement spacing and a progressive phase shift across the array. In this case the field

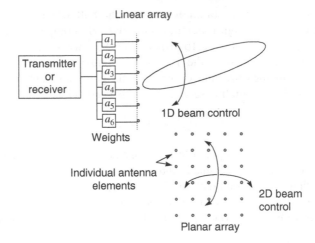

Figure 4.13: Array types: Each array element transmits a signal which is a version of the same signal, apart from a weighting coefficient a_i

pattern of the total array is equivalent to the pattern of the individual elements multiplied by an *array factor*, whose normalised value is

$$F_a = \frac{\sin(n\psi/2)}{(n\psi/2)} \text{ where } \psi = \frac{2\pi d}{\lambda}\cos\phi + \alpha \qquad (4.32)$$

Here n is the number of elements, d is the interelement spacing and α is the phase shift between adjacent elements. The weight applied to the ith element is then $a_i = e^{j(i-1)\alpha}$. The peak gain is $20 \log n$ dB greater than that for a single element, so in principle the gain may be increased to any desired level. The weights may be created by splitting the signal through a phasing network, consisting of lengths of transmission line of increasing length for each element. Figure 4.14 shows the radiation pattern which can be obtained from a 10-element uniform linear array.

4.5.4 Parasitic Elements: Uda–Yagi Antennas

Another array-based approach to enhancing the directivity of dipole antennas is to use *parasitic elements*. Parasitic elements are mounted close to the driven dipole and are not connected directly to the source. Instead, the radiation field of the driven element induces currents in the parasitics, causing them to radiate in turn. If the length and position of the parasitic elements are chosen appropriately, then the radiation from the parasitics and the driven element add constructively in one direction, producing an increase in directivity. The classic form of such an antenna is the *Uda–Yagi*, or simply *Yagi* antenna, illustrated in Figure 4.15 and widely used as a television reception antenna.

Typically, the driven element is made a little shorter than $\lambda/2$, to permit a good match to 50Ω. Elements in the radiation direction, called *directors*, are made a little shorter than the driver element, and an element very close to $\lambda/2$ is placed behind and called the *reflector*. Increasing the number of directors increases the gain, although the improvement diminishes

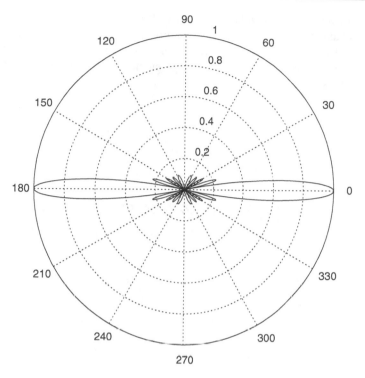

Figure 4.14: Array factor for a 10-element uniform linear array: $\alpha = 0$ and $d = \lambda/2$

according to how far the director is from the driven element. A 4-director Yagi can have a gain of up to around 12 dBi.

4.5.5 Reflector Antennas

Reflector antennas rely on the application of *image theory*, which may be described as follows. If an antenna carrying a current is placed adjacent to a perfectly conducting plane, the *ground plane*, then the combined system has the same fields above the plane as if an image

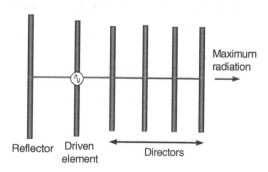

Figure 4.15: The Uda–Yagi antenna

of the antenna were present below the plane. The image carries a current of equal magnitude to the real antenna but in the opposite direction and is located an equal distance from the plane as the real antenna but on the other side, as shown in Figure 4.16. This statement is a consequence of Snell's law of reflection, given the Fresnel reflection coefficients for a perfect conductor, as described in Chapter 3.

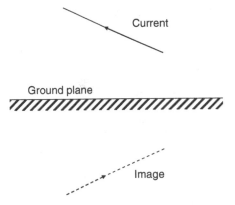

Figure 4.16: Illustrating image theory

4.5.6 Monopole Antennas

The *monopole* antenna results from applying image theory to the dipole. If a conducting plane is placed below a single element of length $L/2$ carrying a current, then the combined system acts essentially identically to a dipole of length L except that the radiation takes place only in the space above the plane, so the directivity is doubled and the radiation resistance is halved (Figure 4.17). The quarter-wave monopole ($L/2 = \lambda/4$) thus approximates the half-wave dipole and is a very useful configuration in practice for mobile antennas, where the conducting plane is the car body or handset case.

4.5.7 Corner Reflectors

The monopole principle can be extended by using two reflectors. This results in multiple images and a correspondingly increased gain. This is known as the corner reflector antenna.

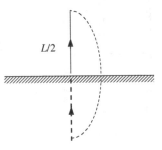

Figure 4.17: Image theory applied to the monopole antenna

In Figure 4.18 the gain is up to 12 dBi. Variations on this type of antenna are often used in cellular base station sites.

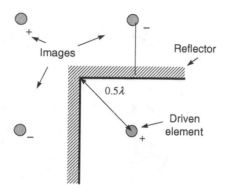

Figure 4.18: The corner reflector antenna

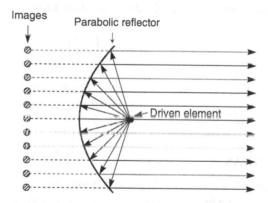

Figure 4.19: A parabolic reflector (dish) antenna

4.5.8 Parabolic Reflector Antennas

The parabolic (dish) antenna Figure 4.19 extends the reflector antenna concept to curved reflectors. In this case the number of images is effectively infinite and the locations of the images are such as to produce a parallel beam from the reflector, provided that the driven element is placed at the focus of the parabola.

The antenna then acts like an infinite uniform array and the array factor can therefore be found from Eq. (4.32) by setting $\alpha = 0, nd = D$, where D is the diameter of the dish, and letting $n \to \infty$. This yields

$$P = \sin\left(\frac{\pi D \cos\phi}{\lambda}\right) \bigg/ \left(\frac{\pi D \cos\phi}{\lambda}\right) \qquad (4.33)$$

The result is shown in Figure 4.20 for various values of D. The gain can be increased essentially arbitrarily by enlarging the size of the dish; this makes it appropriate for long-range communications applications, such as satellite systems.

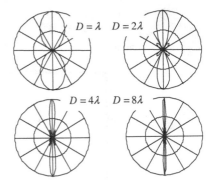

Figure 4.20: Array factors for parabolic dish antennas of various diameters

In general, the gain of a parabolic dish antenna is given by

$$G = \eta \left(\frac{\pi D}{\lambda} \right)^2 \tag{4.34}$$

where η is the antenna efficiency as a fraction, D is the diameter of the dish [m] and λ is the wavelength [m]. The efficiency is dependent on the material and accuracy of the dish construction and the details of the feed system.

4.6 HORN ANTENNAS

The horn antenna is a natural evolution of the idea that any antenna represents a region of transition between guided and propagating waves (Figure 4.1). Horn antennas are highly suitable for frequencies (typically several gigahertz and above) where waveguides are the standard feed method, as they consist essentially of a waveguide whose end walls are flared outwards to form a megaphone-like structure (Figure 4.21). In the case illustrated, the aperture is maintained as a rectangle, but circular and elliptical versions are also possible. The dimensions of the aperture are chosen to select an appropriate resonant mode, giving rise to a controlled field distribution over the aperture. The best patterns (narrow main lobe, low side lobes) are produced by making the length of the horn large compared to the aperture

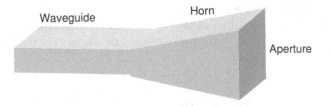

Figure 4.21: The rectangular horn antenna

width, but this must be chosen as a compromise with the overall volume occupied. A common application of horn antennas is as the feed element for parabolic dish antennas in satellite systems (Chapter 7); see [Stutzman, 81] for more details.

4.7 LOOP ANTENNAS

The loop antenna is a simple loop of wire of radius a; it is small enough in comparison to a wavelength that the current I can be assumed constant around its circumference (Figure 4.22). The resulting radiation pattern is

$$H_\theta = -\frac{I}{r}\left(\frac{ka}{2}\right)^2 e^{-jkr}\sin\theta \tag{4.35}$$

Note that this has exactly the shape of the Hertzian dipole pattern in Eq. (4.3), except that the electric and magnetic fields are reversed in their roles. As with the Hertzian dipole, the loop is relatively inefficient. In practice, loops are usually applied as compact receiving antennas, e.g. in pagers. The loop need not be circular, with approximately the same fields being produced provided the area enclosed by the loop is held constant.

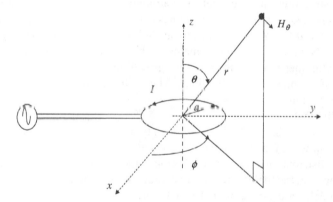

Figure 4.22: The loop antenna

4.8 HELICAL ANTENNAS

The helical antenna (Figure 4.23) can be considered as a vertical array of loops, at least for the case when the diameter of the helix is small compared to a wavelength. The result is *normal mode* radiation with higher gain than a single loop, providing an omnidirectional antenna with compact size and reasonable efficiency, but rather narrow bandwidth. It is commonly used for hand-portable mobile applications where it is more desirable to reduce the length of the antenna below that of a quarter-wave monopole.

In the case where the diameter is around one wavelength or greater, the mode of radiation changes completely to the axial mode, where the operation of the antenna is similar to that of a

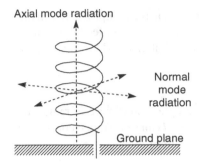

Figure 4.23: The helical antenna

Yagi, but with circular polarisation. This mode is commonly used for satellite communications, particularly at lower frequencies where a dish would be impractically large.

4.9 PATCH ANTENNAS

Patch antennas, as seen in Figure 4.24, are based upon printed circuit technology to create flat radiating structures on top of dielectric, ground-plane-backed substrates. The appeal of such structures is in allowing compact antennas with low manufacturing cost and high reliability. It is in practice difficult to achieve this at the same time as acceptably high bandwidth and efficiency. Nevertheless, improvements in the properties of the dielectric materials and in design techniques have led to enormous growth in their popularity and there are now a large number of commercial applications. Many shapes of patch are possible, with varying applications, but the most popular are rectangular (pictured), circular and thin strips (i.e. printed dipoles).

In the rectangular patch, the length L is typically up to half of the free space wavelength. The incident wave fed into the feed line sets up a strong resonance within the patch, leading to a specific distribution of fields in the region of the dielectric immediately beneath the patch, in which the electric fields are approximately perpendicular to the patch surface and the magnetic fields are parallel to it. The fields around the edges of the patch create the radiation, with contributions from the edges adding as if they constituted a four-element array. The resultant radiation pattern can thus be varied over a wide range by altering the length L and

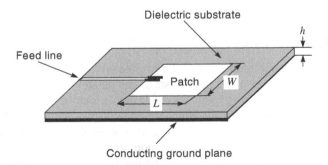

Figure 4.24: Microstrip patch antenna

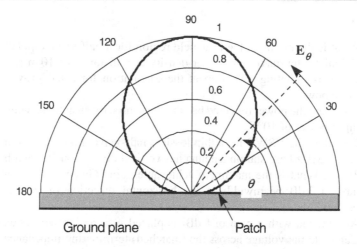

Figure 4.25: Typical patch antenna radiation pattern

width W, but a typical pattern is shown in Figure 4.25. In this case the polarisation is approximately linear in the θ direction, but patches can be created with circular polarisation by altering the patch shape and/or the feed arrangements. A major application of patch antennas is in arrays, where all of the elements, plus the feed and matching networks, can be created in a single printed structure. The necessary dimensions can be calculated approximately by assuming that the fields encounter a relative dielectric constant of $(1 + \varepsilon_r)/2$ due to the combination of fields in the air and in the dielectric substrate.

4.10 CONCLUSION

The fundamental parameters of antennas have been described here, together with a description of several of the most important antenna types. This should be sufficient to provide an appreciation of the trade-offs inherent in antenna design and the meaning of manufacturers' specifications. There are many other antenna types, but the set described here should be sufficient to understand the general operation of antennas used in wireless communication systems. Applications of antennas in specific system types will be described in subsequent chapters. For detailed methods for antenna design, consult books such as [Lee, 84] or [Stutzman 81].

REFERENCES

[Balanis, 97] C. A. Balanis, *Antenna Theory: analysis and design*, 2nd edn, John Wiley & Sons, Ltd, Chichester, ISBN 0-471-59268-4, 1997.

[Kraus, 01] J. D. Kraus and R. Marfehka, *Antennas*, 2nd edn, McGraw Hill education, United States of America, ISBN 0-07-232103-2, 2001.

[Lee, 84] K. F. Lee, *Principles of Antenna Theory*, John Wiley & Sons Ltd, Chichester, ISBN 0-471-90167-9, 1984.

[Stutzman, 81] W. L. Stutzman and G. A. *Thiele, Antenna theory and design*, John Wiley & Sons Ltd, Chichester, ISBN 0-471-04458, 1981.

PROBLEMS

4.1   What is the radius of the near-field region for a half-wave dipole?

4.2   Calculate the maximum power density at a distance of 10 m from a Hertzian dipole which is radiating 1 W. Repeat the calculation for a half-wave dipole radiating the same power.

4.3   Calculate the power reflected back to the source by an antenna with a VSWR of 2 and an input power of 10 W.

4.4   Calculate the VSWR for a thin lossless half-wave dipole fed from a 50 Ω source.

4.5   A 900 MHz base station antenna has vertical dimension 2 m and horizontal dimension 35 cm. What is the maximum achievable gain? Given that the antenna is required to have a 3 dB beamwidth in the horizontal direction of 90°, estimate the vertical beamwidth.

4.6   An antenna with a gain of 8 dBi is placed in a region with power density 1 W m^{-2}. Calculate the voltage across the (matched) terminating impedance of the antenna.

4.7   Calculate the polarisation mismatch loss for two linearly polarised antennas in free space at an angle of 20° to each other.

4.8   Calculate the polarisation mismatch loss for a linearly polarised antenna receiving a circularly polarised wave. Show how two linear antennas operated together can be used to reduce this value to unity (i.e. no loss).

4.9   What is the uplink and downlink gain of a 2.7 m diameter parabolic dish antenna at Ku-band (14/12 GHz), if it is to be used in a satellite ground station? Assume an antenna efficiency of 65%. *Hint:* the frequency bands for satellite communications are often specified as f_1/f_2, where f_1 represents the uplink frequency (from Earth to satellite) and f_2 is the downlink frequency (from satellite to Earth).

4.10 Explain why a helical antenna is often used for satellite links. What would the minimum diameter of such an antenna be to make it suitable for L-band satellite applications?

4.11 A satellite communications system is operated in the C-band (6 GHz in the uplink, 4 GHz in the downlink) to provide satellite TV (SATV) services to customers. You have been hired by a SATV company as the telecommunications expert and have been asked to select an antenna which could be used by a customer at home to receive the satellite signal. Determine

(a) The type of antenna required, the operating frequency band and the reasons for choosing this antenna.

(b) Assuming that link budget calculations show that a minimum gain of 38 dBi is required, calculate the dimensions of the antenna.

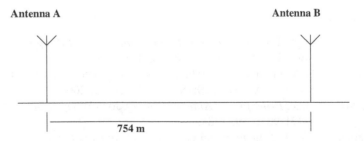

Figure 4.26: Antennas in free-space at 1800 MHz for Problem 4.12

(c) If the customer is unhappy with the dimensions of this antenna (it may not fit in the physical space allocated for the antenna), suggest ways in which this size could be reduced.

4.12 Two directional antennas are aligned facing each other in the boresight direction at 1800 MHz in free-space conditions, as shown in Figure 4.26. Antenna A is a parabolic dish with 65% radiation efficiency, and antenna B is a horn antenna which has a gain of 15 dBi.

Determine

(a) The radiation pattern cuts for azimuth and elevation angles for antenna A, assuming a vertical beamwidth of 45° and a horizontal beamwidth of 55°. You can assume that some side and back lobes for both azimuth and elevation planes are significant.
(b) The directivity for antenna A, in decibels.
(c) The power received at antenna B, given that antenna A has an input power of 1 W. State clearly any assumptions made.
(d) The distance at which antenna B is considered to be in the far-field, if this antenna has a diagonal distance in its mouth of 10 cm.
(e) If antenna B is used as a transmit antenna, what is the power received at antenna A given an input power of 1 W? Explain your answer.

5 Basic Propagation Models

'You see, wire telegraph is a kind of a very, very long cat. You pull his tail in New York and his head is meowing in Los Angeles. Do you understand this? And radio operates exactly the same way: you send signals here, they receive them there. The only difference is that there is no cat'.

Albert Einstein

5.1 INTRODUCTION

In this chapter, we establish how the basic parameters of antennas can be used, together with an understanding of propagation mechanisms, to calculate the range of a wireless communication system. The models introduced here are of an approximate and idealised nature, but will be useful for approximate calculations and for illustrating the principles to be applied in later chapters.

5.2 DEFINITION OF PATH LOSS

The path loss between a pair of antennas is the ratio of the transmitted power to the received power, usually expressed in decibels. It includes all of the possible elements of loss associated with interactions between the propagating wave and any objects between the transmit and the receive antennas. In the case of channels with large amounts of fast fading, such as mobile channels, the path loss applies to the power averaged over several fading cycles (the local median path loss). This path loss is hard to measure directly, since various losses and gains in the radio system also have to be considered. These are best accounted for by constructing a *link budget*, which is usually the first step in the analysis of a wireless communication system (described in detail in Section 5.7). In order to define the path loss properly, the losses and gains in the system must be considered. The elements of a simple wireless link are shown in Figure 5.1.

The power appearing at the receiver input terminals, P_R, can then be expressed as

$$P_R = \frac{P_T G_T G_R}{L_T L L_R} \tag{5.1}$$

where the parameters are defined in Figure 5.1, with all gains G and losses L expressed as power ratios and powers expressed in watts. The antenna gains are expressed with reference to an isotropic antenna, which radiates the power delivered to it equally in all directions. The values used are those corresponding to the direction of the other antenna and may not necessarily be the maximum values. The effective isotropic radiated power (EIRP) is then

---

Antennas and Propagation for Wireless Communication Systems Second Edition   Simon R. Saunders and Alejandro Aragón-Zavala
© 2007 John Wiley & Sons, Ltd

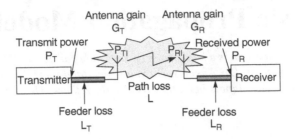

Figure 5.1: Elements of a wireless communication system

given by

$$\text{EIRP} = \frac{P_T G_T}{L_T} = P_{TI} \tag{5.2}$$

where P_{TI} is the effective isotropic transmit power. Similarly, the effective isotropic received power is P_{RI}, where

$$P_{RI} = \frac{P_R L_R}{G_R} \tag{5.3}$$

The advantage of expressing the powers in terms of EIRP is that the path loss, L, can then be expressed independently of system parameters by defining it as the ratio between the transmitted and the received EIRP, or the loss that would be experienced in an idealised system where the feeder losses were zero and the antennas were isotropic radiators ($G_{R,T} = 1, L_{R,T} = 1$):

$$\text{Path loss,} \ \ L = \frac{P_{TI}}{P_{RI}} = \frac{P_T G_T G_R}{P_R L_T L_R} \tag{5.4}$$

The propagation loss, thus defined, can be used to describe the propagation medium essentially independently of the system gains and losses. Note, however, that the propagation loss may change somewhat if the shape of the antenna radiation patterns is altered, since the relative contribution of component waves reaching the receiver from different directions changes. This effect may safely be ignored for many systems, however, particularly for systems with highly directional antennas and little off-path scattering.

The main goal of propagation modelling is to predict L as accurately as possible, allowing the range of a radio system to be determined before installation. The maximum range of the system occurs when the received power drops below a level which provides just acceptable communication quality. This level is often known as the *receiver sensitivity*. The value of L for which this power level is received is the *maximum acceptable path loss*. It is usual to express the path loss in decibels, so that

$$L_{\text{dB}} = 10 \log \left(\frac{P_{TI}}{P_{RI}} \right) \tag{5.5}$$

Note that, as a consequence of the reciprocity theorem, the definition of the path loss is unaltered by swapping the roles of the transmit and the receive antennas, provided that the frequency is maintained and the medium does not vary with time. However, the maximum acceptable propagation loss may be different in the two directions, as the applicable losses and sensitivities may be different. For example, a base station receiver is usually designed to be more sensitive than the mobile, to compensate for the reduced transmit power available from the mobile.

Example 5.1

A base station transmits a power of 10 W into a feeder cable with a loss of 10 dB. The transmit antenna has a gain of 12 dBd in the direction of a mobile receiver, with antenna gain 0 dBd and feeder loss 2 dB. The mobile receiver has a sensitivity of -104 dBm.

(a) Determine the effective isotropic radiated power.
(b) Determine the maximum acceptable path loss.

Solution

All quantities must be expressed in consistent units, usually [dBi] and [dBW]:

Quantity	Value in original units	Value in consistent units
P_T	10 W	10 dBW
G_T	12 dBd	14.15 dBi
G_R	0 dBd	2.15 dBi
P_R	-104 dBm	-134 dBW
L_T	10 dB	10 dB
L_R	2 dBd	2 dBi

(a) From Eq. (5.2) the EIRP is

$$P_{TI} = P_T + G_T - L_T$$
$$= 10 + 14.15 - 10 = 14.15 \, \text{dBW} = 26 \, \text{W}$$

Note that the radiated power may also be expressed using a half-wave dipole rather than an isotropic antenna as reference. This is simply the *effective radiated power*, ERP.

$$\text{ERP [dBW]} = \text{EIRP [dBW]} - 2.15 \text{ [dBi]} \qquad (5.6)$$

(a) From Eq. (5.4) the maximum acceptable path loss is

$$L = P_T + G_T + G_R - P_R - L_T - L_R$$
$$= 10 + 14.15 + 2.15 - (-134) - 10 - 2 = 148.3 \, \text{dB}$$

In a more realistic system than Figure 5.1, other losses would need to be included in the calculation, such as those due to connectors, combiners and filters.

5.3 A BRIEF NOTE ON DECIBELS

The decibel is an important unit in propagation studies but is often confused. These notes may help.

The bel is a logarithmic unit of power ratios, where 1 bel corresponds to an increase in power by a factor of 10 relative to some reference power, P_{ref}. Thus a change to a power P from P_{ref} may be expressed in bels as

$$\text{Power ratio in bels} = \log\left(\frac{P}{P_{ref}}\right) \tag{5.7}$$

Note that these units are only consistent if it is clear what the reference power is. The bel is too large a unit for convenience, so the decibel [dB], one-tenth of a bel, is almost always used instead,

$$\text{Power ratio in decibels} = 10\log\left(\frac{P}{P_{ref}}\right) \tag{5.8}$$

So the loss due to a feeder cable, or the gain of an amplifier, can be expressed simply in decibels, since it is clear that we are expressing the ratio of the output to the input power. Equation (5.8) may also be used to express a ratio of voltages (or field strengths) provided that they appear across the same impedance (or in a medium with the same wave impedance). Then

$$\text{Voltage ratio in decibels} = 20\log\left(\frac{V}{V_{ref}}\right) \tag{5.9}$$

Similarly, field strength ratios can be calculated as

$$\text{Field strength ratio in decibels} = 20\log\left(\frac{E}{E_{ref}}\right) \text{ or } 20\log\left(\frac{H}{H_{ref}}\right) \tag{5.10}$$

Table 5.1: Examples of the application of decibels

Unit	Reference power	Application
dBW	1 W	Absolute power
dBm	1 mW	Absolute power $P\,[\text{dBW}] = P\,[\text{dBm}] - 30$
dBμV	1 μV e.m.f.	Absolute voltage, typically at the input terminals of a receiver (dBμV = dBm + 107 for a 50 Ω load).
dB	Any	Gain or loss of a network, e.g. amplifiers, feeders or attenuators
dBμV m^{-1}	1 μVm^{-1}	Electric field strength
dBi	Power radiated by an isotropic reference antenna	Gain of an antenna
dBd	Power radiated by a half-wave dipole	Gain of an antenna 0 dBd = 2.15 dBi

In other cases, it is conventional to express a power in decibels relative to some standard reference power. This reference is usually indicated by appending an appropriate letter. Examples which are useful in propagation studies are shown in Table 5.1. Thus the power produced by a transmitter can be expressed in [dBm] or [dBW], while the feeder loss is in [dB]. The gain of the transmitter antenna is in [dBi] or [dBd], while the propagation path loss will be expressed in [dB] *between isotropic antennas*. Common errors include expressing an antenna gain in [dB] with no reference or calculating a voltage ratio in [dB] using a factor of 10 rather than 20.

5.4 NOISE MODELLING

In Example 5.1 the effect of the various system elements on the received power was calculated. It is also necessary, however, to calculate the impact of noise on the system, since it is eventually the ratio of signal power to noise power (SNR) which will determine the system performance.

The major noise contributions will usually come from the receiver itself, although external noise contributions may also be significant in systems such as fixed satellite links (Chapter 7). In any case, the total noise associated with the system can be calculated by assuming that the system consists of a *two-port network*, with a single input and a single output as shown in Figure 5.2. The network is characterised by a gain G, being the ratio of the signal power at the output to the signal power at its input, and by a noise factor F. The noise factor is the ratio between the output noise power of the element divided by G (i.e. referred to the input) and the input noise.

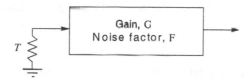

Figure 5.2: A noisy two-port network representing a complete system

The noise power available at the input of the network from a resistor with an absolute temperature of T K is [Connor, 82]

$$P_N = kTB \qquad (5.11)$$

where k is Boltzmann's constant $= 1.379 \times 10^{-23}$ W Hz^{-1} K^{-1}, T is the absolute temperature of the input noise source [K], B is the effective noise bandwidth of the system [Hz][1]. It is assumed that the network is impedance matched to the resistance. The noise factor is then

$$F = \frac{N_{\text{out}}}{N_{\text{in}}} = \frac{N_{\text{out}}}{kTB} \qquad (5.12)$$

---

[1]This is the bandwidth of an ideal rectangular filter which would produce the same power as the real filter.

where N_{out} is the output noise power of the element *referred to the input*, i.e. the actual noise output power divided by G.

F depends on the design and physical construction of the network. Its value in decibels is the *noise figure* of the network,

$$F_{dB} = 10 \log F \qquad (5.13)$$

The numerical value of the noise power [dBW] can be expressed approximately as

$$N_{out}|_{dBW} = F_{dB} - 204 + 10 \log B \qquad (5.14)$$

where $T = 290\,\text{K}$ (23°C) is assumed. Equivalently, using [dBm]

$$N_{out}|_{dBm} = F_{dB} - 174 + 10 \log B \qquad (5.15)$$

An alternative approach is to characterise the network by an *equivalent input noise temperature T_e*. This is the temperature of a noise source which, when placed at the input of the network, yields the same output noise as if the network were noiseless, i.e.

$$N_{out} = kTB + kT_eB = k(T + T_e)B \qquad (5.16)$$

Hence

$$F = \frac{N_{out}}{N_{in}} = \frac{k(T + T_e)B}{kTB} = 1 + \frac{T_e}{T} \qquad (5.17)$$

Usually, B will simply be the intermediate frequency (IF) bandwidth of the receiver.

Example 5.2

A receiver in a digital mobile communication system has a noise bandwidth of 200 kHz and requires that its input SNR should be at least 10 dB when the input signal is -104 dBm. (a) What is the maximum permitted value of the receiver noise figure? (b) What is the equivalent input noise temperature of such a receiver?

Solution

(a) The overall SNR, expressed in [dB], is

$$SNR = P_s - N$$

where P_s is the input signal power [dBW] and N is the noise power of the receiver referred to its input [dBW].

From Eq. (5.12),

$$N = 10 \log(FkTB)$$

so

$$SNR = P_s - N = P_s - F_{dB} - 10 \log(kTB)$$

Rearranging,

$$F_{dB} = P_s - SNR - 10\log(kTB)$$
$$= (-104 - 30) - 10 - 10\log(1.38 \times 10^{-23} \times 290 \times 200 \times 10^3)$$
$$= 7.0\,dB$$

(b) From Eq. (5.17)

$$T_e = T(F - 1) = 290(10^{7.0/10} - 1) = 1163\,K$$

using 290 K as a reference value.

A complete system can be characterised by a cascade of two-port elements, where the ith element has gain G_i and noise factor F_i (Figure 5.3). Each element could consist of an individual module within a receiver, such as an amplifier or filter, or one of the elements within the channel such as the antenna, feeder or some source of external noise.

The gain G of the complete network is then simply given by

$$G = G_1 \times G_2 \times \cdots \times G_N \tag{5.18}$$

while the overall noise factor is given by

$$F = F_1 + \frac{F_2 - 1}{G_1} + \frac{F_3 - 1}{G_1 G_2} + \cdots + \frac{F_N - 1}{G_1 G_2 \times \cdots \times G_{N-1}} \tag{5.19}$$

Equivalently, the overall effective noise temperature of the network, T_e, can be written in terms of the effective noise temperatures of the individual elements as

$$T_e = T_{e1} + \frac{T_{e2}}{G_1} + \frac{T_{e3}}{G_1 G_2} + \cdots + \frac{T_{eN}}{G_1 G_2 \times \cdots \times G_{N-1}} \tag{5.20}$$

It is important to note from Eqs. (5.19) and (5.20) that the noise from the first element adds directly to the noise of the complete network, while subsequent contributions are divided by the gains of the earlier elements. It is therefore important that the first element in the series has a low noise factor and a high gain, since this will dominate the noise in the whole system. As a result, receiver systems often have a separate *low noise amplifier* (LNA) placed close to the antenna, often at the top of the mast and sometimes directly attached to the feed of a dish antenna in order to overcome the impact of the feeder loss.

Figure 5.3: A cascade of two-port elements

An important special case of a two-port element is an attenuator, which has passive components only and a gain $G = 1/L$, where L is the attenuator insertion loss. This might be a feeder cable linking an antenna and a receiver. In this case, assuming the attenuator is itself at reference temperature T, Eq. (5.12) shows

$$F = \frac{N_{\text{out}}}{N_{\text{in}}} = \frac{N_{\text{out}}}{kTB} = \frac{kTB}{G}\frac{1}{kTB} = L \qquad (5.21)$$

More detail is available in [Connor, 82].

Example 5.3

A receiver is made up of three main elements: a preamplifier, a mixer and an IF amplifier with noise figures 3 dB, 6 dB and 10 dB, respectively. If the overall gain of the receiver is 30 dB and the IF amplifier gain is 10 dB, determine the minimum gain of the preamplifier to achieve an overall noise figure of no more than 5 dB. If its gain is set to this minimum, what would the system noise figure become if the noise figure of the IF amplifier is increased to 20 dB?

Solution

The receiver is modelled as a three-element network, with individual noise factors $F_1 = 2$, $F_2 = 4$ and $F_3 = 10$, where subscripts 1, 2 and 3 represent the preamplifier, mixer and IF amplifier, respectively. Since the overall gain is 30 dB, we have $G_1 G_2 G_3 = 1000$ and $G_3 = 10$, so $G_1 G_2 = 100$.
From Eq. (5.19), the overall noise factor is

$$F = F_1 + \frac{F_2 - 1}{G_1} + \frac{F_3 - 1}{G_1 G_2}$$

Thus

$$3.2 = 2.0 + \frac{4.0 - 1}{G_1} + \frac{10.0 - 1}{100}$$

Rearranging,:

$$G_1 = \frac{3.0}{3.2 - 2.0 - \dfrac{9.0}{100}} = 2.7 = 4.3\,\text{dB minimum}$$

If the noise figure of the IF amplifier is now increased to 20 dB, i.e. $F_3 = 100$, we have

$$F = F_1 + \frac{F_2 - 1}{G_1} + \frac{F_3 - 1}{G_1 G_2} = 2.0 + \frac{4.0 - 1}{2.7} + \frac{100 - 1}{100} = 4.1 = 6.1\,\text{dB}$$

This is clearly a very small increase on the previous figure, since the result is highly insensitive to the noise figure of the final element in the cascade.

5.5 FREE SPACE LOSS

In Chapter 4, the Friis transmission formula was derived as

$$\frac{P_r}{P_t} = G_a G_b \left(\frac{\lambda}{4\pi r}\right)^2 \tag{5.22}$$

where G_a and G_b are the gains of the terminal antennas, r is the distance between the antennas and λ is the wavelength. This can be visualised as arising from the spherical spreading of power over the surface of a sphere of radius r centred at the antenna (Figure 5.4). Since power is spread over the surface area of the sphere, which increases as r^2, the available power at a receiver antenna of fixed aperture decreases in proportion to r^2.

Equation (5.22) can be rearranged into the same form as Eq. (5.4) in order to express it as a propagation loss in free space,

$$L_F = \frac{P_t G_a G_b}{P_r} = \left(\frac{4\pi r}{\lambda}\right)^2 = \left(\frac{4\pi r f}{c}\right)^2 \tag{5.23}$$

This expression defines L_F, *the free space loss*. Note especially the square law dependence on both frequency and distance.

Expressing the free space loss in decibels, with frequency in megahertz and distance R in kilometres, we obtain

$$L_{F(\text{dB})} = 32.4 + 20 \log R + 20 \log f_{\text{MHz}} \tag{5.24}$$

Thus, the free space loss increases by 6 dB for each doubling in either frequency or distance (or 20 dB per decade of either).

Only in highly anomalous propagation conditions, the loss between two antennas can be any less than its free space value. An example of an exception to this is when propagation is confined to some guided structure, such as a waveguide. In most cases, due to other sources of loss, the received power will be considerably smaller. The free space value of the loss is then

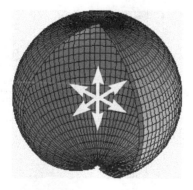

Figure 5.4: Isotropic radiation causing free space loss

used as a basic reference, and the loss experienced in excess of this value (in decibels) is referred to as the excess loss, L_{ex}. Hence

$$L = L_F + L_{ex} \qquad (5.25)$$

Example 5.4

The communication system described in Example 5.1 is operated under free space propagation conditions at 900 MHz. Determine its maximum range.

Solution

Substituting in Eq. (5.23) and rearranging for $\log R$ using the maximum acceptable propagation loss of 148.3 dB, we have

$$\log R = \frac{L_{F(\text{dB})} - 32.4 - 20 \log f_{\text{MHz}}}{20} = \frac{148.3 - 32.4 - 20 \log 900}{20}$$
$$\approx 2.84$$

Hence $R \approx 10^{2.84} = 693 \, \text{km}$

This is an impractically large range for any mobile communications system. In practice, other factors will reduce the range substantially, so more practical models must be considered. Nevertheless, the free space loss is usually considered as a practical minimum to the path loss for a given distance.

5.6 PLANE EARTH LOSS

Another fundamental propagation situation is illustrated in Figure 5.5. Here the base and mobile station antennas are situated above a flat reflecting ground (plane earth), at heights h_b and h_m, respectively, so that propagation takes place via both a direct path between the antennas and a reflection from the ground. These two paths sum at the receiver with a phase difference related to the difference in length between the two paths. A simple way to

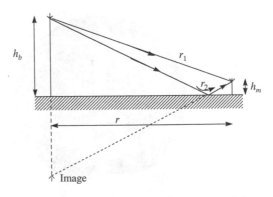

Figure 5.5: Physical situation for plane earth loss

analyse the situation is to make use of image theory, which considers the reflected ray as having come from an image of the transmitter in the ground, just as if the ground were a mirror. It is then easy to see that the path lengths of the direct and reflected rays (r_1 and r_2, respectively) are given by

$$r_1 = \sqrt{(h_b - h_m)^2 + r^2}$$
$$r_2 = \sqrt{(h_b + h_m)^2 + r^2}$$

(5.26)

The path length difference is then

$$(r_2 - r_1) = r \left[\sqrt{\left(\frac{h_b + h_m}{r}\right)^2 + 1} - \sqrt{\left(\frac{h_b - h_m}{r}\right)^2 + 1} \right]$$

(5.27)

Assuming the antenna heights are small compared with the total path length ($h_b, h_m \ll r$), then the following approximation applies from the binomial theorem,

$$(1 + x)^n \approx 1 + nx \text{ for } x \ll 1$$

(5.28)

The result is

$$(r_2 - r_1) \approx \frac{2h_m h_b}{r}$$

(5.29)

Since the path length is large compared with the antenna heights, the arriving amplitudes of the waves are identical apart from the reflection loss, R. The overall amplitude of the result (electric or magnetic field strength) is then

$$A_{\text{total}} = A_{\text{direct}} + A_{\text{reflected}} = A_{\text{direct}} \left| 1 + R \exp\left(jk \frac{2h_m h_b}{r} \right) \right|$$

(5.30)

where k is the free space wavenumber.

Since the power is proportional to the amplitude squared, we can write

$$\frac{P_r}{P_{\text{direct}}} = \left(\frac{A_{\text{total}}}{A_{\text{direct}}} \right)^2 = \left| 1 + R \exp\left(jk \frac{2h_m h_b}{r} \right) \right|^2$$

(5.31)

where P_r is the received power.

The direct path is itself subject to free space loss, so it can be expressed in terms of the transmitted power as

$$P_{direct} = P_T \left(\frac{\lambda}{4\pi r} \right)^2$$

(5.32)

so the path loss can be expressed as

$$\frac{P_r}{P_T} = \left(\frac{\lambda}{4\pi r} \right)^2 \left| 1 + R \exp\left(jk \frac{2h_m h_b}{r} \right) \right|^2$$

(5.33)

Since the angle of incidence with the ground is close to grazing, the magnitude of the reflection coefficient will be close to one, whatever its conductivity or roughness, as shown in Chapter 3. If we further assume that the signal always undergoes a phase change of 180°, then $R \approx -1$ and the result can be expressed as

$$\frac{P_r}{P_T} = 2\left(\frac{\lambda}{4\pi r}\right)^2 \left[1 - \cos\left(k\frac{2h_m h_b}{r}\right)\right] \tag{5.34}$$

This expression is shown as the solid line in Figure 5.6. For comparison, the free space loss is shown as a dotted line. For small distances, the influence of the interference between the two paths is visible as the combined signal undergoes distinct peaks and troughs. As the distance is increased, however, the loss monotonically increases. The equation of the asymptotic dashed line can be found by considering that for small angles, $\cos\theta \approx 1 - \theta^2/2$, so the previous assumption that $h_b, h_m \ll r$ leads to

$$\frac{P_r}{P_T} \approx \left(\frac{\lambda}{4\pi r}k\frac{2h_m h_b}{r}\right)^2 \approx \frac{h_m^2 h_b^2}{r^4} \tag{5.35}$$

Expressing this in decibels,

$$L_{\text{PEL}} = 40\log r - 20\log h_m - 20\log h_b \tag{5.36}$$

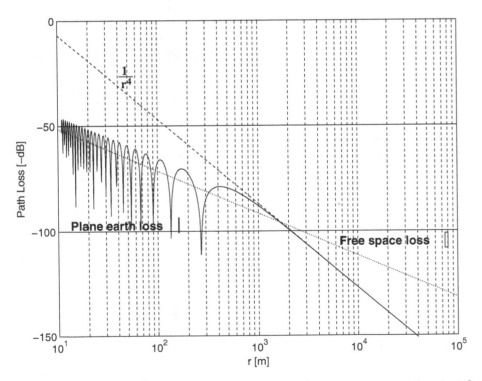

Figure 5.6: Plane earth loss (–), free space loss (...) and approximate plane earth loss (– –) from Eq. (5.36). Here $h_m = 1.5\,\text{m}$, $h_b = 30\,\text{m}$, $f = 900\,\text{MHz}$

This is the usual form of the *plane earth loss*. Note that the plane earth loss increases far more rapidly than the free space loss and that it is independent of carrier frequency. The loss now increases by 12 dB per doubling of distance or by 40 dB per decade.

The plane earth loss is rarely an accurate model of real-world propagation when taken in isolation. It only holds for long distances and for cases where the amplitude and phase of the reflected wave is very close to the idealised -1. In practice, loss is almost never independent of frequency. However, it is sometimes used as a reference case as will be seen in Chapter 8.

Example 5.5

Calculate the maximum range of the communication system in Example 5.1, assuming $h_m = 1.5$ m, $h_b = 30$ m, $f = 900$ MHz and that propagation takes place over a plane earth. How does this range change if the base station antenna height is doubled?

Solution

Assuming that the range is large enough to use the simple form of the plane earth model (5.36), then

$$\log r = \frac{L_{PEL} + 20 \log h_m + 20 \log h_b}{40} = \frac{148.3 + 3.5 + 29.5}{40} \approx 4.53$$

Hence $r = 34$ km, a substantial reduction from the free space case described in Example 5.4. If the antenna height is doubled, the range may be increased by a factor of $\sqrt{2}$ for the same propagation loss. Hence $r = 48$ km.

5.7 LINK BUDGETS

A calculation of signal powers, noise powers and/or signal-to-noise ratios for a complete communication link is a *link budget*, and it is a useful approach to the basic design of a complete communication system.[2] Such calculations are usually fairly simple, but they can give very revealing information as to the system performance, provided appropriately accurate assumptions are made in calculating the individual elements of the link budget.

Essentially, the link budget is simply an application of the principles already explained in this chapter. The maximum acceptable path loss is usually split into two components, one of which is given by the distance-dependent path loss model (such as the free space or plane earth models of Sections 5.5 and 5.6) plus a *fade margin*, which is included to allow the system some resilience against the practical effects of signal fading beyond the value predicted by the model. Thus

$$\begin{array}{lll} \text{Maximum acceptable} & = & \text{Predicted} & + & \text{Fade} \\ \text{propagation loss [dB]} & & \text{loss} & & \text{margin} \end{array} \qquad (5.37)$$

---

[2]Note that the process of calculating the maximum acceptable path loss resembles that of calculating a bank balance to determine the money available to spend, justifying the term *link budget*. A system where the desired range causes the loss to exceed the MAPL is thus analogous to an overdrawn account.

The greater the fade margin, the greater the reliability and quality of the system, but this will constrain the maximum system range. Later chapters will give details of how the fade margin may be chosen.

A sample link budget for the downlink (base station to mobile) of an imaginary, but representative, terrestrial mobile system is shown in Table 5.2. The output in this case is the maximum acceptable propagation loss, but it may be that in practice the designer starts with the propagation loss, knowing how this corresponds to the desired range of the system, and then uses it to calculate some other parameter such as the effective isotropic transmit power or antenna height. Notice that how the units initially used to specify the parameters are rarely the most convenient to work with in practice, and how the units have been converted to consistent units [dBi, dBW, dB] in the last two columns.

In the case of the uplink, the mobile transmit power would typically be only 1 W or less, which would reduce the maximum acceptable propagation loss and limit the range of the system to less than that calculated in Table 5.2. The system would then be *uplink limited*.

Table 5.2: Sample downlink budget for terrestrial mobile communications

	Quantity	Value	Units	Value	Units
(a)	Base station transmit power	10	W	10	dBW
(b)	Base station feeder loss	10	dB	10	dB
(c)	Base station antenna gain	6	dBd	8.2	dBi
(d)	Effective isotropic transmit power $(a - b + c)$			8.2	dBW
(e)	Maximum acceptable propagation loss			L	dB
(f)	Mobile antenna gain	-1	dBd	1.2	dBi
(g)	Body and matching loss	6	dB	6	dB
(h)	Signal power at receiver terminals $(d - e + f - g)$			$3.4 - L$	dBW
(i)	Receiver noise Bandwidth	200	kHz	53	dBHz
(j)	Receiver noise figure	7	dB	7	dB
(k)	$10 \log_{10}(kT)$ with $k = 1.38 \times 10^{-23}\,\mathrm{W\,Hz^{-1}\,K^{-1}}$, $T = 290\,\mathrm{K}$			-204	$\mathrm{dBW\,Hz^{-1}}$
(l)	Receiver noise power referred to input $(i + j + k)$			-144	dBW
(m)	Required signal-to-noise ratio	9	dB	9	dB
(n)	Required input signal power $(l + m)$			-135	dBW
(o)	Maximum acceptable propagation loss (by solving h = n)			138.4	dB
(p)	Fade margin			15	dB
(q)	Predicted cell radius (using plane earth loss with $h_m = 1.5\,\mathrm{m}$ and $h_b = 15\,\mathrm{m}$)			5.8	km

The two links are usually balanced in practice by improvements in the base station receiver, such as using a lower noise figure, or by using some *diversity gain* (Chapter 16).

5.8 CONCLUSION

This chapter has defined the important steps in the prediction of the useful range of a wireless communication system.

- Construction of a link budget, which identifies the maximum acceptable propagation loss of the system, with due regards to the key system parameters, including signal and noise powers and by selecting an appropriate fade margin.
- Use of a propagation model to predict the corresponding maximum range.

In the following chapters, more sophisticated propagation models will be used to estimate the system range and to predict the fade margin.

REFERENCE

[Connor, 82] F. R. Connor, *Noise*, 2nd edn, Edward Arnold, London, ISBN 0-7131-3459-3, 1982.

PROBLEMS

5.1 A base station antenna with a 9 dBd gain is supplied with 10 W of input power. What is the effective isotropic radiated power?

5.2 Demonstrate that free space loss increases by 6 dB each time the distance between transmitter and receiver is doubled.

5.3 The speech quality for a particular mobile communication system is just acceptable when the received power at the terminals of the mobile receiver is −104 dBm. Find the maximum acceptable propagation loss for the system, given that the transmit power at the base station is 30 W, base station feeder losses are 15 dB, the base station antenna gain is 6 dBi, the mobile antenna gain is 0 dBi and the mobile feeder losses are 2 dB. Express the field strength at the receiver antenna in dBμV m^{-1}.

5.4 See Problem 4.12 (Chapter 4): Calculate the received power (in dBm) at antenna B if antenna A is transmitting at 60 W.

5.5 Calculate the maximum range of the system described in 5.3 using (a) the free space loss and (b) the plane earth loss, assuming a frequency of 2 GHz and antenna heights of 15 m and 1.5 m.

5.6 A 1 V RMS sinusoidal source is applied across the terminals of a half-wave dipole, with source and load impedances perfectly matched. An identical dipole is placed 10 m from the first. What is the maximum power available at the terminals of the receive antenna and under what conditions is this power produced? State any assumptions made in the calculation.

5.7 A satellite is operated at C-band (6 GHz in the uplink and 4 GHz in the downlink) for video broadcasting. Calculate the free-space loss experienced at this frequency, if the satellite is in geostationary orbit (GEO), located at 36 000 km above ground level. What is the minimum EIRP needed to provide adequate reception, assuming a receiver sensitivity of -120 dBm and effective antenna gain of 20 dBi?

5.8 A one-way microwave link operating at 10 GHz has these parameters:

Transmitter power 13 dBW
Transmitter feeder loss 5 dB
Transmitter antenna gain 18 dBd
Receiver antenna gain 10 dBi
Receiver feeder loss 3 dB
Receiver noise bandwidth 500 kHz
Receiver noise figure 3 dB

The receiver operates satisfactorily when the signal-to-noise ratio at its input is at least 10 dB. Calculate the maximum acceptable path loss.

5.9 A transmit antenna produces an EIRP of 1 W and is received by an antenna with an effective aperture of 1 m^2 at a distance of 1 km. The frequency is 100 MHz. Calculate the received power and the path loss. Assuming this is the maximum acceptable path loss for the system, how does the maximum system range change at 1 GHz and 10 GHz, assuming all other parameters remain constant?

6 Terrestrial Fixed Links

*'It seems hardly worthwhile to expend much effort on the building of long-distance
relay chains. Even the local networks which will soon be under construction may
have a working life of only 20–30 years'.*

Arthur C. Clarke

6.1 INTRODUCTION

It is frequently useful to provide wireless connectivity between two fixed stations separated by
a large distance on the Earth. This may act as the main connectivity between these stations or
can act as a fall-back link for a cabled system. In this chapter, the principles established in
earlier chapters will be applied to show how propagation in practical terrestrial fixed links
may be predicted. These links are radio systems involving a pair of stations mounted on masts
and separated by tens or even hundreds of kilometres. They are commonly used for very high
data rate systems, such as telephone trunked lines, and are intended to have very high
reliability. The stations are typically mounted on masts of many tens of metres above the
Earth's surface, usually located in positions free from any local clutter, including trees and
buildings. Highly directional antennas, usually parabolic dishes or Uda–Yagis, are typically
applied in order to permit a generous fade margin.

6.2 PATH PROFILES

It is usually assumed in fixed link propagation prediction that propagation takes place
predominantly across and through obstacles on the *great circle* or *geodesic* path between
the terminals. This path represents the shortest distance between the two terminals, measured
across the Earth's surface. Since the Earth is approximately spherical, the great circle path
will not be the same as the straight-line path drawn between the terminals on a conventional
map, but the difference is usually small. An example is shown in Figure 6.1, where the
great circle path between two terminal locations is shown on a contour map. The great
circle path length between two points with latitudes θ_1 and θ_2 and longitudes λ_1 and λ_2 is
given by

$$r = R\cos^{-1}[\cos\theta_1\cos\theta_2\cos(\lambda_1 - \lambda_2) + \sin\theta_1\sin\theta_2] \qquad (6.1)$$

assuming the Earth is spherical, with a radius of around $R = 6375$ km. For really accurate
work, particularly over very long paths, it is necessary to account for departures of the
Earth's shape from spherical. This will not be considered in this chapter, but see [OS, 06]
for details.

Antennas and Propagation for Wireless Communication Systems Second Edition   Simon R. Saunders and
Alejandro Aragón-Zavala
© 2007 John Wiley & Sons, Ltd

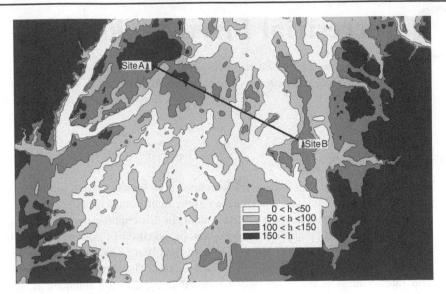

Figure 6.1: A great circle path between two antenna sites; contours are marked with heights in metres above mean sea level. Data is for the Seattle area using the USGS 3 and 1 arcsecond digital terrain models translated into Vertical Mapper format (reproduced by permission of Ericsson)

In practice, there may be some propagation effects which take place off the path; for example, there may be significant reflecting or scattering surfaces which are illuminated by the transmit antenna. These will usually be much smaller in amplitude than the direct path, however, an appropriate choice of an antenna with high directivity can minimise illumination of such areas. Similarly, diffraction associated with the direct path will not be confined to the great circle path, but will be affected by objects within a volume associated with the Fresnel zones around the direct ray, as explained in Chapter 3. At VHF frequencies and higher, the width of these zones is relatively small, so it is usually assumed that the objects along the great circle path are representative.

These assumptions permit the great circle path to be used to construct a *path profile*. This is a section through the Earth along the great circle path, showing the terrain heights and the terminal heights. An example corresponding to the path marked in Figure 6.1 is shown in Figure 6.2.

In the past, terrain data has been collected using conventional land-based surveying techniques. Now methods involving aerial photography, laser range-finding and satellite imaging are common. Height resolutions as low as 1 m are available from commercial data providers. Interpolation techniques are used to reconstruct a continuous path profile from a finite set of data points between the terminals of the path. Grid intervals of $50\,\text{m} \times 50\,\text{m}$ or $10\,\text{m} \times 10\,\text{m}$ are common for prediction work.

Also, a baseline is plotted in Figure 6.2, the *Earth bulge*, which shows the height of the Earth at mean sea level, given by

$$b \approx \frac{r_1 r_2}{2R} \qquad (6.2)$$

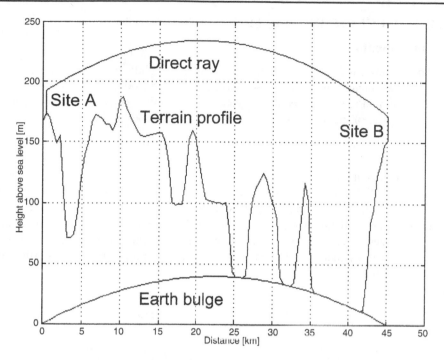

Figure 6.2: Path profile corresponding to the path shown in Figure 6.1

where the approximation holds provided that $r_1, r_2 \ll R$. The Earth bulge is illustrated in Figure 6.3. This Earth bulge has been used to uplift the height of the terrain in Figure 6.2 to correctly represent the obstructing effect of the Earth's curvature. Also, the horizon distances d_{r1} and d_{r2} are shown, which are the greatest distances on the Earth's surface that are visible from antennas of height h_1 and h_2, respectively. Again assuming $d_1, d_2 \ll R$, the horizon distances are

$$d_{r1} = \sqrt{2Rh_1} \quad \text{and} \quad d_{r2} = \sqrt{2Rh_2} \tag{6.3}$$

When the terrain variations are slight, the Earth bulge represents the basic obstruction to long-range terrestrial systems. More usually, the terminals are placed on elevated terrain to greatly extend the effective antenna height and thereby to increase the horizon distance.

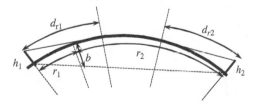

Figure 6.3: Geometry of elevated antennas on spherical Earth surface

6.3 TROPOSPHERIC REFRACTION

6.3.1 Fundamentals

The refractive index n of the Earth's atmosphere is slightly greater than the free space value of 1, with a typical value at the Earth's surface of around 1.0003. Since the value is so close to 1, it is common to express the refractive index in N units, which is the difference between the actual value of the refractive index and unity in parts per million:

$$N = (n - 1) \times 10^6 \tag{6.4}$$

This equation defines the *atmospheric refractivity* N. Thus, the surface value of $N = N_S \approx 300 \, N$ units. N varies with pressure, with temperature and with the water vapour pressure of the atmosphere. Although all of these quantities vary with both location and height, the dominant variation is vertical with height above the Earth's surface, with N value reducing towards zero (i.e. n value coming close to 1) as the height is increased. The variation is approximately exponential within the first few tens of kilometres of the Earth's atmosphere – the *troposphere*, so:

$$N = N_S \exp\left(-\frac{h}{H}\right) \tag{6.5}$$

where h is the height above mean sea level, and $N_S \approx 315$ and $H = 7.35 \, \text{km}$ are standard reference values [ITU, 453]. An environment with a good approximation to the behaviour described by Eq. (6.5) is a *standard atmosphere*.

This refractive index variation with height causes the phase velocity of radio waves to be slightly slower closer to the Earth's surface, such that the ray paths are not straight, but tend to curve slightly towards the ground. This curvature can be calculated from Snell's law of refraction (3.2). Figure 6.4 defines the geometry of a simple case, where the atmosphere is divided into two layers of differing refractive index, with the transition from $n = n_1$ to $n = n_2$ occurring at a height h above the surface. A ray is launched at an angle α to the surface and is incident on the transition between layers at an angle θ_1 to the normal.

Applying Snell's law of refraction, with the assumption that the refraction occurs as if the wave and the interface were locally plane, we obtain

$$\frac{\sin \theta_1}{\sin \theta_2} = \frac{n_2}{n_1} \tag{6.6}$$

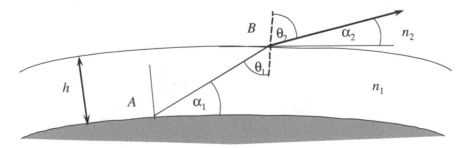

Figure 6.4: Refraction from a vertically stratified atmospheric layer

In the triangle ABX, where X is the centre of the Earth, the angles and side lengths can be related using the sine law as

$$\frac{R+h}{\sin(\alpha_1 + \pi/2)} = \frac{R}{\sin \theta_1} \tag{6.7}$$

Substituting this expression into Eq. (6.6) gives

$$\frac{\cos \alpha_1}{\sin \theta_2} = \frac{R+h}{R}\left(\frac{n_2}{n_1}\right) = \frac{\cos \alpha_1}{\cos \alpha_2} \tag{6.8}$$

This is known as *Bouger's law*. It can be applied to thin layers (small h) to trace rays approximately through regions with any refractive index profile. More accurately, the radius of curvature of the ray, ρ, at any point is given in terms of the rate of change of n with height,

$$\frac{1}{\rho} = -\frac{\cos \alpha}{n}\frac{dn}{dh} \tag{6.9}$$

where α is the elevation angle of the ray at the point [ITU, 834]. The resulting ray curvature is visible in the ray illustrated in Figure 6.2 and in more detail in Figure 6.6. The curvature is not normally as great as that of the Earth's surface, but is nevertheless sufficient to cause the ray to bend around the geometrical horizon calculated in Eq. (6.3), extending the overall range somewhat.

For small heights, the standard atmosphere of Eq. (6.5) can be approximated as linear, as shown in Figure 6.5, according to the following equation:

$$N \approx N_s - \frac{N_s}{H}h \tag{6.10}$$

The refractivity near the ground thus has a nearly constant gradient of about -43 N units per kilometre. In this case, when the elevation angle of the ray is close to zero, Eq. (6.9) shows that the curvature of the ray is constant, so the ray path is an arc of a circle.

A common way to represent this is to increase the radius of the Earth to a new effective value, so that the ray now appears to follow a straight path. The Earth bulge in Eq. (6.2) and the horizontal distances in Eq. (6.3) are recalculated with R replaced by a new value R_{eff}:

$$R_{eff} = k_e R \tag{6.11}$$

where k_e is the *effective Earth radius factor*, given by

$$k_e = \frac{1}{R\frac{dn}{dh} + 1} \tag{6.12}$$

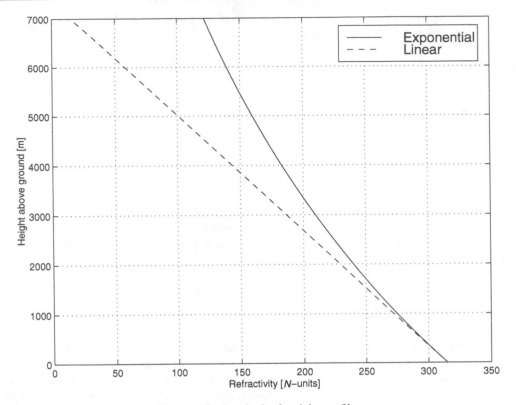

Figure 6.5: Standard refractivity profiles

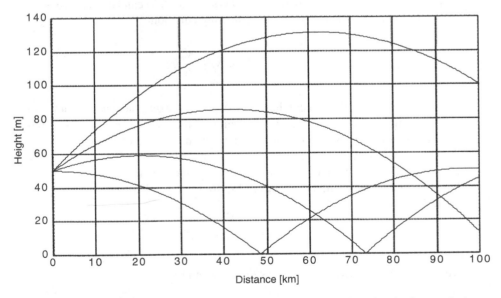

Figure 6.6: Rays launched from a height of 50 m above the Earth's surface in the standard atmosphere of Eq. (6.5). Angles of elevation are 0°, 0.05°, 0.1° and 0.15°. Snell's law of reflection has been applied when rays hit the Earth's surface

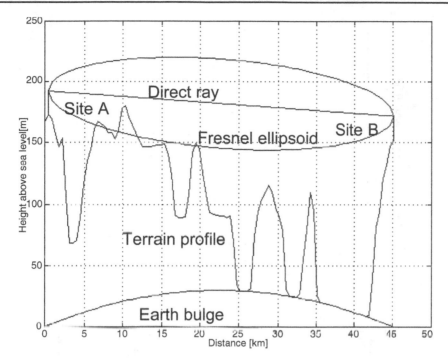

Figure 6.7: Path profile as Figure 6.2 but with Earth radius corrected to account for atmospheric refractive index gradient. The Fresnel ellipsoid represents 0.6 times the first Fresnel zone at 900 MHz

The median value for k_e is taken to be 4/3, so the effective radius for 50% of the time is $(4 \times 6375)/3 = 8500$ km. Figure 6.7 shows the path profile of Figure 6.2 redrawn using Eq. (6.11). The Earth bulge and terrain profile is reduced and the ray path can now be drawn as straight, without changing the obstruction of the Fresnel ellipsoid by the terrain.

6.3.2 Time Variability

The standard atmosphere, which yields the value $k_e = 4/3$, is the median value only, exceeded for 50% of the time in a typical year. In practice, meteorological conditions may cause large deviations. The most severe deviations occur when the refractive index decreases with height much more rapidly than normal, reducing k_e and hence reducing the distance to the horizon as given by Eq. (6.3). Such conditions are known as *superrefractive*. The value of k_e exceeded for 99.9% of the worst month in temperate climates is given in Figure 6.8. In order to minimise the obstruction loss associated with these conditions, the antenna heights should be planned to provide clearance of 0.6 times the first Fresnel zone for a value of k_e corresponding to the desired availability of the link.

6.3.3 Ducting and Multipath

If a significant reflection from the ground or the sea is also present, a two-path system is created, causing multipath interference at the receiver similar to that analysed in the plane

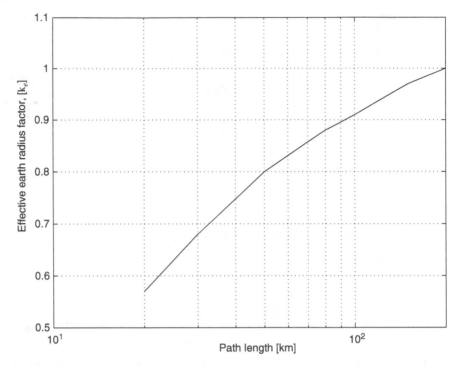

Figure 6.8: Value of effective Earth radius factor exceeded for 99.9% of the worst month in continental temperate climates [ITU, 530]

earth model of Chapter 5. This may cause severe degradations in the signal level if the interference is destructive. This situation may be avoided on stable paths by adjusting the antenna heights to provide constructive interference in normal atmospheric conditions, but the time variability of k_e causes variations in the path length differences, resulting in time variations (fading) of the received signal which must be allowed for by providing an adequate fade margin.

Another way in which multipath fading can occur is by *ducting*. This typically occurs in unusual atmospheric conditions where the temperature increases with height over some limited range, creating a *temperature inversion* and *subrefractive* conditions. Here the refractive index gradient may be very steep over some height interval, creating the possibility that a ray launched from the transmitter with a large elevation angle may return to the receiver, when it would normally be lost into space (see Figure 6.9 for an extreme example). This ray combines with the normal path to produce multipath fading. In some cases, there may be a large number of rays simultaneously interfering at the receiver, and this may lead to the severe case of *Rayleigh fading* as detailed in Chapter 10. In hot climates, and over the sea, ducting conditions may be present almost all of the time. In temperate climates, ducting usually occurs in the early evening as the Sun sets, leading to rapid cooling of the ground. The effects of ducting may be minimised by either making the antenna heights very different or applying *vertical space diversity* as described in Chapter 16.

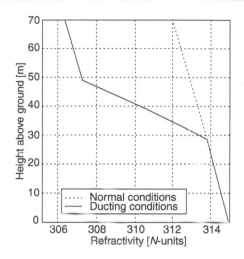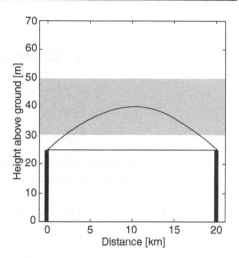

Figure 6.9: Example refractivity profile and two corresponding ray paths during ducting conditions: the inversion layer occurs between 30 m and 50 m above ground level

6.4 OBSTRUCTION LOSS

In Figure 6.7, the 0.6 times first Fresnel zone ellipsoid contains three terrain peaks, which will create some *obstruction loss* in excess of the free space loss. In many practical fixed links, particularly those operated at high microwave frequencies (tens of gigahertz), the available link margin is too small to allow any obstruction loss and the link must be planned to ensure that the path is completely unobstructed. For links operated at lower frequencies, however, the Fresnel zone radius may be too large to completely avoid obstruction. The designer must then rely on being able to accurately predict the obstruction loss.

The usual approach to calculation is to represent peaks in the terrain by a series of equivalent absorbing knife-edges. This makes the implicit assumption that the terrain peak is sharp enough to be represented by a knife-edge and that the peak is wide enough transverse to the path to neglect any propagation from around the sides of the hill. If only a single equivalent knife-edge exists within 0.6 times the first Fresnel zone, then the obstruction loss can be predicted directly using the methods in Chapter 3. If there are several edges, then either the approximate methods in Section 6.5 or the more accurate approach in Section 6.6 can be applied. Objects with shapes other than knife-edges are considered in Sections 6.7 and 6.8.

Example 6.1

A microwave link operating at 10 GHz with a path length of 30 km has a maximum acceptable path loss of 169 dB. The transmitter antenna is mounted at a height of 20 m above ground level, while the height of the receiver antenna is to be determined. The ground is level apart from a hill of height 80 m, located 10 km away from the transmitter antenna.

(a) Calculate the total path loss assuming the receiver antenna is mounted at a height of 20 m above ground level.

(b) Calculate the height of the receiver antenna for the path loss to be just equal to the maximum acceptable value.

Solution

(a) From Eq. (5.22) the free space loss is

$$L_{F(\text{dB})} = 32.4 + 20\log R + 20\log f_{\text{MHz}}$$
$$= 32.4 + 20\log 30 + 20\log 10\,000$$
$$\approx 142\,\text{dB}$$

Assume that the hill can be modelled as a perfectly absorbing, infinitely thin knife-edge of infinite extent transverse to the propagation direction. This leads to the geometry shown in the diagram.
Using Eq. (3.28) the Fresnel diffraction parameter is then given by

$$v = h\sqrt{\frac{2(d_1 + d_2)}{\lambda\, d_1 d_2}}$$
$$= 60\sqrt{\frac{2 \times 30 \times 10^3}{3 \times 10^{-2} \times 10 \times 10^3 \times 20 \times 10^3}} = 6$$

Hence the approximate form of the knife-edge attenuation (3.26) can be applied,

$$L_{ke} \approx -20\log\frac{0.225}{v} = 28.5\,\text{dB}$$

The total path loss is therefore $142 + 28.5 = 170.5\,\text{dB}$, which is in excess of the maximum acceptable limit and so the system must be redesigned.

(b) In order to reduce the obstruction loss to an acceptable level, the excess loss must be no greater than $169 - 142 = 27\,\text{dB}$. Hence, using the approximate attenuation again

$$L_{ke} \approx -20\log\frac{0.225}{v} = 27\,\text{dB}$$

Rearranging this for v yields

$$v = \frac{0.225}{10^{-27/20}} = 5$$

This value is large enough to justify continued use of the approximation. Then rearranging the expression for the diffraction parameter yields the excess knife-edge height as

$$h = v\sqrt{\frac{\lambda d_1 d_2}{2(d_1 + d_2)}} = v \times 10 = 50\,\text{m}$$

The new receiver height can then be calculated by simple geometry as 50 m.

6.5 APPROXIMATE MULTIPLE KNIFE-EDGE DIFFRACTION

If more than one terrain obstruction exists within the Fresnel ellipsoid, a multiple diffraction problem must be solved. It is not correct to simply add the obstruction losses from each edge individually, since each edge disturbs the field, producing wave fronts incident on the next edge which are not plane, contravening the assumptions in the single-edge theory. The most widely used approach to predict multiple knife-edge diffraction is to use an approximate method which simplifies the path geometry, using simple geometrical constructions, to calculate a total diffraction loss in terms of combinations of single-edge diffractions between adjacent edges. Examples are [Bullington, 77], [Epstein, 53], [Deygout, 66] and [Giovanelli, 84]. Only the last two methods will be described here, since both have been widely used and give reasonable results in practical terrain diffraction problems.

6.5.1 The Deygout Method

The method by [Deygout, 66] is simple to apply and explain and can give reasonable results under restrictive circumstances.

First, the Fresnel diffraction parameter v for a single knife-edge is expressed using Eq. (3.28),

$$v(d_a, d_b, h) = h\sqrt{\frac{2(d_a + d_b)}{\lambda d_a d_b}} \tag{6.13}$$

where d_a is the distance from the source to the edge, d_b is the distance from the edge to the field point and h is the excess height of the edge.

Consider the situation involving three edges illustrated in Figure 6.10. The v parameter for each of the edges is first calculated as if the edge were present alone, i.e.

$$\begin{aligned}
v_1 &= v(d_1, d_2 + d_3 + d_4, h_1) \\
v_2 &= v(d_1 + d_2, d_3 + d_4, h_2) \\
v_3 &= v(d_1 + d_2 + d_3, d_4, h_3)
\end{aligned} \tag{6.14}$$

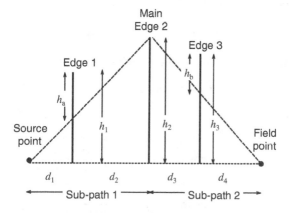

Figure 6.10: Geometry for the Deygout model

The edge with the largest positive value of v is designated the *main edge*, in this case edge 2. The single-edge loss (in decibels) for this edge is calculated from the Fresnel theory as

$$L_{main} = L_{ke}(v_2) \tag{6.15}$$

where L_{ke} is given by Eq. (3.23).

The main edge is then used to split the path into two sub-paths. The point at the top of the main edge is treated as a field point with respect to the original source, and as a source for the original field point. New v parameters are then calculated for the two path segments with respect to these new paths,

$$\begin{aligned} v_1' &= v(d_1, d_2, h_a) \\ v_3' &= v(d_3, d_4, h_b) \end{aligned} \tag{6.16}$$

The total excess loss is then calculated by combining the loss from the main edge and from each of the two sub-paths,

$$L_{ex} = L_{ke}(v_1') + L_{ke}(v_2) + L_{ke}(v_3') \tag{6.17}$$

This is the excess loss and must be combined with the free space loss to give the total loss.

In this example, the two sub-paths contain only a single edge. If multiple edges are present, the whole method may be reapplied to the sub-path and this is continued until each sub-path contains only a single edge or no edge. In this way, diffraction over any number of edges may be predicted.

6.5.2 The Causebrook Correction

The Deygout method tends to overestimate the true path loss when there are a large number of edges, or when pairs of edges are very close together. In order to reduce this problem [Causebrook, 71] proposed an approximate correction derived from the exact analysis of the two-edge solution [Millington, 62]. If we express Eq. (6.17) in decibels as

$$L_{ex} = L_1' + L_2 + L_3' \tag{6.18}$$

then the corrected form is given by

$$L_{ex}^{corr} = L_1' + L_2 + L_3' - C_1 - C_2 \tag{6.19}$$

The correction factors C_1 and C_2 are given by

$$C_1 = (6 - L_2 + L_1) \cos \alpha_1 \quad \text{and} \quad C_2 = (6 - L_2 + L_3) \cos \alpha_3 \tag{6.20}$$

where

$$\begin{aligned} \cos \alpha_1 &= \sqrt{\frac{d_1(d_3 + d_4)}{(d_1 + d_2)(d_2 + d_3 + d_4)}} \\ \cos \alpha_3 &= \sqrt{\frac{(d_1 + d_2)d_4}{(d_1 + d_2 + d_3)(d_3 + d_4)}} \end{aligned} \tag{6.21}$$

and L_1 and L_3 are the losses due to edges 1 and 3, respectively, as if they existed on their own between the original source and field points. In this method, only the edges that lie above the relative line-of-sight paths are taken into account.

6.5.3 The Giovanelli Method

Another development of the Deygout method has also been proposed [Giovanelli, 84]. The geometry for the method is shown in Figure 6.11. Only two edges are shown here, but the method is straightforward to extend to multiple edges.

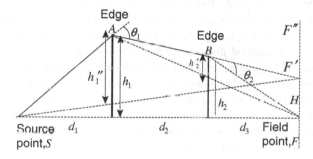

Figure 6.11: Geometry for the Giovanelli method

The main edge is again identified, just as given in Section 6.5.1. Assume edge A to be the main edge in this case. Then a reference field point F' is found by projecting AB onto FF'' and h_1'' (an effective excess height for edge A above SF') is defined by

$$h_1'' = h_1 - \frac{d_1 H}{d_1 + d_2 + d_3} \tag{6.22}$$

where $H = h_2 + md_3$ and $m = (h_2 - h_1)/d_2$. The effective height for the secondary edge B is then given by

$$h_2' = h_2 - \frac{d_3 h_1}{d_2 + d_3} \tag{6.23}$$

The excess loss is now given by the following expression:

$$L_{ex} = L_{ke}(v(d_1, d_2 + d_3, h_1'')) + L_{ke}(v(d_2, d_3, h_2'')) \tag{6.24}$$

The results are not unconditionally reciprocal (i.e. swapping source and field points may give slightly different results), which is also the case for Deygout. The method is conveniently extended to more than two edges by recursively applying the above procedure.

6.5.4 Test Cases

The Deygout, Causebrook and Giovanneli methods are compared here for three test cases. As an accurate reference we use the Vogler method, itself described in Section 6.6.

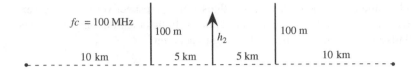

Figure 6.12: Multiple diffraction: test case 1

The first test case (Figure 6.12) shows a three-edge arrangement which might be encountered in a terrain diffraction calculation. The edges obstruct the line-of-sight path and h_2 is varied over a wide range. Two cases of grazing incidence occur when $h_2 = 100$ m and $h_2 = 150$ m. The results for this case are shown in Figure 6.13. All of the models perform reasonably well when h_2 is large, since the diffraction loss is then dominated by a single edge. The Causebrook method does not predict the oscillatory behaviour of the field, whereas Giovanelli and Deygout achieve this well with only a small error at the peaks and troughs of the variation. Around the region of grazing incidence, the errors are large, peaking at around 8 dB for Causebrook and Deygout and around 4 dB for Giovanelli. For lower heights, where the test case reduces to a two-edge situation, Deygout has an offset of around 3 dB, whereas Causebrook estimates the loss very accurately. The Giovanelli method has a small offset of around 0.5 dB and overestimates the extent of the fluctuations.

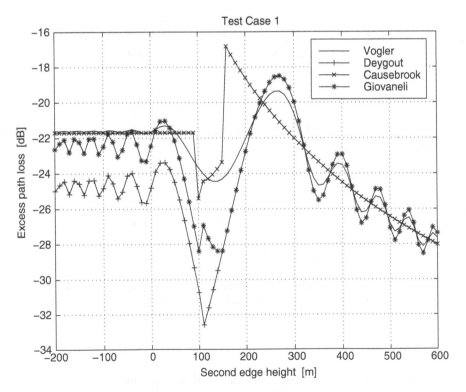

Figure 6.13: Comparison of models for test case 1

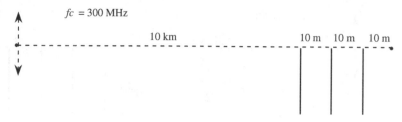

Figure 6.14: Multiple diffraction: test case 2

Test case 2 (Figure 6.14) is more severe. Three edges and the field point are collinear, and the transmitter is varied in height by 100 m at either sides of grazing incidence. The results of this case are shown in Figure 6.15. The Deygout and Giovanelli methods consistently overestimate the loss by around 7 dB. The Causebrook correction does considerably better, but underestimates the loss by around 5 dB when the transmitter height is positive.

Finally, test case 3 (Figure 6.16) examines the grazing incidence case as the number of edges n is varied with results as shown in Figure 6.17. Here the Giovanelli and Deygout approaches predict 6 dB loss per edge, resulting in a serious overestimate of loss for $n > 1$

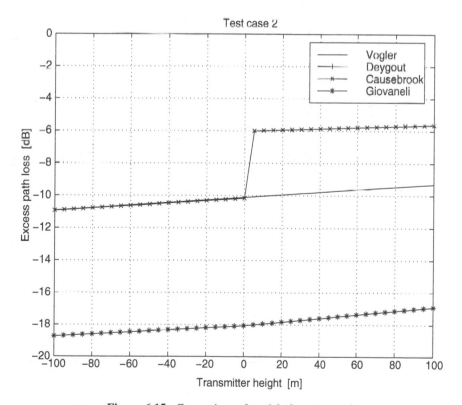

Figure 6.15: Comparison of models for test case 2

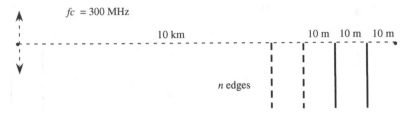

Figure 6.16: Multiple diffraction: test case 3

(Figure 6.17). The Causebrook method does rather better, although the slope is different to the accurate case, which could cause significant errors for large numbers of edges.

It is concluded that the approximate methods presented in this section are inadequate for situations involving large number of edges, particularly when a series of edges is collinear, although the ease of implementation and relative simplicity of calculation make them attractive for many terrain diffraction problems. The following section describes a more accurate but more complicated method.

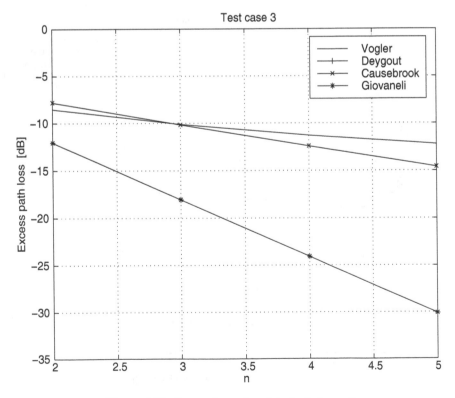

Figure 6.17: Comparison of models for test case 3

6.6 THE MULTIPLE-EDGE DIFFRACTION INTEGRAL

For cases involving large number of obstructions, particularly, close to grazing incidence, the last section showed that methods based on single-edge diffraction are unreliable.

In 1963, [Furutsu, 63] published complete expressions for propagation over various configurations of inhomogeneous terrain. One of the cases considered was a two-dimensional representation of propagation over multiple cascaded circular cylinders; this includes the multiple knife-edge case when the radius of curvature becomes vanishingly small. The results were given in the form of a residue series, which is slow to converge in the knife-edge case. This series is transformed by [Vogler, 82] into a multiple integral representation of the diffraction loss. The resulting *multiple-edge diffraction integral* is a very general and useful tool and is presented and briefly discussed here.

Vogler expresses the excess diffraction loss (as a ratio between the received field strengths with and without the edges present) due to n knife-edges as

$$A_n = \sqrt{L_{ex}} = C_n \pi^{-n/2} e^{\sigma_n} I_n \tag{6.25}$$

where

$$I_n = \int\limits_{x_n=\beta_n}^{\infty} \cdots \int\limits_{x_1=\beta_1}^{\infty} \exp\left(2f - \sum_{m=1}^{n} x_m^2\right) dx_1 \ldots dx_n \tag{6.26}$$

with

$$f = \sum_{m=1}^{n-1} \alpha_m (x_m - \beta_m)(x_{m+1} - \beta_{m+1}) \quad \text{for } n \geq 2 \tag{6.27}$$

where

$$\alpha_m = \sqrt{\frac{d_m d_{m+2}}{(d_m + d_{m+1})(d_{m+1} + d_{m+2})}} \tag{6.28}$$

$$\beta_m = \theta_m \sqrt{\frac{jk d_m d_{m+1}}{2(d_m + d_{m+1})}} \tag{6.29}$$

$$\sigma_n = \beta_1^2 + \cdots + \beta_n^2 \tag{6.30}$$

$$C_n = \sqrt{\frac{d_2 d_3 \times \ldots \times d_n d_T}{(d_1 + d_2)(d_2 + d_3) \times \ldots \times (d_n + d_{n+1})}} \tag{6.31}$$

$$d_T = d_1 + \ldots + d_{n+1} \tag{6.32}$$

and the geometrical parameters are defined in Figure 6.18 (note that θ_1 and θ_3 are positive in this diagram, θ_2 is negative).

This expression is valid in general, subject only to the usual assumptions of the same Huygens–Fresnel theory which is used to derive the single knife-edge result (3.23) in Chapter 3. It therefore potentially constitutes an extremely powerful tool, provided that efficient

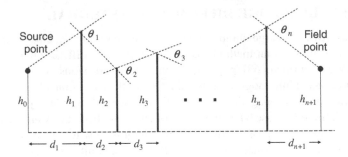

Figure 6.18: Geometry for the multiple-edge integral

methods of evaluating the integrals can be found. This problem is non-trivial, since some of the characteristics of the integral present special computational difficulties. Additionally, the high dimensionality of the integral when n is large makes computation slow.

The β_m parameters can be expressed in terms of the Fresnel zone radius as follows:

$$\beta_m = \theta_m \sqrt{\pi} \frac{r_1}{\lambda} \tag{6.33}$$

where r_1 is the radius of the first Fresnel zone for a ray joining the $(m-1)$ and the $(m+1)$ edges. The β_m terms are therefore analogous to the Fresnel diffraction parameter v, as if the $(m-1)$ edge was a source and $(m+1)$ was a field point. The α_m terms express the coupling between adjacent edges. When there is little coupling $(r_{m+1} \gg r_m, r_{m+2})$, the α_m term is negligible so the m and $(m+1)$ integrals can be separated into the product of two independent terms. In such cases, the approximate methods described in Section 6.5 should give good results.

Computation of the multiple-edge integral (6.26) can be achieved via a method described by [Vogler, 82] which transforms the integral into an infinite series. This takes a significant amount of computation time, rising rapidly with the number of edges. Additionally, the series only works in the form given for up to around 10 knife-edges. Nevertheless, this method represents a useful reference for comparison of other techniques and can be rendered practical for real predictions by the use of advanced computational techniques and evaluation methods different to that proposed by Vogler; see for example [Saunders, 94] and [Tzaras, 01]. A similar approach was used for computing the reference results in the test cases of the previous section.

6.6.1 Slope-UTD Multiple-Edge Diffraction Model

A new method for performing multiple-edge diffraction modelling has been developed which approximates the full multiple-edge diffraction integral with far better accuracy than the methods detailed in Section 6.5, while providing far shorter computational time than the Vogler method. This method takes the uniform theory of diffraction (UTD) introduced in Section 3.5.3 as a starting point. In principle, UTD diffraction coefficients, such as Eq. (3.34), can simply be cascaded for multiple diffracting edges. Used on its own, however, the UTD produces inaccurate results when edges are illuminated by diffracting fields within the transition regions of previous edges because the spatial variation of the field is too large for the basic plane-wave assumptions of UTD to hold. Instead, the diffraction coefficient is supplemented by an additional 'slope' diffraction coefficient which accounts for the spatial rate of change of the diffracted field. This approach was first introduced in the case of large numbers of edges by [Andersen, 97] and was subsequently refined by [Tzaras, 01] to improve the accuracy of the method when the number of edges increases or the edges have unequal heights.

According to the slope-UTD theory, the diffracted field for a single absorbing knife-edge is given by

$$E_d = \left[E_i D(\alpha) + \frac{\partial E_i}{\partial n} d_S \right] A(s) e^{-jks} \tag{6.34}$$

The geometry for this situation is shown in Figure 6.19. In Eq. (6.34), E_d is the diffracted field, α is the angle above the shadow boundary $(\phi - \phi')$, ϕ is the diffraction angle with respect to the transmitter side of the edge, ϕ' is the incident angle with respect to the same side, n' is the normal to the wave direction and $A(s)$ is the spreading factor,

$$A(s) = \sqrt{\frac{s_0}{s(s + s_0)}} \tag{6.35}$$

The amplitude diffraction coefficient $D(\alpha)$ is given by

$$D(\alpha) = -\frac{e^{-j\pi/4}}{2\sqrt{2\pi k} \cos(\alpha/2)} F[2kL \cos^2(\alpha/2)] \tag{6.36}$$

whereas the slope diffraction coefficient is

$$d_s(\alpha) = \frac{1}{jk} \frac{\partial D(\alpha)}{\partial \alpha} \tag{6.37}$$

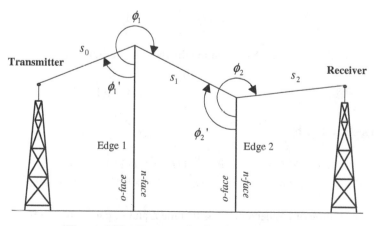

Figure 6.19: Geometry for the slope-UTD model

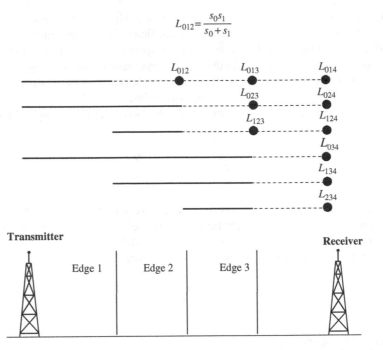

$$L_{012} = \frac{s_0 s_1}{s_0 + s_1}$$

Figure 6.20: Path geometry for three knife edges with equal heights and spacing

The term $F(x)$ is known as the transition function,

$$F(x) = 2j\sqrt{x}e^{jx} \int_{\sqrt{x}}^{\infty} e^{-ju^2} du \qquad (6.38)$$

and the derivative of this function is

$$F'(x) = j[F(x) - 1] + \frac{F(x)}{2x} \qquad (6.39)$$

From Eqs. (6.37) and (6.38), the slope diffraction coefficient is then given by

$$d_s(\alpha) = -\frac{e^{-j\pi/4}}{\sqrt{2\pi k}} L_s \sin(\alpha/2)(1 - F(x)) \qquad (6.40)$$

and its derivative by

$$\frac{\partial d_s(\alpha)}{\partial n} = -\frac{1}{2s} \frac{e^{-j\pi/4}}{\sqrt{2\pi k}} \left\{ \begin{array}{l} L_s \cos(\alpha/2)[1 - F(x)] \\ +4L_s^2 k \sin^2(\alpha/2) \cos(\alpha/2) F'(x) \end{array} \right\} \qquad (6.41)$$

The contribution of [Andersen, 97] is that the L and L_s parameters in Eqs. (6.36) and (6.40) are calculated according to some continuity equations, which ensure continuity of the diffracted field and its slope along the shadow boundaries of the edges. The work in [Tzaras, 01] has built upon this approach by observing that the continuity equations vary for each ray independently, producing values of $D(\alpha)$ and $d_s(\alpha)$ which are different for every ray, even if the diffracted edge and the receiver point are the same. Although such an approach results in a little extra complexity, this method produces a more accurate result compared to [Andersen, 97]. Furthermore, the approach in [Tzaras, 01] is more physically correct, since both the magnitude and the phase of the signal are taken into account when the L and L_s values are calculated. The resulting produces accurate and reliable output which, in most cases, has almost the same accuracy as the Vogler solution [Vogler, 82] but with far less computational complexity.

To illustrate the calculation of the L and L_s parameters using different continuity equations for each ray, consider the geometry of Figure 6.20, where three edges of equal heights and spacing are shown. It is assumed that the slope of the field is zero only when it originates from the transmitter. Hence, denoting the total field at edge m by E_m and the field at edge m due to edge n as E_{nm}, the total field on the second edge is

$$E_2 = E_{02} + E_{01}D_1(\alpha_{012})A_1(s_1)e^{-jks_1} \tag{6.42}$$

In Eq. (6.42), E_{02} is the incident field from the source, E_{01} is the incident field from the source at edge 1 and α_{012} comprises the transmitter, the first and the second edge. The continuity equation for evaluating the L parameter that appears in $D_1(\alpha)$ is

$$0.5E_{02} = E_{01}D_1(\alpha_{012})A_1(s_1)e^{-jks_1} \tag{6.43}$$

For $\alpha = \pi$, the amplitude diffraction coefficient is given by

$$D(\pi) = 0.5\sqrt{L} \tag{6.44}$$

Hence, following Eq. (6.43),

$$\frac{0.5}{s_0 + s_1}e^{-jk(s_0+s_1)} = \frac{0.5}{s_0}e^{-jks_0}\sqrt{L_{012}}\sqrt{\frac{s_0}{s_1(s_0 + s_1)}}e^{-jks_1} \tag{6.45}$$

from which

$$L_{012} = \frac{s_0 s_1}{s_0 + s_1} \tag{6.46}$$

The total field at edge 2 is then

$$E_2 = \frac{0.5}{s_0 + s_1}e^{-jk(s_0+s_1)} \tag{6.47}$$

When the field at edge 3 and beyond needs to be calculated, the L parameters for the second edge need to be evaluated. The continuity equations for amplitude diffraction are

$$0.5E_{03} = E_{01}D_1(\alpha_{013})A_1(s_1 + s_2)e^{-jk(s_1+s_2)} \tag{6.48}$$

$$0.5E_{03} = E_{02}D_2(\alpha_{023})A_2(s_2)e^{-jks_2} \tag{6.49}$$

$$0.5E_{13} = E_{12}D_2(\alpha_{123})A_2(s_2)e^{-jks_2} \tag{6.50}$$

From Eqs. (6.48), (6.49) and (6.50), we have

$$L_{013} = \frac{s_0(s_1 + s_2)}{s_0 + s_1 + s_2} \tag{6.51}$$

$$L_{023} = \frac{(s_0 + s_1)s_2}{s_0 + s_1 + s_2} \tag{6.52}$$

$$L_{123} = \frac{(s_0 + s_1)s_2}{s_0 + s_1 + s_2} \tag{6.53}$$

Finally, for the slope term

$$\frac{\partial E_2}{\partial n'} = \frac{e^{j\pi/4}}{\sqrt{2\pi}}\sqrt{k}\sqrt{\frac{s_0}{s_1}}\frac{1}{(s_0 + s_1)^{3/2}}e^{-jks_1} \tag{6.54}$$

The L_s parameter is then given by

$$L_s = \left(\frac{s_0 + s_1}{s_0 + s_1 + s_2}\right)^{2/3}\left(\frac{s_1}{s_1 + s_2}\right)^{1/3}s_2 \tag{6.55}$$

The solution proposed by the slope-UTD model can be extended for more than three edges with far better accuracy than [Andersen, 97]. The case for unequal heights and spacing has also been solved in [Tzaras, 01] and a significant improvement in accuracy is reported.

6.6.2 Test Case: Comparison of Multiple Models

In [Tzaras, 00], a comparison of five different multiple diffraction methods is made (Deygout, Causebrook, Giovanelli, Vogler and slope-UTD), in real outdoor environments using measured results originally from broadcasting applications in various parts of the UK. The frequency varied between 40 MHz and 900 MHz depending on the site.

Figure 6.21 shows the probability of the received signal being less than the diffraction loss for one of the sites studied, containing 1390 records, as detailed in [Tzaras, 00]. The slope-UTD and Vogler methods perform much better than the others. For the total number of sites, the Causebrook solution had a standard deviation of error of around 8 dB with a mean of −3 dB, whereas the slope-UTD and Vogler methods have a standard deviation of error of 7.5 dB with a mean error of around −2 dB.

When implementing diffraction predictions starting from real terrain height data, it is necessary to represent the path profile, which may consist of many hundreds of height points, with a specific number of edges. This process can introduce uncertainty into the results. One approach is to choose points which have the smallest Fresnel zone clearance, as if single knife-edge diffraction were occurring, but this is also an approximation. Figure 6.22 shows how the standard deviation of the error behaves as the number of edges selected from the profile increases. From these results, the Deygout and Giovanelli methods are found to be very sensitive to the number of edges selected for prediction (5 per path profile, in this case). These methods cannot thus be regarded as reliable, since it is difficult to know how many edges should be considered for accurate estimation, demonstrating their improved physical basis. Only the Vogler and slope-UTD produced a decrease in standard deviation of the prediction error. [Tzaras, 00]

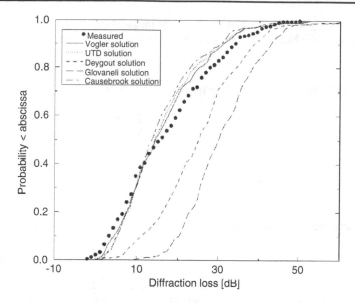

Figure 6.21: First-order statistics for digital broadcasting prediction comparisons [Tzaras, 00]

also demonstrates that the computational time required for the slope-UTD is much less than that for Vogler, with almost identical accuracy. Further comparisons of the slope-UTD method with detailed measurements are available in [Lee, 02].

6.7 DIFFRACTION OVER OBJECTS OF FINITE SIZE

When the terrain profile has significant curvature at its peak, the assumption that its width can be assumed infinitesimal may be inaccurate, so knife-edge diffraction calculations are no longer appropriate. An alternative approach is to treat the obstacle as a cylinder of finite radius. One straightforward method of doing this is given in [ITU, 526]. As shown in Figure 6.22, the cylinder is

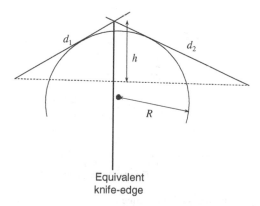

Figure 6.22: Geometry for single-cylinder diffraction

first replaced by an equivalent knife-edge at the intersection of the rays which start at the source and field points and which are tangential to the cylinder. The loss for this edge is calculated as

$$L_1 = L_{ke}(v(d_1, d_2, h)) \tag{6.56}$$

An extra decibel loss term is then introduced,

$$L_c(d_1, d_2, h, R) = (8.2 + 12n)\, m^{0.73+0.27[1-\exp(-1.43n)]} \tag{6.57}$$

where

$$m = \frac{R(d_1 + d_2)/d_1 d_2}{(\pi R/\lambda)^{1/3}} \quad \text{and} \quad n = \frac{h}{R}\left(\frac{\pi R}{\lambda}\right)^{2/3} \tag{6.58}$$

The overall excess loss is then

$$L_{ex} = 10 \log L_{ke}(v(d_1, d_2, h)) + L_c(d_1, d_2, h, R) \tag{6.59}$$

The additional cylinder loss L_c is illustrated in Figure 6.23 for a particular example.

This loss may be extended to multiple cylinders either using an approximate technique [ITU, 526] or by a more exact method, which is closely related to the multiple diffraction integral [Sharples, 89].

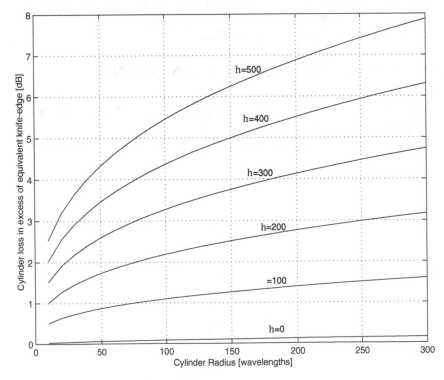

Figure 6.23: Loss in excess of knife-edge due to cylinder for the case where $d_1 = d_2 = 15$ km and $f = 300$ MHz for various heights of the equivalent knife-edge

In cases where the terrain is very smooth, or where the main part of the path profile exists over the sea, diffraction over the spherical earth must be accounted for. This may again be performed in an exact [ITU, 368] or approximate fashion [ITU, 526].

6.8 OTHER METHODS FOR PREDICTING TERRAIN DIFFRACTION

The methods outlined so far have all depended on reducing the surface to *canonical objects* such as edges, cylinders and spheres. This approach allows rapid computation and good accuracy, but it may sometimes be desirable to take into account the actual shape of the terrain, avoiding the need to select edges and potentially increasing the accuracy. This is practical only if sufficiently closely sampled information on the terrain heights, constitutive parameters and surface roughness are available. Two such methods, the *integral equation* and *parabolic equation* methods, are presented in this section.

Beyond these, there are also situations in which a significant contribution to the field comes from terrain which is not present along the great circle path. This may occur if the receiver is situated in a valley or bowl, so that it receives a strong field reflected or scattered from the terrain. Prediction of such cases requires very much greater accuracy and resolution for the terrain data and much more computation time.

6.8.1 The Integral Equation Model

The integral equation propagation model is a 'full-wave' method, originally developed by [Hufford, 52] in an attempt to predict field strength over irregular terrain from a geographical database, hence taking into account the actual shape of the terrain. Here we present a version of the integral equation model from [Hviid, 95] which is an evolution of the Hufford approach designed to avoid numerical instabilities, which the latter method exhibits at high frequencies. It makes the assumptions that the surface has to be a smooth, perfect magnetic conductor and that there are no variations of the surface transverse to the direction of propagation. The perfect magnetic conductivity assumption is equivalent to a reflection coefficient of -1, so this assumption is also reasonable for vertical polarisation near grazing incidence.

The geometry for the model is shown in Figure 6.24. Fields from the source point p_0 are scattered at points along the surface, denoted p'. The total field received at the observation point p is a combination of the incident field directly from the source and the fields scattered from all points along the surface. In principle, scattering occurs from points before the source and after the observation point, but in practice these 'backscatter' contributions can reasonably be assumed negligible. Sidescatter contributions from outside the x-z plane are also neglected.

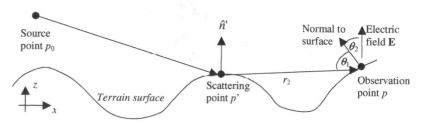

Figure 6.24: Geometry for integral equation method

The incident field induces currents on the surface of the terrain which radiate to produce the scattered field. The scattered field is calculated by summing scattered contributions from all elements of current along the surface. In principle, this requires an integral along the surface to sum infinitesimal surface current elements, hence an integral equation results. In practice, however, the integral can be replaced with a finite summation over surface elements provided these are of size Δx_m less than one half-wavelength.

The electric field is expressed in terms of an equivalent source current M^s flowing normal to the x-z plane of magnitude

$$M^s = |\mathbf{E}| \sin \theta_2 \tag{6.60}$$

The surface current M_n^s at a point at $x = n\Delta x_m$ from the source point is then given by the summation shown in Eq. (6.61). The first term represents the field incident from the source and the second term is the summation over the surface currents from source to observation points,

$$M_n^s = TM_{i,n}^s + \frac{T}{4\pi} \sum_{m=0}^{n-1} M_m^s f(n, m) \Delta x_m \tag{6.61}$$

T is a constant defined as 2 if p is on the surface and 1 elsewhere. The terms in this expression are defined as

$$f(n, m) = (r_2 \cos \theta_2) \frac{jk}{R_2} \sqrt{\lambda \frac{R_1 R_2}{R_1 + R_2}} e^{-j(kR_2 + \pi/4)} \frac{\Delta l_m}{\Delta x_m} \tag{6.62}$$

Here Δl_m is an increment of distance measured along the surface (rather than along the x axis), k is the wave number and

$$R_1 = \sqrt{x_m^2 + (z_m - h_1)^2} \tag{6.63}$$

$$R_2 = \sqrt{(x_n - x_m)^2 + (h_2 - z_m)^2} \tag{6.64}$$

where x_n is the x coordinate of the field point, x_m is that of the current point in the summation and h_1, z_m and h_2 represent the heights of the source point, current point and observation point, respectively.

The principle of the algorithm in Eq. (6.61) is illustrated in Figure 6.25.

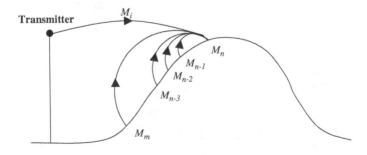

Figure 6.25: Illustration of the integral equation algorithm

[Hviid, 95] reports the use of the integral equation for path loss predictions at frequencies from 145 MHz to 1.9 GHz, for different path profiles and height variations in the order of 20–50 m. To verify such predictions, field strength measurements were conducted using a $\lambda/4$ monopole antenna on top of a van over five profiles of irregular terrain. The standard deviation of the path loss prediction error was found to increase smoothly with frequency from 3 dB at 144 MHz to 9 dB at 1900 MHz.

Since the integral equation method depends only on simple summations, it is relatively straightforward to calculate and has the advantage of working directly with the terrain data. This avoids the need to approximate the terrain with canonical objects such as knife-edges, wedges or cylinders. However, the method is slow since the computation effort increases quadratically with problem size and frequency. Similarly, calculations must start again with every new observation point.

6.8.2 The Parabolic Equation Method

The parabolic equation method is an alternative full-wave approach for predictions of variations in signal level, first introduced into radiowave propagation by [Leontovich, 46]. Numerical methods for solution of the equation were not available at that time, and only since the late 1980s has the application to tropospheric propagation been revived. The success is due to the application of an efficient numerical algorithm as specified in [Craig, 88]. The main motivation for the solution of the parabolic equation came from an analogous problem in underwater acoustics [Dinapoli, 79].

Starting from Maxwell's curl equations (2.1) and combining them, under very general conditions it is possible to derive the *scalar wave equation* which describes how either field component (E or H) varies spatially with range x and height z on the x-z plane. Denoting either field component as $\psi(x, z)$, the scalar wave equation is

$$\frac{\partial^2 \psi}{\partial z^2} + \frac{\partial^2 \psi}{\partial x^2} + k^2 n^2 \psi = 0 \tag{6.65}$$

where k is the free-space wave number and n is the refractive index, which can depend on range x as well as height z, in contrast to the simplistic assumptions in Section 6.3. Since the main interest is in the variations of the field on scales that are large compared to a wavelength, the rapid phase variation of the wave with x can be removed, so ψ can be expressed in terms of an attenuation function $u(x, z)$,

$$\psi(x, z) \approx u(x, z) \frac{e^{jkx}}{\sqrt{x}} \tag{6.66}$$

For field points, many wavelengths from the source and assuming relatively slowly varying fields, u satisfies a parabolic equation of the form [Craig, 88]

$$\frac{\partial^2 u}{\partial z^2} + 2jk \frac{\partial u}{\partial x} + k^2 (m^2 - 1)u = 0 \tag{6.67}$$

To account for the curvature of the Earth, $m(x, z)$ is the modified refractive index, which can be computed as follows (taking R as the Earth radius):

$$m(x, z) = n(x, z) + z/R \tag{6.68}$$

Equation (6.67) is then a parabolic equation (PE). It can be simplified by factorisation into a forward and backward travelling wave component by making a *paraxial* approximation which assumes that all the energy propagates close to a preferred direction. It is further assumed that only the forward travelling wave in the positive x direction is of interest, yielding

$$\frac{\partial u}{\partial x} = \frac{j}{2k_0} \frac{\partial^2 u}{\partial z^2} \tag{6.69}$$

This particular form of the PE is known as the *narrow angle parabolic equation* and is valid for propagation angles within about 15° from the paraxial direction [Levy, 00]. To make solution of the PE manageable, further assumptions are necessary: the z direction must be truncated to make the computation finite, but in a way which avoids spurious fields arising from sudden truncations of the field. One approach is to use an absorbing boundary condition which smoothly reduces the field using an appropriate windowing function. Horizontal or vertical polarisation and the details of the terrain shape and constitutive parameters are introduced by implementing appropriate boundary conditions at the lower end of the computational domain [Dockery, 91].

Finally, the PE is solved by discretising and adopting an appropriate numerical technique for solving the differentials. One approach is via a discrete Fourier transform as in [Craig, 91]. This approach is computationally efficient but it lacks flexibility for boundary modelling. Instead, finite difference methods are more appropriate for complicated boundaries. The *implicit finite difference* method takes the field values at discretised points on a given vertical line and uses these to determine all the field values on an adjacent line. The differentials in the equation are replaced by spatial differences, so that fields on a given vertical line depend only on those in the previous line. In this way, field values are marched along starting close to the source and moving in the direction of increasing x until the whole region of interest is covered. The discretisation interval need not be at half-wavelengths due to the removal of the fast varying phase term. For example, if we assume the maximum propagation angle is 15°, then the maximum range and height step sizes are 14.6λ and 3.8 λ, respectively, producing efficient calculations. This whole process is illustrated in Figure 6.26.

Unlike the integral equation, the PE method can account for detailed spatial variations in the refractive index of the atmosphere [Hviid, 95]. Another advantage of the PE is that the field over a complete plane is determined in one run of the simulation, so the variation in field range of heights and distances is calculated in a single step. In principle, PE can also be applied to fully 3D problems although complexity increases dramatically. It is worth emphasizing that the PE is valid in those areas where ray tracing methods break down, i.e. near caustics and focal points. See [Levy, 00] for a full description of PE and alternative solution techniques and applications.

As an example of results from the PE method, see Figure 6.27, which displays the path loss as predicted by the PE in decibels as the shading across height and distance along the profile. The source is a Gaussian source with a beamwidth of 2° at an elevation angle of 0°. The plot displays up to 160 m in height, but the total height simulated was 240 m, as an absorbing layer was applied in the top third of the calculation domain. The terrain at the lower boundary was simulated as a perfect electric conductor. The frequency is 970 MHz, with a vertical step size of one wavelength and horizontal step size of ten wavelengths.

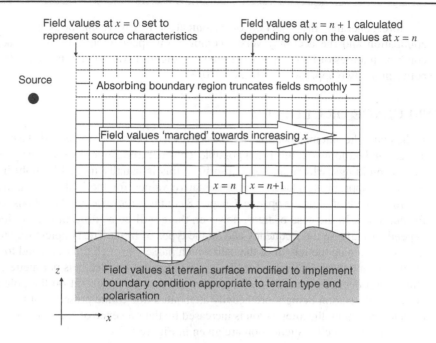

Figure 6.26: Illustration of parabolic equation method calculation

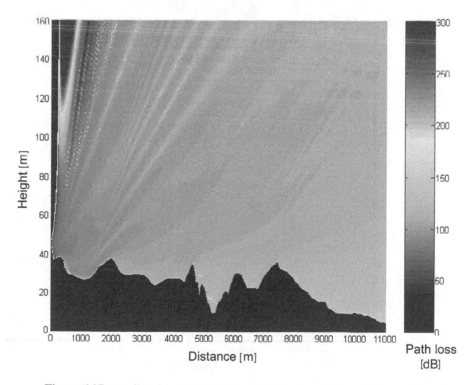

Figure 6.27: Predicted path loss over irregular terrain using parabolic equation

More generally, the selection of canonical or full-wave methods is dependent on the application and the accuracy versus computation speed required. Hybrid methods which combine features of both approaches are one potential way to obtain a suitable compromise between the two approaches [Owadally, 03].

6.9 INFLUENCE OF CLUTTER

In the main, the antenna locations for fixed links will be chosen to be well clear of *clutter* such as trees and buildings. If this is not possible, considerable clutter loss may be produced. In the case of buildings, which are good absorbers, this is usually treated by uplifting the terrain heights at appropriate places by representative values for the building heights, effectively assuming the buildings are opaque. Chapter 8 contains more detailed methods for accounting for buildings. In the case of trees, however, there will be significant penetration which will depend strongly on the frequency and on the types of trees. The simplest way to account for this is to estimate the length of the path which passes through the trees, and to multiply this length by an appropriate value for the *specific attenuation* [decibels per metre]. The specific attenuation depends strongly on frequency, and to a lesser extent on the polarisation, with vertical polarisation being more heavily attenuated due to the presence of tree trunks. To an even lesser extent, the attenuation is increased by the presence of leaves on the trees. Typical values for the specific attenuation are given in Figure 6.28.

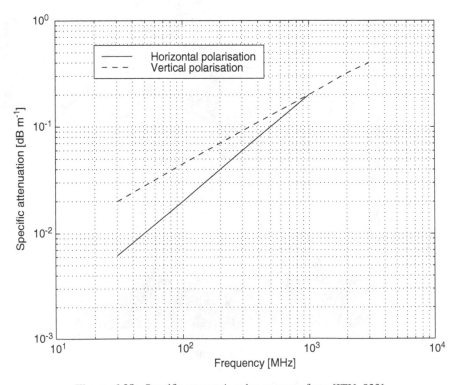

Figure 6.28: Specific attenuation due to trees, from [ITU, 833]

It is often found in practice, however, that the specific attenuation is a function of the path length, so more accurate models account for the path length explicitly. One empirical approach is the *modified exponential decay model* [ITU, 236],

$$L = \eta f_c^\nu d^\gamma \qquad (6.70)$$

with f_c in [MHz] and d in [km], where $\eta = 0.187$, $\nu = 0.284$ and $\gamma = 0.588$ are parameters of the model, valid for frequencies between 200 MHz and 95 GHz. Other approaches, based directly on the relevant propagation mechanisms, can also be applied, e.g. [Tamir, 77].

6.10 CONCLUSION

The prediction of propagation over fixed links is a mature art and reasonably accurate predictions may be made in most cases, although the accuracy will depend on the data available and the time available for computation, since the more exact methods can be very computationally intensive. Here are the main steps in predicting the propagation loss:

1. Locate the positions and heights of the antennas.
2. Construct the great circle path between the antennas.
3. Derive the terrain path profile; this step can be done by reading contour heights from conventional maps, although it is increasingly common to use digital terrain maps.
4. Uplift the terrain profile by representative heights for any known buildings along the path.
5. Select a value for the effective Earth radius factor appropriate to the percentage of time being designed for; modify the path profile by this value.
6. Calculate the free space loss for the path.
7. If any obstructions exist within 0.6 times the first Fresnel zone, calculate diffraction over these obstructions, using canonical or full-wave methods depending on the application, and add the resulting excess loss to the link budget.
8. Compute the path length which passes through trees and add the corresponding extra loss.

For systems which require very high availability, the time variability of the signal due to multipath propagation from ducting and reflections must also be accounted for.

REFERENCES

[Andersen, 97] J. B. Andersen, UTD Multiple-edge transition zone diffraction, *IEEE Transactions on Antennas and Propagation*, 45 (7), 1093–1097, 1997.

[Bullington, 77] K. Bullington, Radio propagation for vehicular communications, *IEEE Transactions On Vehicular Technology*, 26 (4), 295–308, 1977.

[Causebrook, 71] J. H. Causebrook and B. Davies, Tropospheric radio wave propagation over irregular terrain: the computation of field strength for UHF broadcasting, *BBC Research Report* 43, 1971.

[Craig, 88] K. H. Craig, Propagation modelling in the troposphere: parabolic equation method, *Electronics Letters*, 24 (18), 1136–1139, 1988.

[Craig, 91] K. H. Craig and M. F. Levy, Parabolic equation modelling of the effects of multipath and ducting on radar systems, *IEEE Proceedings, Part F*, 138 (2), 153–162, 1991.

[Deygout, 66] J. Deygout, Multiple knife-edge diffraction of microwaves, *IEEE Transactions on Antennas and Propagation*, 14 (4), 480–489, 1966.

[Dinapoli, 79] F. R. Dinapoli and R. L. Davenport, Numerical models of underwater acoustics propagation, *Ocean Accoustics, Topics in Current Physics, Vol. 8*, Springer Verlag, New York, pp. 79–157, 1979.

[Dockery, 91] G. D. Dockery and J. R. Kuttler, Theoretical description of the parabolic approximation/Fourier split-step method of representing electromagnetic propagation in the troposphere, *Radio Science*, (2), 381–393, 1991.

[Epstein, 53] J. Epstein and D. W. Peterson, An experimental study of wave propagation at 850 MC, *Proceedings of Institute of Radio Engineers*, 595– 611, 1953.

[Furutsu, 63] K. Furutsu, On the theory of radio wave propagation over inhomogeneous earth, *Journal of Research of the National Bureau of Standards*, 67D (1), 39–62, 1963.

[Giovanelli, 84] C. L. Giovanelli, An analysis of simplified solutions for multiple knife-edge diffraction, *IEEE Transactions on Antennas and Propagation*, 32 (3), 297–301, 1984.

[Hata, 80] M. Hata, Empirical formulae for propagation loss in land mobile radio services, *IEEE Transactions On Vehicular Technology*, VT-29, 317–325, 1980.

[Hufford, 52] G. A. Hufford, An integral equation approach to the problem of wave propagation over irregular terrain, *Quarterly Journal of Applied Mathematics*, 9 (4), 391–404, 1952.

[Hviid, 52] J. T. Hviid, J. Bach Andersen, J. Toftgard and J. Boger, Terrain based propagation model for rural area – An integral equation approach, *IEEE Transactions on Antennas and Propagation*, (43), 41–46, 1995.

[ITU, 236] International Telecommunication Union, ITU-R Report 236-6, *Influence of Terrain Irregularities and Vegetation on Tropospheric Propagation*, Vol. V: *Propagation in non-ionised media*, CCIR XVth Plenary Assembly, Dubrovnik, 1986.

[ITU, 833] International Telecommunication Union, ITU-R Recommendation PN.833-1, *Attenuation in vegetation*, Geneva, 1994.

[ITU, 638] International Telecommunication Union, Recommendation P.638-7, *Ground-wave propagation curves for frequencies between 10 kHz and 30 MHz*, Geneva, 1997e.

[ITU, 453] International Telecommunication Union, ITU-R Recommendation P.453-6, *The radio refractive index: its formula and refractivity data*, Geneva, 1997a.

[ITU, 526] International Telecommunication Union, ITU-R Recommendation P.526-5, *Propagation by diffraction*, Geneva, 1997d.

[ITU, 530] International Telecommunication Union, ITU-R Recommendation P.530-7, *Propagation data and prediction methods required for the design of terrestrial line-of-sight systems*, Geneva, 1997c.

[ITU, 834] International Telecommunication Union, ITU-R Recommendation P.834-2, *Effects of tropospheric refraction on radiowave propagation*, Geneva, 1997b.

[Lee, 02] M. B. R. Lee, S. R. Saunders, C. Tzaras, E. Montiel and J. Scrivens, *The Digiplan Project: Coverage Prediction For Digital Broadcast Services*, International Broadcasting Convention, IBC2002, 12th–16th September 2002.

[Leontovich, 46] M. A. Leontovich and V. A. Fock, Solution of the problem of propagation of electromagnetic waves along the Earth's surface by the method of parabolic equations, *The Journal of Physiology, USSR*, 10 (1), 13–24, 1946.

[Levy, 00] M. F. Levy, Parabolic equation methods for electromagnetic wave propagation, *IET Electromagnetic Waves Series 45*, 2000, ISBN 0-85-2967640.

[Millington, 62] G. Millington, R. Hewitt and F. S. Immirzi, Double knife-edge diffraction in field strength predictions, *IEE Monograph*, 509E, 419–29, 1962.

[OS, 06] A guide to coordinate systems in Great Britain, Ordnance Survey, 2006. Available from www.ordsvy.gov.uk

[Ott, 70] R. H. Ott and L.A. Berry, An alternative integral equation for propagation over irregular terrain, *Radio Science*, 5 (5), 767–771, 1970.

[Owadally, 03] A. S. Owadally and S. R. Saunders, Performance analysis of modified terrain propagation models against their reference models in terms of speed and accuracy, *International Conference on Antennas and Propagation 2003, Vol. 1*, 2003, pp. 43–46.

[Saunders, 94] S. R. Saunders and F. R. Bonar, Prediction of mobile radio wave propagation over buildings of irregular heights and spacings, *IEEE Transactions on Antennas and Propagation*, 42 (2), 137–144, February 1994.

[Sharples, 89] P. A. Sharples and M. J. Mehler, Cascaded cylinder model for predicting terrain diffraction loss at microwave frequencies, *Proceedings of IEE*, 136H (4), 331–337, 1989.

[Tamir, 77] T. Tamir, Radio wave propagation along mixed paths in forest environments, *IEEE Transactions on Antennas and Propagation*, 25 (4), 471–477, 1977.

[Tzaras, 00] C. Tzaras and S. R. Saunders, Comparison of multiple-diffraction models for digital broadcasting coverage prediction, *IEEE Transactions on Broadcasting*, 46 (3), 221–226, 2000.

[Tzaras, 01] C. Tzaras and S. R. Saunders, An improved heuristic UTD solution for multiple-edge transition zone diffraction, *IEEE Transactions on Antennas and Propagation*, 49 (12), 1678–1682, 2001.

[Vogler, 82] L. E. Vogler, An attenuation function for multiple knife-edge diffraction, *Radio Science*, 17 (6), 1541–1546, 1982.

PROBLEMS

6.1 An omnidirectional antenna is placed 10 m above the Earth's surface.

(a) What is the distance to the horizon, assuming a smooth Earth and median refractivity?

(b) How would the antenna height have to be increased in order to double the visible coverage area?

(c) What would be the horizon distance with a 10 m antenna height under *super-refractive* conditions with $k_e = 1.5$?

6.2 A ray is launched at an angle of 1° to the Earth's surface. The refractivity is constant at 300 N units between the ground and a height of 50 m, when it abruptly decreases to 250 N units. What is the new elevation angle of the ray?

6.3 The refractivity in a certain region decreases by 80 N units per kilometre of height. What is the effective Earth radius?

6.4 Calculate the total attenuation for a 4 km path, with equal height antennas, and three knife-edges with excess heights of 50 m, 100 m and 50 m (in that order) equally spaced along the path at 100 MHz, using (a) the Deygout method and (b) the Causebrook method.

6.5 In Problem 6.4, how would the total loss change if the central knife-edge were replaced by a cylinder with the same effective height but a radius of 5 m?

6.6 When a diffracting obstacle is below the direct path between a transmitter and a receiver, the excess diffraction loss is lower at higher frequencies. Why? Does this imply that the total loss is lower?

7 Satellite Fixed Links

'A hundred years ago, the electric telegraph made possible - indeed, inevitable - the United States of America. The communications satellite will make equally inevitable a United Nations of Earth; let us hope that the transition period will not be equally bloody'.

Arthur C. Clarke

7.1 INTRODUCTION

The use of satellites as relay stations between earthbound stations permits high reliability communication over distances of many hundreds of kilometers, where the Earth bulge would completely obstruct the path for a terrestrial link, even with very high antennas. Fixed earth stations for communication with satellites typically involve very large-aperture dish antennas, carefully sited to avoid local obstructions. Communication then takes place usually with geostationary satellites located in orbits some 36 000 km above the Earth and orbiting the Earth at the same angular speed as the Earth's rotation. As a result they appear in the same position in sky at all times, and three of them can serve all points on the Earth with a view of the sky [Clarke, 45]. The antenna is therefore broadly fixed in pointing direction, although small corrections are usually required for 'station keeping', in order to track orbital perturbations. In these systems, the dominant component of the propagation loss is simply the free space loss. Beyond this, the radiowave propagation effects can be divided into three components as illustrated in Figure 7.1:

- *Ionospheric*, involving interactions between the layers of charged particles around the Earth, the Earth's magnetic field and the radio waves.

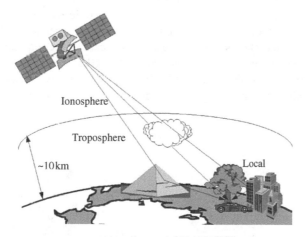

Figure 7.1: Three contributions to Earth-satellite propagation effects

Antennas and Propagation for Wireless Communication Systems Second Edition   Simon R. Saunders and Alejandro Aragón-Zavala
© 2007 John Wiley & Sons, Ltd

- *Tropospheric*, involving interactions between the waves and the lower layer of the Earth's atmosphere, including the effects of the gases composing the air and hydrometeors such as rain.
- *Local*, involving interactions between the waves and features of the environment in the vicinity of the earth station such as terrain, trees and buildings. These effects may be significant in providing some site shielding of fixed earth stations from terrestrial inter-ferers, but are most important in mobile satellite systems, where the direct path may frequently be wholly or partially obscured. Local effects will be treated in Chapter 14 in the context of mobile systems.

7.2 TROPOSPHERIC EFFECTS

7.2.1 Attenuation

The troposphere consists of a mixture of particles, having a wide range of sizes and characteristics, from the molecules in atmospheric gases to raindrops and hail. The total loss (in decibels) resulting from a wave passing through such a medium made up of many small particles is composed of two additive contributions from absorption and scattering processes,

$$L_{\text{total}} = L_{ab} + L_{sc} \tag{7.1}$$

Absorption is the result of conversion from radio frequency energy to thermal energy within an attenuating particle, such as a molecule of gas or a raindrop (Figure 7.2).

Scattering results from redirection of the radiowaves into various directions, so that only a fraction of the incident energy is transmitted onwards in the direction of the receiver (Figure 7.3). The scattering process is strongly frequency-dependent, since wavelengths which are long compared to the particle size will be only weakly scattered.

The main scattering particles of interest to satellite systems are hydrometeors, including raindrops, fog and clouds. In these cases, the scattering component of attenuation is only significant to systems operating above around 10 GHz. The absorption component also rises with frequency, although not so rapidly. In this chapter the hydrometeor effects will be limited to rain, as rain is the most important in determining system reliability. For other effects, see [ITU, 840].

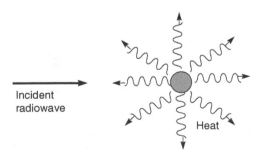

Figure 7.2: Absorption

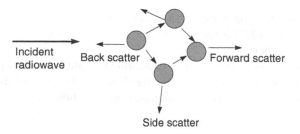

Figure 7.3: Scattering

7.2.2 Rain Attenuation

The attenuation of a wave due to rain increases with the number of raindrops along the path, the size of the drops and the length of the path through the rain (Figure 7.4).

If the density and shape of the raindrops in a given region is constant, then the received power P_r in a given antenna is found to diminish exponentially with distance r through the rain,

$$P_r(r) = P_r(0) \exp(-\alpha r) \tag{7.2}$$

where α is the reciprocal of the distance required for the power to drop to e^{-1} (about 37%) of its initial value. Expressing this as a propagation loss in decibels gives

$$L = 10 \log \frac{P_t}{P_r} - 4.343 \alpha r \tag{7.3}$$

It is usual to calculate the total loss via the specific attenuation in decibels per metre,

$$\gamma = \frac{L}{r} = 4.343 \alpha \tag{7.4}$$

The value of α is given by the following relationship:

$$\alpha = \int_{D=0}^{\infty} N(D) \times C(D) \, dD \tag{7.5}$$

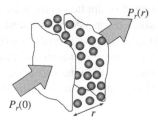

Figure 7.4: Rain path attenuation

where $N(D)$ is the number of drops of diameter D per metre of path length (the *drop size distribution*) and $C(D)$ is the effective attenuation cross-section of a drop [dB m^{-1}] (which depends on frequency).

This formulation is useful when evaluating the attenuation for paths over which the drop size distribution $N(D)$ is not constant. We then take the value of the specific attenuation at a given point on the path, $\gamma(r)$ and integrate over the whole path length r_T to find the total path loss

$$L = \int_0^{r_T} \gamma(r) \ dr \tag{7.6}$$

One way to solve the above equation is to take a particular drop size distribution and integrate the result. Large drop sizes are found to be less probable than smaller ones, and a commonly used distribution obeying this finding is an exponential one [Marshall, 48],

$$N(D) = N_0 \exp\left(-\frac{D}{D_m}\right) \tag{7.7}$$

where N_0 and D_m are parameters, with D_m depending on R, the rainfall rate, measured on the ground in millimetres per hour. Some useful values are

$$\begin{aligned} N_0 &= 8 \times 10^3 \ \text{m}^{-2}\text{mm}^{-1} \\ D_m &= 0.122 R^{0.21} \ \text{mm} \end{aligned} \tag{7.8}$$

As may be expected, the dependence of D_m on R indicates that heavy rain is composed of more large drops, although very small drops are numerous at all rainfall rates. The attenuation cross-section $C(D)$ can be found by theoretical analysis of the attenuation due to an individual spherical raindrop illuminated by a plane wave. For low frequencies, where the drop is small compared to the incident wavelength, most of the attenuation is due to absorption in the drop. In this region, the cross-section rises with frequency according to the *Rayleigh approximation*,

$$C(D) \propto \frac{D^3}{\lambda} \tag{7.9}$$

At higher frequencies, the attenuation increases more slowly, tending eventually towards a constant value known as the *optical limit*. At these frequencies, scattering forms a significant part of the attenuation. In the general case, the cross-section is calculated using the *Mie scattering theory* and this must be applied to treat the situations where the wavelength is of a similar size to the drops since resonance phenomena are produced.

In principle, Eq. (7.5) can now be solved directly and then used to evaluate the total attenuation using Eq. (7.6). A more common and practical approach is to use an empirical model which implicitly combines all of these effects, where γ is assumed to depend only on R, the rainfall rate measured on the ground in millimetres per hour. The usual form of such expressions is

$$\gamma = aR^b \tag{7.10}$$

where a and b depend on frequency and average rain temperature and γ has units [dB km^{-1}]. Table 7.1 shows values for a and b at various frequencies at 20°C for horizontal polarisation [ITU, 838]. A more complete set of curves of γ versus f is shown in Figure 7.5.

Table 7.1: Parameters for empirical rain attenuation model, extracted from [ITU, 838)

f [GHz]	a	b
1	0.0000387	0.912
10	0.0101	1.276
20	0.0751	1.099
30	0.187	1.021
40	0.350	0.939

The path length r_R used to multiply Eq. (7.10) to find total rain attenuation is the total rainy slant path length as shown in Figure 7.6. All heights in this figure are measured above mean sea level; h_R is the effective rain height, usually the same as the height of the melting layer, at

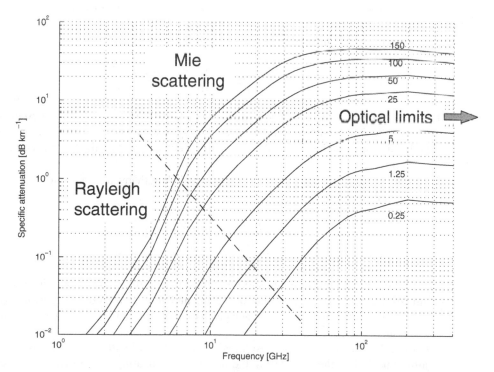

Figure 7.5: Specific rain attenuation: curves marked with rainfall rate, R [mm h^{-1}]

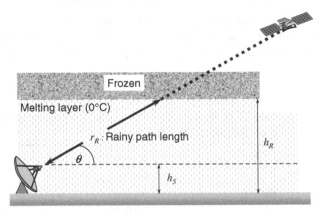

Figure 7.6: Rain attenuation path geometry

which the temperature is 0°C. Representative values for the effective rain height vary according to the latitude ϕ of the earth station [ITU, 618],

$$
h_R \text{ (km)} = \begin{cases}
5 - 0.075\,(\phi - 23) & \text{for} \quad \phi > 23° \text{ Northern Hemisphere} \\
5 & \text{for} \quad 0° \le \phi \le 23° \text{ Northern Hemisphere} \\
5 & \text{for} \quad 0° \ge \phi \ge -21° \text{ Southern Hemisphere} \\
5 + 0.1\,(\phi + 21) & \text{for} -71° \le \phi < -21° \text{ Southern Hemisphere} \\
0 & \text{for} \quad \phi < 71° \text{ Southern Hemisphere}
\end{cases} \quad (7.11)
$$

The rainy path length can then be found geometrically (for $\theta > 5°$) as

$$
r_R = \frac{h_R - h_S}{\sin \theta} \tag{7.12}
$$

For paths in which the elevation angle θ is small, it is necessary to account for the variation in the rain in the horizontal direction. This will tend to reduce the overall rain attenuation due to the finite size of the *rain cells*, which arises from the local structure of rain clouds (Figure 7.7).

This effect can be treated by reducing the path length by a reduction factor s so the attenuation is now given by

$$
L = \gamma\, s r_R = a R^b s r_R \tag{7.13}
$$

Also, rain varies in time over various scales:

- *Seasonal*: The rainfall rate tends to be highest during the summer months in temperate climates, but has two peaks during the spring and autumn periods in tropical regions.
- *Annual*: The values of rainfall rate at a given time of year may be significantly different in each year.
- *Diurnal*: Rainfall tends to be most intense during the early afternoon. This is because the Earth is heated by the Sun during the day, and this heating sets up convection currents which may lead to thunderstorms and hence rain.

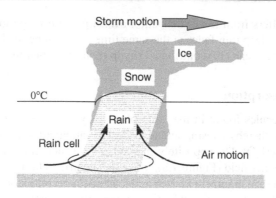

Figure 7.7: Structure of a typical mature thunderstorm

It is important to realise that it is not the total *amount* of rain which falls during a given year which matters, but rather the period of time for which the rainfall *rate* exceeds a certain value, at which the rain attenuation exceeds the system fade margin and hence causes outage. Thus frequent light rain will produce far less outage than the same amount of rain falling in occasional heavy storms.

All of the above temporal variations are usually accounted for by using Eq. (7.13) to predict the rain attenuation not exceeded for 0.01% of the time ($L_{0.01}$) in terms of $R_{0.01}$, the rainfall rate exceeded 0.01% of the time in an average year (i.e. around 53 min), and then correcting this attenuation according to the percentage level of actual interest. Thus

$$L_{0.01} = aR_{0.01}^b s_{0.01} r_R \qquad (7.14)$$

and the following empirical relation for $s_{0.01}$ is used:

$$s_{0.01} = \cfrac{1}{1 + \cfrac{r_R \sin \theta}{35 \exp(-0.015 R_{0.01})}} \qquad (7.15)$$

The attenuation can then be corrected to the relevant time percentage P using

$$L_P = L_{0.01} \times 0.12 P^{-(0.546 + 0.043 \log P)} \qquad (7.16)$$

where P is between 0.001 and 1% [ITU, 618].

The reference rainfall rate $R_{0.01}$ is strongly dependent on the geographical location. For most of Europe it is around 30 mm h^{-1}, except for some Mediterranean regions where it may be as high as 50 mm h^{-1}. In equatorial regions it may reach as high as 160 mm h^{-1}.

The methodology for predicting rainfall attenuation explained above is fully defined in [ITU, 618] and its impact on the system performance is examined in [Maral, 93].

The effect of rain fading may be reduced by applying *site diversity*, where two earth stations are constructed so that the paths to the satellite are separated by greater than the extent of a typical rain cell. The signal is then switched between the earth stations, according to which one suffers from least attenuation over a given time period. The probability of both links suffering deep rain fades at the same time can then be made very small.

Clouds and fog can also contribute to hydrometeor attenuation – see [ITU, 840] for details.

7.2.3 Gaseous Absorption

Gaseous molecules found in the atmosphere may absorb energy from radiowaves passing through them, thereby causing attenuation. This attenuation is greatest for polar molecules such as water (H_2O). The oppositely charged ends of such molecules cause them to align with an applied electric field (Figure 7.8). Since the electric field in the wave changes in direction twice per period, realignment of such molecules occurs continuously, so a significant loss may result. At higher frequencies this realignment occurs faster, so the absorption loss has an overall tendency to increase with frequency.

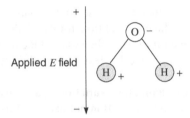

Figure 7.8: Alignment of polar molecules with an applied electric field

Non-polar molecules, such as oxygen (O_2) may also absorb electromagnetic energy due to the existence of magnetic moments. Each type of absorbing molecule tends to contribute both a general background level of absorption, which rises with frequency, and several resonance peaks, each corresponding to different *modes* of vibration (lateral vibration, longitudinal vibration, flexing of interatomic bonds).

In normal atmospheric conditions, only O_2 and H_2O contribute significantly to absorption, although other atmospheric gases may be significant in very dry air at above 70 GHz. The main resonance peaks of O_2 and H_2O are given in Table 7.2.

The oxygen peak at around 60 GHz is actually a complex set of a large number of closely spaced peaks which contribute significant attenuation, preventing the use of the band 57–64 GHz for practical satellite communications.[1]

Table 7.2: Peaks of atmospheric absorption

Gas	Resonance frequencies [GHz]		
Water vapour (H_2O)	22.3	183.3	323.8
Oxygen (O_2)	60.0	118.74	

---

[1]This property makes the 60 GHz range useful for short range terrestrial communications with limited interference between links.

The specific attenuation [dB km^{-1}] for water vapour (γ_w) and for oxygen (γ_0) is given in Figure 7.9 for a standard set of atmospheric conditions.

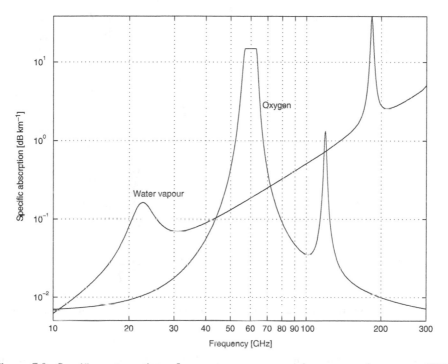

Figure 7.9: Specific attenuation for water vapour and oxygen (pressure = 1013 mb, temperature = 15°C, water vapour content = 7.5 g m^{-3}). Considerable variation appears within the peak around 60 GHz but it is not shown here. Calculated using equations from ITU, 676

The total atmospheric attenuation L_a for a particular path is then found by integrating the total specific attenuation over the total path length r_T:

$$L_a = \int_0^{r_T} \gamma_a(l)\ dl = \int_0^{r_T} \{\gamma_w(l) + \gamma_0(l)\}\ dl [\text{dB}] \qquad (7.17)$$

This integration has been performed for the total zenith ($\theta = 90°$) attenuation in Figure 7.10 by assuming an exponential decrease in gas density with height and by using equivalent water vapour and dry air heights of around 2 and 6 km, respectively. The attenuation for an inclined path with an elevation angle $\theta > 10°$ can then be found from the zenith attenuation L_z as

$$L_a = \frac{L_z}{\sin \theta} \qquad (7.18)$$

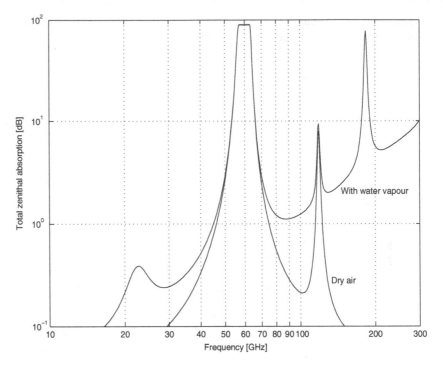

Figure 7.10: Total one-way zenith attenuation in dry air and including water vapour

For lower elevation angles, the effect of Earth curvature must be taken into account; see [ITU, 618] for more details.

Note that atmospheric attenuation results in an effective upper frequency limit for practical Earth–space communications.

7.2.4 Tropospheric Refraction

Chapter 6 described the effects of tropospheric refraction on the curvature of rays. Since the variation in refractive index is mostly vertical, rays launched and received with the relatively high elevation angles used in satellite fixed links will be mostly unaffected. Nevertheless, the bending is sufficient that it must be accounted for when calculating ideal antenna pointing angles (Figure 7.11).

Variations in the refractive index gradient with time may lead to some loss in effective antenna gain due to misalignment or may be corrected by using automatic steering.

7.2.5 Tropospheric Scintillation

When the wind blows, the mainly horizontal layers of equal refractive index in the troposphere tend to become mixed due to turbulence, leading to rapid refractive index variations over small distances or *scale sizes* and over short time intervals. Scintillation also occurs at optical frequencies, where it is more commonly known as the *twinkling of stars*. Waves travelling through these rapid variations of index therefore vary in amplitude and phase. This

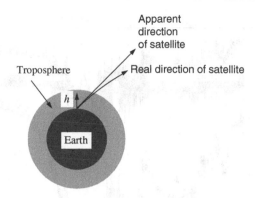

Figure 7.11: Ray bending due to tropospheric refractivity

is *dry tropospheric scintillation*. Another source of tropospheric scintillation is rain; rain leads to a *wet* component of variation, which tends to occur at slower rates than the dry effects.

Scintillation is not an absorptive effect in that the mean level of the signal is essentially unchanged. The effect is strongly frequency-dependent in that shorter wavelengths will encounter more severe variations resulting from a given scale size. The scale size can be determined by monitoring the scintillation on two nearby paths and examining the cross-correlation between the scintillation on the paths. If the effects are closely correlated, then the scale size is large compared with the path spacing. Figure 7.12(a) shows an example of the signal measured simultaneously at three frequencies during a scintillation event. It is clear that there is some absorption taking place, but this changes relatively slowly. In order to extract the scintillation component, the data is filtered with a high-pass filter having a cut-off frequency of around 0.01 Hz, yielding the results shown in Figure 7.12(b). The magnitude of the scintillation is measured by its standard deviation, or *intensity* (in decibels), measured usually over 1 min intervals as shown in Figure 7.13. Notice the close similarity between the curves at the three frequencies.

The distribution of the fluctuation (in decibels) is approximately a Gaussian distribution, whose standard deviation is the intensity. The physics of the air masses in the troposphere leads to a well-defined roll-off of the spectrum, reducing at the rate of $f^{-8/3}$ at frequencies above around 0.3 Hz [Tatarski, 61]. This is evident in Figure 7.14. The scintillation intensity σ_{pre} may be predicted from an ITU-R model [ITU, 618] as follows:

$$\sigma_{\text{pre}} = \frac{\sigma_{\text{ref}} f^{7/12} g(D)}{(\sin\theta)^{1.2}} [\text{dB}] \qquad (7.19)$$

where f is the carrier frequency, θ is the elevation angle, σ_{ref} depends on the weather conditions (temperature, atmospheric pressure and water vapour pressure) and $g(D)$ accounts for averaging of the scintillation across the aperture of the antenna, which leads to a reduction in the scintillation intensity for large-aperture diameter D.

Scintillation is most noticeable in warm, humid climates and is greatest during summer days. One way to reduce the effect of scintillation is to use an antenna with a wide aperture,

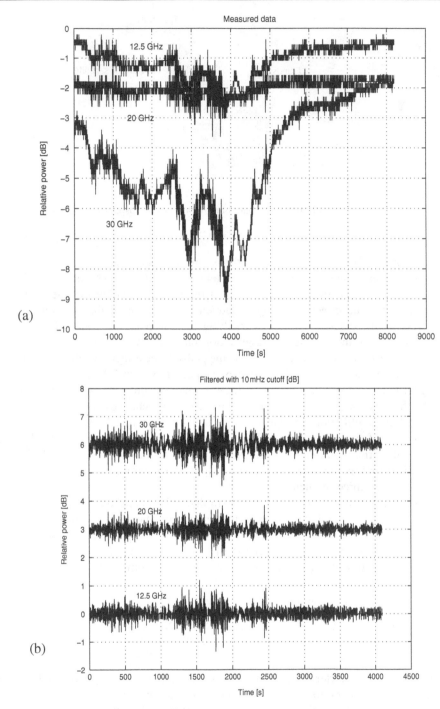

Figure 7.12: Measured signals at 12.5, 20 and 30 GHz during a scintillation event: (a) raw measurements, (b) after high-pass filtering. The filtered signals are offset by 3 dB for clarity, and all actually have 0 dB mean. Details of the measurement set-up are given in [Howell, 92] and further data analysis is described in [Belloul, 98]

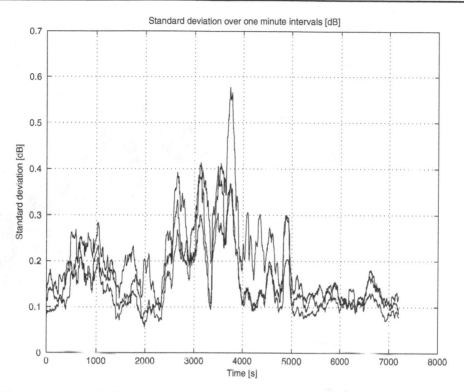

Figure 7.13: Scintillation intensity, calculated from Figure 7.12. The curves increase in order of increasing frequency

because this produces averaging of the scintillation across the slightly different paths taken to each point across the aperture.

Another approach is to use spatial diversity, where the signals from two antennas are combined to reduce the overall fade depth (Chapter 16 gives more details). Best results for a given antenna separation are produced using vertically separated antennas due to the tendency for horizontal stratification of the troposphere.

7.2.6 Depolarisation

The polarisation state of a wave passing through an anisotropic medium such as a rain cloud is altered, such that a purely vertical polarised wave may emerge with some horizontal components, or an RHCP wave may emerge with some LHCP component, as shown in Figure 7.15. The extent of this depolarisation may be measured by the terms *cross-polar discrimination* (XPD) or *cross-polar isolation* (XPI). These are defined (in decibels) by the following field ratios:

$$\text{XPD} = 20 \log \frac{E_{ac}}{E_{ax}} \tag{7.20}$$

$$\text{XPI} = 20 \log \frac{E_{ac}}{E_{bx}} \tag{7.21}$$

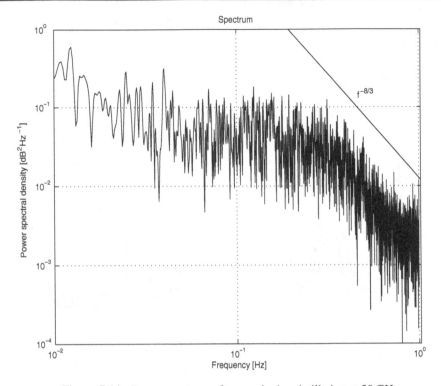

Figure 7.14: Power spectrum of tropospheric scintillation at 20 GHz

where the E terms are the electric fields defined as shown in Figure 7.16. Essentially, XPD expresses how much of a signal in a given polarisation is scattered into the opposite polarisation by the medium alone, while XPI shows how much two signals of opposite polarisations transmitted simultaneously will interfere with each other at the receiver. Raindrops are a major source of tropospheric depolarisation and their shape may be approximated by an oblate spheroid. In still air, the drops tend to fall with their major axis parallel to the ground. If there is a horizontal wind component, however, the axis tilts through a *canting angle* (Figure 7.17). All of the drops in a given rain cloud will be subject to similar forces, so there is an overall imbalance in the composition of the cloud.

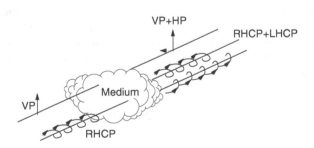

Figure 7.15: Depolarisation in an anisotropic medium

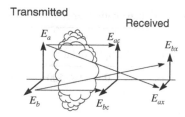

Figure 7.16: Field definitions for XPD and XPI

A wave passing through such raindrops will tend to have the component of the electric field parallel to the major axis attenuated by more than the orthogonal polarisation and will therefore emerge depolarised. The typical shape of a raindrop depends on its size, as shown in Table 7.3, where D is the diameter of a sphere with the same volume as the raindrop.

Depolarisation is strongly correlated with rain attenuation, and standard models of depolarisation use this fact to predict the XPD directly from the attenuation. One such model [NASA, 83] takes the form

$$\mathrm{XPD} = a - b \log L \tag{7.22}$$

where L is the rain attenuation [dB] and a and b are constants. Representative values for these constants are $a = 35.8$ and $b = 13.4$, resulting in the curve as shown in Figure 7.18. This is a reasonably accurate approach for frequencies above 10 GHz. Depolarisation may also have other causes:

- hydrometeors other than rain, particularly, needle-shaped ice crystals;
- tropospheric scintillation;
- ionospheric scintillation (Section 7.3.4).

7.2.7 Sky Noise

When a receiving antenna is pointed at a satellite in the sky, it receives noise from a variety of sources which add to the receiver internal noise, degrading the performance of the receiver. These effects are accounted for by calculating an effective noise temperature for the antenna

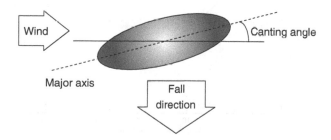

Figure 7.17: Raindrop asymmetry

Table 7.3: Variation of raindrop shape with diameter (after [Allnutt, 89])

Shape	Spherical	Spheroidal	Flattened spheroidal	Pruppacher and Pitter [Pruppacher, 71]
Diameter D [μm]	$D \leq 170$	$170 < D \leq 500$	$500 < D \leq 2000$	$D > 2000$
Appearance	●	◖◗	⬭	⬭

from all sky sources, then combining this with the receiver effects to obtain an overall system noise temperature, following the principles of Section 5.4.

The extraterrestrial sources of noise are shown in Figure 7.19. The cosmic background radiation D is independent of frequency and appears equally everywhere in the sky at a temperature of 2.7 K. The Sun's noise temperature varies with the sunspot cycle and with frequency in the approximate range 10^2 to 10^5 K, but its effect is localised to a $0.5°$ angular diameter (A). The Moon has virtually the same angular extent, but a considerably lower noise temperature (B). Finally, the galactic noise temperature of radio stars and nebulas varies considerably across the sky and decreases rapidly with frequency. The range of this variation is indicated in region C. The total cosmic noise temperature is denoted T_C. For more details of noise sources see [ITU, 372].

When an atmospheric absorptive process takes place, the absorption increases the effective noise temperature according to the total absorption. If the total absorption due to rain and gaseous attenuation is A and the physical temperature of the rain medium is T_m, then the overall antenna noise temperature is

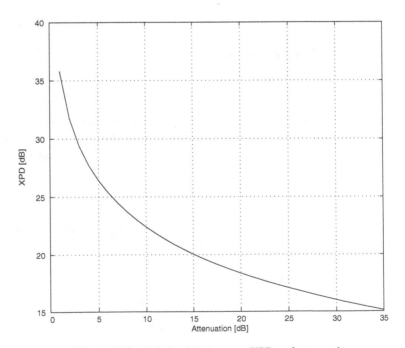

Figure 7.18: Relationship between XPD and attenuation

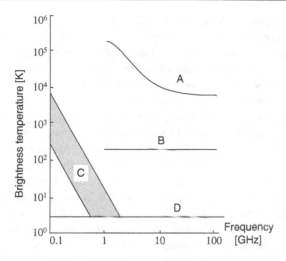

Figure 7.19: Extraterrestrial noise sources

$$T_A = T_m(1 - 10^{-A/10}) + T_c \times 10^{-A/10} \tag{7.23}$$

where, typically, $T_m = 280$ K for clouds and 260 K for rain. The total system noise temperature can then be calculated as

$$T_{sys} = T_R + T_f(1 - 10^{-L_f/10}) + T_A \times 10^{-L_f/10} \tag{7.24}$$

where T_R is the equivalent noise temperature of the receiver, T_f is the physical temperature of the feeder and L_f is the feeder loss as illustrated in Figure 7.20.

7.3 IONOSPHERIC EFFECTS

The ionosphere is a region of ionised plasma (a gas consisting mostly of charged particles) which surrounds the Earth at a distance ranging from around 50 to 2000 km above its surface.

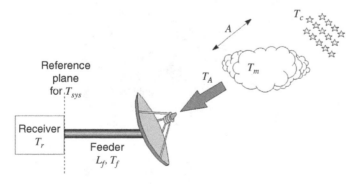

Figure 7.20: Definitions for system noise calculations

The ions are trapped in the Earth's magnetic field and may have come originally from either the *solar wind* or ionisation of atmospheric particles by the Sun's electromagnetic radiation. Since the Sun's radiation penetrates deeper into the Earth's atmosphere at zenith, the ionosphere extends closest to the Earth around the equator and is more intense on the daylight side. Figure 7.21 shows how the ionosphere separates into four distinct layers (D, E, F1 and F2) during the day and how these layers coalesce during the night into the E and F layers. The intensity of the solar wind is closely correlated with the sunspot density, which in turn varies with a period of approximately 11 years, so Earth–space communication paths can experience gross variations in properties which are only apparent on this timescale.

Radiowaves passing though plasma regions set the charged particles within them into oscillation, causing them to reradiate in all directions. The forward-going scattered wave is phase advanced by $\pi/2$ radians and combines with the original wave, causing the resultant wave to be further advanced in regions with a higher density of charged particles.

The key parameter relating the structure of the ionosphere to its effect on radio communications is thus the electron concentration N, measured in free electrons per cubic metre. The variation of N versus height above the Earth for a typical day and night is shown in Figure 7.22 [Budden, 61]. The electron content of the ionosphere changes the effective refractive index encountered by waves transmitted from the Earth, changing their direction by increasing the wave velocity. Given the right conditions (frequency, elevation angle, electron content), the wave may fail to escape from the Earth and may appear to be 'reflected' back to Earth, although the process is actually refraction (Figure 7.23).

The path of a radio wave is affected by any free charges in the medium through which it is travelling. The refractive index is governed by the electron concentration and the magnetic field of the medium and the frequency and polarisation of the transmitted wave as mentioned earlier. These lead to some important properties for waves propagating in the ionosphere

- The refractive index is proportional to the electron concentration.
- The refractive index is inversely proportional to the frequency of the transmitted wave.

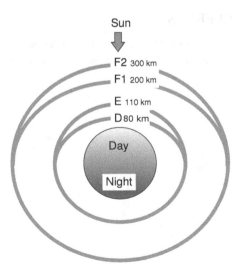

Figure 7.21: Ionospheric layers and typical heights above the Earth's surface

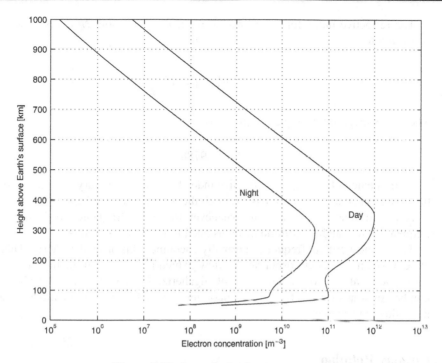

Figure 7.22: Ionospheric electron concentration

- There are two possible ray paths depending on the sense of polarisation of the transmitted wave. The two rays are referred to as the *ordinary* and *extraordinary* components. The ordinary wave is that for which its electric field vector is parallel to the Earth's magnetic field; for the extraordinary wave, its electric field vector is perpendicular to the Earth's magnetic field vector.

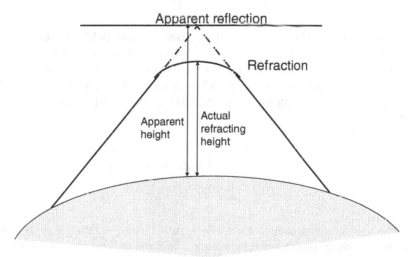

Figure 7.23: Apparent reflection from the ionosphere

The refractive index for an ordinary wave depends on both N and the wave frequency f according to

$$n_0^2 = 1 - \left(\frac{f_c}{f}\right)^2 \tag{7.25}$$

where f_c is the critical frequency, given by

$$f_c = 8.9788\sqrt{N} \text{ (Hz)} \tag{7.26}$$

At frequencies below f_c the refractive index becomes imaginary, i.e. the wave is exponentially attenuated and does not pass through the ionosphere.

Apparent reflection can occur whenever the wave frequency is below f_c, so useful frequencies for satellite communications need to be well above this.

The greatest critical frequency normally encountered is around 12 MHz. This is the other extreme of an overall atmospheric 'window' which is bounded at the high-frequency end by atmospheric absorption at hundreds of gigahertz. Even well above 12 MHz, however, a number of ionospheric effects are important in satellite communications as described in the following sections.

7.3.1 Faraday Rotation

A linearly polarised wave becomes rotated during its passage through the ionosphere due to the combined effects of the free electrons and the Earth's magnetic field. The angle associated with this rotation depends on the frequency and the total number of electrons encountered along the path according to

$$\phi = \frac{2.36 \times 10^{20}}{f^2} B_{av} N_T \tag{7.27}$$

where f is in hertz and $B_{av} = \mu H_{av}$ is the average magnetic field of the Earth [weber per square metre]; a typical value is $B_{av} = 7 \times 10^{-21}$ Wb m^{-2}. The parameter N_T in this expression can be pictured as the total number of electrons contained in a vertical column of cross-sectional area 1 m^2 and length equal to the path length, i.e. the *total electron content*, N_T, sometimes called the *vertical total electron content* (VTEC). It is a key parameter in several ionospheric effects and is given by

$$N_T = \int_0^{r_T} N \, dr \text{ [electrons m}^{-2}] \tag{7.28}$$

The total electron content for a zenith path varies over the range 10^{16} to 10^{19} electrons [m^{-2}], with the peak taking place during the day time. This phenomenon is called *Faraday rotation*. If linearly polarised waves are used, extra path loss will result due to polarisation mismatch between the antennas. Here are some ways to minimise this extra path loss

1. Use of circular polarisation;
2. Physical or electrical variation of receive antenna polarisation;
3. Aligning the antennas to compensate for an average value of the rotation, provided that the resulting mismatch loss is acceptable.

7.3.2 Group Delay

The ionospheric refractivity encountered by a radiowave means the resulting phase shift differs from the expected phase shift based on the physical path length. This can be considered as a change Δr [m] in the apparent path length.

$$\Delta r = \frac{40.3}{f^2} N_T \tag{7.29}$$

Typical values for a 4 GHz zenith path system are between 0.25 and 25 m. The change in path length can equivalently be considered as a time delay t [s],

$$t = \frac{40.3}{cf^2} N_T \tag{7.30}$$

This variation in apparent path length with electron density causes uncertainties for positioning systems such as GPS (Global Positioning System) or Galileo which depend on measurements of relative path delay from multiple satellites to determine the receiver position. This can be overcome by making measurements at multiple frequencies to estimate and remove the offset.

7.3.3 Dispersion

The change in effective path length arising from the group delay described above would not be problematic in itself if it were applied equally to all frequencies. However, the delay is frequency dependent, so a transmitted pulse occupying a wide bandwidth will be smeared in time when it arrives at the receiver, with the higher frequencies arriving earliest. The dispersion is defined as the rate of change of the delay with respect to frequency,

$$\frac{dt}{df} = -\frac{80.6}{f^3} N_T \, [\text{s Hz}^{-1}] \tag{7.31}$$

The differential delay associated with opposite extremes of a signal occupying a bandwidth Δf is then

$$\Delta t = -\frac{80.6}{cf^3} \Delta f \, N_T \tag{7.32}$$

This effect limits the maximum signal bandwidths which may be transmitted through the ionosphere without distortion.

Table 7.4: Comparison of ionospheric effects

Effect	1 GHz	3 GHz	10 GHz	30 GHz	Frequency dependence
Faraday rotation [°]	106	12	1.1	0.1	f^{-2}
Propagation delay [μ s]	0.25	0.028	0.003	0.0003	f^{-2}
Dispersion [ps MHz^{-1}]	400	15	0.4	0.015	f^{-3}

7.3.4 Ionospheric Scintillation

There is a wind present in the ionosphere, just as in the troposphere, which causes rapid variations in the local electron density, particularly close to sunset. These density variations cause changes in refraction of an earth satellite wave and hence of the signal levels. Portions of the ionosphere then act like lenses, cause focusing and divergence of the wave and hence lead to signal level variations or *ionospheric scintillation*. The key characteristics of ionospheric scintillation are as follows:

- low-pass power spectrum, roll-off as f^{-3}, corner frequency ≈ 0.1 Hz;
- strong correlation between scintillation occurrence and sunspot cycle;
- size of disturbances proportional to $f_c^{-1.5}$.

7.3.5 Summary of Ionospheric Effects

Table 7.4 summarises the magnitude of various ionospheric effects at various frequencies. The key point to note is that all diminish rapidly with frequency. More detail is available in [ITU, 531].

7.4 SATELLITE EARTH STATION ANTENNAS

The standard antenna type used in satellite earth stations is the parabolic reflector antenna as described in Section 4.5.8. These allow almost arbitrarily large gains to be produced by simply increasing the diameter of the dish, and gains well in excess of 30 dBi are often required in order to provide sufficient fade margin for high-reliability earth satellite links, which often carry signals for broadcast to many users on the Earth. Most high-gain antennas, even for geostationary satellite systems, have to be automatically steered to track irregularities in the satellite orbit and to overcome the effects of wind loading.

There are two basic feed types employed in the design of parabolic reflector antennas, namely, *prime focus* and *Cassegrain* (Figure 7.24). In the prime focus case, the feed antenna, usually a horn antenna, is located directly at the focus of the parabola and illuminates the parabola with circular wavefronts which are converted into plane waves by the reflector curvature. The Cassegrain case, by contrast, has the prime feed located at the apex of the main reflector, and it illuminates a secondary subreflector placed close to the main focus. If the subreflector curvature is chosen to be hyperbolic, then essentially the same far-field radiation pattern is produced as in the prime focus case. The Cassegrain configuration has less feeder loss due to its reduced length, and permits easy access to the feed horn. When used in receive mode, any *spillover* associated with the feed horn receives noise from the relatively low-noise

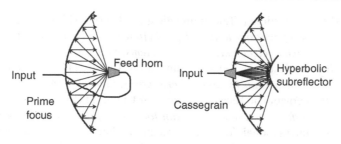

Figure 7.24: Feed configurations for parabolic reflector antennas

sky, whereas the prime focus horn receives spillover noise from the noisy ground. However, the blockage of the aperture by the feed arrangement (and consequent efficiency reduction) is much less in the case of the prime-focus antenna, so the Cassegrain is usually only used for systems requiring beamwidths less than around 1°. See [Stutzman, 81] for more details.

In recent times, small-aperture earth terminals are increasingly implemented using arrays of printed dipoles, which are relatively cheap to manufacture and smaller in volume and wind loading than parabolic reflectors, although the radiation efficiency is usually lower.

7.5 CONCLUSION

Satellite fixed links provide high reliability communications by relying on line-of-sight paths and high-gain antennas, which avoid the obstruction losses encountered in terrestrial links and permit communication over very much greater distances than would be possible when the curvature of the Earth is the limiting factor. Nevertheless, propagation mechanisms due to atmospheric particles are highly significant in creating attenuation and time-variant fading of the signal, which must be characterised in a statistical manner in order to permit low outage probabilities over time periods measured in years.

REFERENCES

[Allnutt, 89] J. A. Allnutt, *Satellite-to-ground radiowave propagation*, Institution of Electrical Engineers, ISBN 0-86341-157-6, 1989.

[Belloul, 98] B. Belloul, S. R. Saunders and B. G. Evans, Prediction of scintillation intensity from sky-noise temperature in earth-satellite links, *Electronics Letters*, 34 (10), 1023–1024, 1998.

[Budden, 61] K. G. Budden, *Radiowaves in the ionosphere*, Cambridge University Press, Cambridge, 1961.

[Clarke, 45] A. C. Clarke, Extra-terrestrial relays – Can rocket stations give world-wide radio coverage?, *Wireless World*, pp. 305–308, 1945.

[Howell, 92] R. G. Howell, R. L. Stuckey and J. W. Harris, The BT laboratories slant-path measurement complex, *BT Technology Journal*, 10 (4), 9–21, 1992.

[ITU, 372] International Telecommunication Union, ITU-R Recommendation PI.372-6, *RadioNnoise*, Geneva, 1994.

[ITU, 531] International Telecommunication Union, ITU-R Recommendation P.531-4, *Ionospheric propagation data and prediction methods required for the design of satellite services and systems*, Geneva, 1997a.

[ITU, 618] International Telecommunication Union, ITU-R Recommendation P.618-5, *Propagation data and prediction methods required for the design of earth-space telecommunication systems*, Geneva, 1997b.

[ITU, 676] International Telecommunication Union, ITU-R Recommendation P.676-3, *Attenuation by atmospheric gases, Geneva*, 1997c.

[ITU, 838] International Telecommunication Union, ITU-R Recommendation 838, *Specific attenuation model for rain for use in prediction methods*, Geneva, 1992.

[ITU, 840] International Telecommunication Union, ITU-R Recommendation P.840-2, *Attenuation due to clouds and fog*, Geneva, 1997d.

[Maral, 93] G. Maral and M. Bousquet, *Satellite Communications Systems – Systems, techniques and technology*, 2nd edn, John Wiley & Sons, Ltd, Chichester, ISBN 0-471-93032-6, 1993.

[Marshall, 48] J. S. Marshall and W. M. K. Palmer, The distribution of raindrops with size, *Journal of Metrology*, 5, 1965–1966, 1948.

[NASA, 83] W. L. Flock, Propagation effects on satellite systems at frequencies below 10 GHz, *NASA Reference Publication 1108*, 1983.

[Pruppacher, 71] H. R. Pruppacher and R. L. Pitter, A semi-empirical determination of the shape of cloud and rain drops, Journal of The Atmospheric Sciences, 28, 86–94, 1971.

[Stutzmen, 81] W. L. Stutzman and G. A. Thiele, *Antenna Theory and Design*, John Wiley & Sons, Inc., New York, ISBN 0-471-04458-X, 1981.

[Tatarski 61] V. I. Tatarski, *Wave propagation in a turbulent medium*, McGraw-Hill, New York, 1961.

PROBLEMS

7.1 Calculate the free space loss for a satellite operating in geostationary orbit, in communication at 10 GHz with an earth station which observes the satellite at zenith.

7.2 Would you expect tropospheric scintillation to be greatest at high or low elevation angles? Why?

7.3 Compute the rain attenuation not exceeded for 0.001% of the time in non-Mediterranean European regions at 30 GHz with an elevation angle of 30°.

7.4 Compute the total atmospheric gas attenuation in Problem 7.3.

7.5 An earth station receive antenna, located at a latitude of 55°, with 3 dB beamwidth of 0.1° at 10 GHz is pointed directly at the Moon with an elevation angle of 45°. Assuming a rainfall rate of 28 mm h^{-1} and feeder loss of 3 dB, calculate the system noise temperature.

7.6 Assuming that the variation of electron density N with height r is given by $N(r) = N_0 e^{k(1-r)}$, where $N_0 = 1 \times 10^{12}$ and $k = 1 \times 10^{-5} \, \text{m}^{-1}$, calculate the following parameters for a zenith path: total electron content, Faraday rotation, propagation delay and dispersion for a 30 GHz wave.

7.7 For a 28 GHz ordinary wave, and given the electron density equation in 7.6, determine the critical frequency and critical angle.

8 Macrocells

8.1 INTRODUCTION

This chapter introduces methods for predicting the path loss encountered in macrocells. Such cells are commonly encountered in cellular telephony, where they are the main means of providing initial network coverage over a wide area. However, the propagation models are also applicable to broadcasting, private mobile radio and fixed wireless access applications including WiMax. For such systems, in principle, the methods introduced in Chapter 6 could be used to predict the loss over every path profile between the base station and every possible user location. However, the data describing the terrain and clutter would be very large and the computational effort involved would often be excessive. Even if such resources were available, the important parameter for the macrocell designer is the overall area covered, rather than the specific field strength at particular locations, so models of a statistical nature are often more appropriate.

The models presented in this chapter treat the path loss associated with a given macrocell as dependent on distance, provided that the environment surrounding the base station is fairly uniform. In consequence, the coverage area predicted by these models for an isolated base station in an area of consistent environment type will be approximated as circular. Although this is clearly inaccurate, it is useful for system dimensioning purposes. Methods will be indicated at the end of this chapter and in Chapter 9 for improving the reality of this picture.

8.2 DEFINITION OF PARAMETERS

The following terms will be used in defining path loss models in this chapter and are illustrated in Figure 8.1:

h_m	mobile station antenna height above local terrain height [m], often taken as 1.5 m
d_m	distance between the mobile and the nearest building [m]
h_0	typical (usually the mean) height of buildings above local terrain height [m]
h_b	base station antenna height above local terrain height [m]
r	great circle distance between base station and mobile [m]
$R = r \times 10^{-3}$	great circle distance between base station and mobile [km]
f	carrier frequency [Hz]
$f_c = f \times 10^{-6}$	carrier frequency [MHz]
λ	free space wavelength [m]

The basic definition of a macrocell is that $h_b > h_0$ Although buildings are not the only obstructions in practice, they are usually by far the most significant at typical macrocellular frequencies. In practice, base station heights are around 15–35 m if a mast is used, or around

Antennas and Propagation for Wireless Communication Systems Second Edition   Simon R. Saunders and Alejandro Aragón-Zavala
© 2007 John Wiley & Sons, Ltd

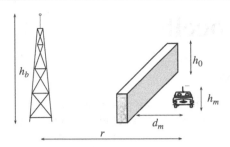

Figure 8.1: Definition of parameters for macrocell propagation models

20 m upwards if deployed on a building rooftop. The effective base station height may be increased dramatically by locating it on a hill overlooking the region to be covered.

8.3 EMPIRICAL PATH LOSS MODELS

The two basic propagation models (free space loss and plane earth loss) examined in Chapter 5 plus the more detailed obstruction loss models described in Chapter 6 account, in principle, for all of the major mechanisms which are encountered in macrocell prediction. However, to use such models would require detailed knowledge of the location, dimension and constitutive parameters of every tree, building and terrain feature in the area to be covered. This is far too complex to be practical and would anyway yield an unnecessary amount of detail, as the system designer is not usually interested in the particular locations being covered, but rather in the overall extent of the coverage area. One appropriate way of accounting for these complex effects is via an empirical model. To create such a model, an extensive set of actual path loss measurements is made, and an appropriate function is fitted to the measurements, with parameters derived for the particular environment, frequency and antenna heights so as to minimise the error between the model and the measurements. Note that each measurement represents an average of a set of samples, the *local mean*, taken over a small area (around 10–50 m), in order to remove the effects of fast fading (Chapter 10), as originally suggested by [Clarke, 68]. See Chapter 19 for further details of measurement procedures. The model can then be used to design systems operated in similar environments to the original measurements. A real example of an empirical model fitted to measurements is shown in Figure 8.2. Methods of accounting for the very large spread of the measurements at a given distance are the subject of Chapter 9.

The simplest useful form for an empirical path loss model is as follows:

$$\frac{P_R}{P_T} = \frac{1}{L} = \frac{k}{r^n} \text{ or in decibels } L = 10n \log r + K \tag{8.1}$$

where P_T and P_R are the effective isotropic transmitted and predicted isotropic received powers as defined in Chapter 5, L is the path loss, r is the distance between the base station and the mobile and $K = -10 \log_{10} k$ and n are constants of the model. Parameter k can be considered as the reciprocal of the propagation loss that would be experienced at one metre range ($r = 1$ m). Models of this form will be referred to as *power law models*. A more convenient form (in decibels) is

$$L(r) = 10n \log(r/r_{\text{ref}}) + L(r_{\text{ref}}) \tag{8.2}$$

where $L(r_{\text{ref}})$ is the predicted loss at a reference distance r_{ref}.

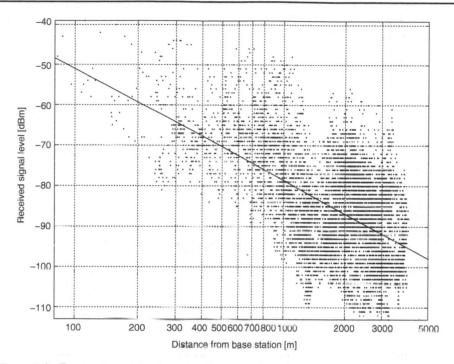

Figure 8.2: Empirical model of macrocell propagation: the dots are measurements taken in a suburban area and the line represents a best-fit empirical model (reproduced by permission of Red-M Services Ltd)

Both the free space loss and the plane earth loss can be expressed in this form. The parameter *n* is known as the *path loss exponent*. It is found by measurement to depend on the system parameters, such as antenna heights and the environment. Chapter 1 showed that the path loss exponent is critical in establishing the coverage and capacity of a cellular system.

8.3.1 Clutter Factor Models

Measurements taken in urban and suburban areas usually find a path loss exponent close to 4, just as in the plane earth loss, but with a greater absolute loss value – i.e. larger *K* in Eq. (8.1). This has led to some models being proposed which consist of the plane earth loss, plus (in decibels) an extra loss component called the clutter factor, as shown in Figure 8.3. The various models differ basically in the values which they assign to *k* and *n* for different frequencies and environments.

Example 8.1

Calculate the range of a macrocell system with a maximum acceptable path loss of 138 dB, assuming $h_m = 1.5\,\text{m}$, $h_b = 30\,\text{m}$, $f_c = 900\,\text{MHz}$ and that path loss can be modelled for this frequency and environment using the plane earth loss plus a clutter factor of 20 dB.

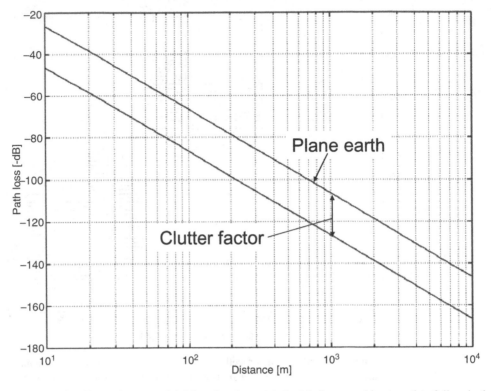

Figure 8.3: Clutter factor model. Note that the y-axis in this figure and in several to follow is the negative of the propagation loss in decibels. This serves to make clear the way in which the received power diminishes with distance.

Solution

The plane earth model of Eq. (5.34) is now modified to the following empirical model:

$$L_{\text{emp}} = 40 \log r - 20 \log h_m - 20 \log h_b + K$$

where K is the clutter factor [decibels]. Equating this to the maximum acceptable path loss and rearranging for $\log r$ yields

$$
\begin{aligned}
\log r &= \frac{L_{\text{emp}} + 20 \log h_m + 20 \log h_b - K}{40} \\
&= \frac{138 + 20 \log 1.5 + 20 \log 30 - 20}{40} \approx 3.78
\end{aligned}
$$

Hence $r = 10^{3.78} \approx 6\,\text{km}$. This shows how the results of Examples 5.4 and 5.5 can be modified to predict more practical ranges.

A good example of a clutter factor model is the method by [Egli, 57], which is based upon a large number of measurements taken around American cities. Egli's overall results were originally presented in nomograph form, but [Delisle, 85] has given an approximation to these results for easier computation:

$$L = 40 \log R + 20 \log f_c - 20 \log h_b + L_m \quad (8.3)$$

where

$$L_m = \begin{cases} 76.3 - 10 \log h_m \text{ for } h_m < 10 \\ 76.3 - 20 \log h_m \text{ for } h_m \geq 10 \end{cases} \quad (8.4)$$

Note that this approximation involves a small discontinuity at $h_m = 10$ m. Although plane earth loss is frequency independent, this model introduces an additional f_c^{-2} received power dependence, which is more representative of the results of real measurements. For very large antenna heights, the loss predicted by (8.3) may be less than the free space value, in which case the free space value should be used.

The mobile antenna characteristic is approximately linear for antennas which clear the surrounding terrain features. Elsewhere there is a square law variation for heights in the range 2–10 m, as for the plane earth loss model. The transition value 10 m presumably corresponds to the mean building height, although no correction is made for other heights. The average effect of polarisation is considered negligible.

8.3.2 The Okumura–Hata Model

This is a fully empirical prediction method [Okumura, 68], based entirely upon an extensive series of measurements made in and around Tokyo city between 200 MHz and 2 GHz. There is no attempt to base the predictions on a physical model such as the plane earth loss. Predictions are made via a series of graphs, the most important of which have since been approximated in a set of formulae by [Hata, 80]. The thoroughness of these two works taken together has made them the most widely quoted macrocell prediction model, often regarded as a standard against which to judge new approaches. The urban values in the model presented below have been standardised for international use in [ITU, 529]

The method involves dividing the prediction area into a series of clutter and terrain categories, namely *open, suburban and urban*. These are summarised as follows:

- *Open area*: Open space, no tall trees or buildings in path, plot of land cleared for 300–400 m ahead, e.g. farmland, rice fields, open fields.
- *Suburban area*: Village or highway scattered with trees and houses, some obstacles near the mobile but not very congested.
- *Urban area*: Built up city or large town with large buildings and houses with two or more storeys, or larger villages with close houses and tall, thickly grown trees.

Okumura takes urban areas as a reference and applies correction factors for conversion to the other classifications. This is a sensible choice, as such areas avoid the large variability present in suburban areas (Chapter 9) and yet include the effects of obstructions better than could be done with open areas. A series of terrain types is also defined for when such information is available. Quasi-smooth terrain is taken as the reference, and correction factors are added for the other types.

Okumura's predictions of median path loss are usually calculated using Hata's approximations as follows [Hata, 80]:

$$
\begin{array}{lll}
\text{Urban areas} & L_{dB} = A + B \log R - E \\
\text{Suburban areas} & L_{dB} = A + B \log R - C & \quad (8.5) \\
\text{Open areas} & L_{dB} = A + B \log R - D
\end{array}
$$

where

$$
\begin{aligned}
A &= 69.55 + 26.16 \log f_c - 13.82 \log h_b \\
B &= 44.9 - 6.55 \log h_b \\
C &= 2(\log(f_c/28))^2 + 5.4 \\
D &= 4.78(\log f_c)^2 - 18.33 \log f_c + 40.94 \qquad (8.6) \\
E &= 3.2(\log(11.75 h_m))^2 - 4.97 \text{ for large cities, } f_c \geq 300\,\text{MHz} \\
E &= 8.29(\log(1.54 h_m))^2 - 1.1 \text{ for large cities, } f_c < 300\,\text{MHz} \\
E &= (1.1 \log f_c - 0.7)h_m - (1.56 \log f_c - 0.8) \text{ for medium to small cities}
\end{aligned}
$$

The model is valid only for $150\,\text{MHz} \leq f_c \leq 1500\,\text{MHz}, 30\,\text{m} \leq h_b \leq 200\,\text{m}$, $1\,\text{m} < h_m < 10\,\text{m}$ and $R > 1\,\text{km}$. The path loss exponent is given by $B/10$, which is a little less than 4, decreasing with increasing base station antenna height.

Base station antenna height h_b is defined as the height above the average ground level in the range 3–10 km from the base station; h_b may therefore vary slightly with the direction of the mobile from the base. The height gain factor varies between 6 dB per octave and 9 dB per octave as the height increases from 30 m to 1 km. Measurements also suggest this factor depends upon range.

Okumura found that mobile antenna height gain is 3 dB per octave up to $h_m = 3$ m and 8 dB per octave beyond. It depends partially upon urban density, apparently as a result of the effect of building heights on the angle-of-arrival of wave energy at the mobile and the consequent shadow loss variation (Chapter 9). Urban areas are therefore subdivided into large cities and medium/small cities, where an area having an average building height in excess of 15 m is defined as a large city.

Other correction factors are included in Okumura's original work for the effects of street orientation (if an area has a large proportion of streets which are either radial or tangential to the propagation direction) and a fine correction for rolling hilly terrain (used if a large proportion of streets are placed at either the peaks or valleys of the terrain undulations). Application of the method involves first finding the basic median field strength in concentric circles around the base station, then amending them according to the terrain and clutter correction graphs.

Okumura's predictions have been found useful in many cases [COST207, 89], particularly in suburban areas. However, other measurements have been in disagreement with these predictions; the reasons for error are often cited as the difference in the characteristics of the area under test with Tokyo. Other authors such as [Kozono, 77] have attempted to modify Okumura's method to include a measure of building density, but such approaches have not found common acceptance.

The Okumura–Hata model, together with related corrections, is probably the single most common model used in designing real systems. Several commercial prediction tools essentially rely on variations of this model, optimised for the particular environments they are catering for, as the basis of their predictions; see www.simonsaunders.com/apbook for a list of such tools.

8.3.3 The COST 231–Hata Model

The Okumura–Hata model for medium to small cities has been extended to cover the band $1500\,\text{MHz} < f_c < 2000\,\text{MHz}$ [COST231, 99].

$$L_{dB} = F + B \log R - E + G \tag{8.7}$$

where

$$F = 46.3 + 33.9 \log f_c - 13.82 \log h_b \tag{8.8}$$

E is as defined in (8.6) for medium to small cities and

$$G = \begin{cases} 0\,\text{dB} & \text{medium-sized cities and suburban areas} \\ 3\,\text{dB} & \text{metropolitan areas} \end{cases} \tag{8.9}$$

8.3.4 The Lee Model

The Lee model is a power law model, with parameters taken from measurements in a number of locations, together with a procedure for calculating an effective base station antenna height which takes account of the variations in the terrain [Lee, 82; Lee, 93]. It can be expressed in the simplified form

$$L = 10n \log R - 20 \log h_{b(\text{eff})} - P_0 - 10 \log h_m + 29 \tag{8.10}$$

where n and P_0 are given by measurements as shown in Table 8.1 and $h_{b(\text{eff})}$ is the effective base station antenna height. The measurements were all made at 900 MHz, and correction factors must be applied for other frequencies, but these do not appear to have been specified in the open literature.

The effective base station height is determined by projecting the slope of the terrain in the near vicinity of the mobile to the base station location. Figure 8.4 shows how this effective height varies for four mobile locations on gently sloping terrain.

Table 8.1: Parameters for the Lee model

Environment		n	P_0
Free space		2	−45
Open area		4.35	−49
Suburban		3.84	−61.7
Urban	Philadelphia	3.68	−70
	Newark	4.31	−64
	Tokyo	3.05	−84
	New York city	4.8	−77

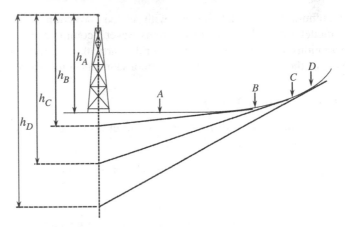

Figure 8.4: Determination of effective base station antenna height for Lee's model

8.3.5 The Ibrahim and Parsons Model

This method [Ibrahim, 83] is based upon a series of field trials around London. The method is not intended as a fully general prediction model, but as a first step towards quantifying urban propagation loss. It integrates well with a previous method [Edwards, 69] for predicting terrain diffraction effects as the same 0.5 km square database is also used. Each square is assigned three parameters, H, U and L, defined as follows.

Terrain height H is defined as the actual height of a peak, basin, plateau or valley found in each square, or the arithmetic mean of the minimum and maximum heights found in the square if it does not contain any such features.

The *degree of urbanisation factor U* is defined as the percentage of building site area within the square which is occupied by buildings having four or more floors. For the 24 test squares in inner London which were analysed, U varied between 2 and 95%, suggesting that this parameter is sensitive enough for the purpose.

Land usage factor L is defined as the percentage of the test area actually occupied by any buildings.

These parameters were selected empirically as having good correlation with the data. Two models were proposed. The fully empirical method shows marginally lower prediction errors but relies on a complex formulation which bears no direct relationship to propagation principles.

The semi-empirical method, as with the Egli clutter factor method, is based upon the plane earth loss together with a clutter factor β, expressed as a function of f_c, L, H and U. Only the semi-empirical method will be examined here, as it has been quoted in later work by the same author [Parsons, 92], and as it forms a better basis for future development. The model is given as

$$L_T = 40 \log r - 20 \log(h_m h_b) + \beta$$

$$\text{where } \beta = 20 + \frac{f_c}{40} + 0.18L - 0.34H + K \tag{8.11}$$

$$\text{and } K = 0.094U - 5.9$$

Data is extracted from the databases compiled by local authorities in the United Kingdom. Although information on U is available only in highly urbanised city centres, K is set to zero elsewhere. Considerably lower accuracy may be expected in such areas as $K = 0$ would correspond to $U = 63\%$. RMS errors calculated from the original data on which the model is based vary from 2.0 to 5.8 dB as frequency is increased from 168 to 900 MHz. A comparison is also shown for some independent data, but error statistics are not given.

The model is of limited use in suburban areas as U will normally be zero, giving no measure of building height distribution.

8.3.6 Environment Categories

In an empirical model, it is crucial to correctly classify the environment in which the system is operating. The models assume the characteristics of the environment to be predicted are sufficiently similar to those where the original measurements were taken that the propagation loss at a given distance will be similar. Good results will therefore be obtained only if the correct classification is chosen. The categories of environment should also be sufficiently numerous that the properties of different locations classed within the same category are not too variable. The decision as to which category an environment fits into is usually purely subjective and may vary between individuals and countries. For example, the Okumura–Hata model uses four categories: large cities, medium-small cities, suburban areas and open areas. Although the original measurements were made in Tokyo, the model relies on other parts of the world having characteristics which are somehow similar to those in Tokyo. Although this is an extremely questionable assumption, it is nevertheless true that the model has been applied to many successful system designs.

Many more detailed schemes exist for qualitative classification of land usage; Table 8.2 shows one example. Schemes often correspond to sources of data, such as satellite remote sensing data which classifies land according to the degree of scattering experienced at various wavelengths. This at least avoids the need for ambiguous judgements to be made. Similarly, the Ibrahim and Parsons model uses a clear numerical approach to classification. Nevertheless, there is no guarantee that there is any one-to-one mapping between the propagation characteristics and such measures of land usage. In order to find more appropriate parameters,

Table 8.2: British Telecom land usage categories [Huish, 88]

Category	Description
0	Rivers, lakes and seas
1	Open rural areas, e.g. fields and heathlands with few trees
2	Rural areas, similar to the above, but with some wooded areas, e.g. parkland
3	Wooded or forested rural areas
4	Hilly or mountainous rural areas
5	Suburban areas, low-density dwellings and modern industrial estates
6	Suburban areas, higher density dwellings, e.g. council estates
7	Urban areas with buildings of up to four storeys, but with some open space between
8	Higher density urban areas in which some buildings have more than four storeys
9	Dense urban areas in which most of the buildings have more than four storeys and some can be classed as skyscrapers; this category is restricted to the centres of a few large cities

the growing tendency in macrocellular propagation is towards models which have a physical basis, and these are examined in the next section.

8.4 PHYSICAL MODELS

Although empirical models have been extensively applied with good results, they suffer from a number of disadvantages:

- They can only be used over parameter ranges included in the original measurement set.
- Environments must be classified subjectively according to categories such as 'urban', which have different meanings in different countries.
- They provide no physical insight into the mechanisms by which propagation occurs.

The last point is particularly significant, as empirical models are unable to account for factors such as an unusually large building or hill which may greatly modify propagation in particular locations. Although the plane earth model has a path loss exponent close to that observed in actual measurements (i.e. 4), the simple physical situation it describes is rarely applicable in practice. The mobile is almost always operated (at least in macrocells) in situations where it does not have a line-of-sight path to either the base station or to the ground reflection point, so the two-ray situation which the plane earth model relies upon is hardly ever applicable. In order to find a more satisfactory physical propagation model, the remaining models in this chapter examine diffraction as a potential mechanism.

8.4.1 The Allsebrook and Parsons Model

Although this model [Allsebrook, 77] is based upon a series of measurements, it may be regarded as an early attempt to provide a physical basis for urban prediction models.

Measurements were made in three British cities (Bradford, Bath and Birmingham) at 86, 167 and 441 MHz. These cities cover a wide range of terrain and building classifications. A 40 dB/decade range dependence is again forced, as would be expected for plane earth loss. This results in an Egli type model with a maximum RMS error of 8.3 dB at 441 MHz. (Note that a least-squares curve fit at this frequency results in a range dependence of only 24 dB/decade). A clutter factor β is introduced to account for excess loss relative to the plane earth calculation.

The frequency dependence of the measured clutter factor is compared with an approximation to the excess loss expected from a 10 m absorbing knife-edge, placed 30 m away from a 2 m high mobile antenna. The predictions compare reasonably well with the mean values of β at 86 and 167 MHz, but considerably underestimate it at 441 MHz. The knife-edge calculation is used as a generalised means of calculating diffraction from the rooftop of the building adjacent to the mobile, with a UHF correction factor γ included to force agreement with the measured values above 200 MHz. It is suggested that this deviation is the result of building width being more significant at the higher frequencies, but this is not confirmed by any analysis.

Allsebrook and Parsons' 'flat city' model can be expressed as

$$L_T = L_P + L_B + \gamma$$

$$\text{where } L_B = 20 \log \left(\frac{h_0 - h_m}{548 \sqrt{(d_m \times 10^{-3})/f_c}} \right) \tag{8.12}$$

and L_P is the plane earth loss. For ease of computation the prediction curve for γ can be replaced by the following quadratic approximation:

$$\gamma = -2.03 - 6.67f_c + 8.1 \times 10^{-5}f_c^2 \tag{8.13}$$

Note that, in the calculation of L_P here, the effective antenna heights are those of the base station and the *building*, giving an overall physical model which may be represented by Figure 8.5.

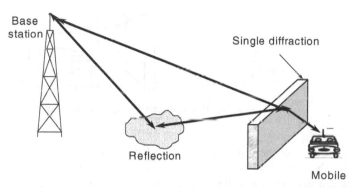

Figure 8.5: Physical interpretation of Allsebrook and Parsons model

A published discussion of this model [Delisle, 85] finds that the correction factor γ is necessary in open areas as well as in (sub)urban areas, although the physical cause suggested for γ cannot apply in line-of-sight situations. Additionally, the quoted value of γ is too large in all situations, casting doubt upon the model's generality. The model is physically valid only in terms of the final building diffraction; the use of the plane earth calculation suggests the existence of a specular ground reflection, which is highly unlikely in a built-up area. Despite this, the model may be considered an improvement over empirical methods as it was the first to make any allowance for the geometry of the specific path being considered.

8.4.2 The Ikegami Model

This model attempts to produce an entirely deterministic prediction of field strengths at specified points [Ikegami, 91]. Using a detailed map of building heights, shapes and positions, ray paths between the transmitter and receiver are traced, with the restriction that only single reflections from walls are accounted for. Diffraction is calculated using a single edge approximation at the building nearest the mobile, and wall reflection loss is assumed to be fixed at a constant value. The two rays (reflected and diffracted) are power summed, resulting in the following approximate model:

$$\begin{aligned} L_E = 10\log f_c + 10\log(\sin\phi) + 20\log(h_0 - h_m) \\ - 10\log w - 10\log\left(1 + \frac{3}{L_r^2}\right) - 5.8 \end{aligned} \tag{8.14}$$

where ϕ is the angle between the street and the direct line from base to mobile and $L_r = 0.25$ is the reflection loss. The analysis assumes that the mobile is in the centre of the street. The model therefore represents the situation illustrated in Figure 8.6. It further assumes that the elevation angle of the base station from the top of the knife-edge is negligible in comparison to the diffraction angle down to the mobile level, so there is no dependence on base station height.

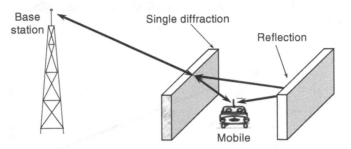

Figure 8.6: Physical interpretation of Ikegami model

A comparison of the results of this model with measurements at 200, 400 and 600 MHz shows that the general trend of variations along a street is accounted successfully. The predictions suggest that field strength is broadly independent of a mobile's position across the street. This is confirmed by the mean values of a large number of measurements, although the spread of values is rather high. Acceptable agreement is also obtained for variations with street angle and width.

Although it accounts reasonably well for 'close-in' variations in field strength, it is a flawed assumption that base station antenna height does not affect propagation. The same assumption means that the free space path loss exponent is assumed, so the model tends to underestimate loss at large distances. Similarly, the variation with frequency is underestimated compared with measurements.

8.4.3 Rooftop Diffraction

When a macrocell system is operated in a built-up area with reasonably flat terrain, the dominant mode of propagation is multiple diffraction over the building rooftops. Diffraction can occur around the sides of individual buildings, but this tends to become highly attenuated over reasonable distances as many interactions with individual buildings are involved.

The diffraction angle over most of the rooftops is small for typical base station heights and distances, usually less than $1°$. In these cases the diffraction is largely unaffected by the particular shape of the obstacles, so it is appropriate to represent the buildings by equivalent knife-edges. The one exception to this is diffraction from the 'final building' at which the wave is diffracted from rooftop level down to the street-level antenna of the mobile (Figure 8.7). It is usual to separate these processes into multiple diffraction across the first $(n-1)$ buildings, treated as knife-edges, and a final building which can be treated either as a knife-edge or as some more complex shape for which the diffraction coefficient is known.

The small diffraction angles encountered have two negative consequences for prediction of these effects. Firstly, a large number of building rooftops may appear within the first Fresnel zone, all contributing to the propagation loss. Secondly, the near-grazing incidence

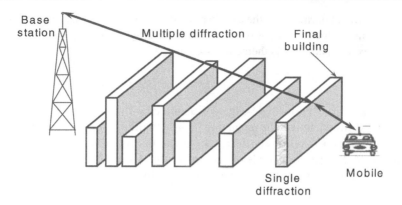

Figure 8.7: Multiple diffraction over building rooftops

implies that the approximate models described in Chapter 6 will fail, leading to very inaccurate predictions. The full multiple edge integral must instead be applied (Section 6.6), which could lead to very long computation times, particularly as it is desired to predict the base station coverage over a wide area, which would require a large number of individual path profiles. Special methods have been developed to enable reasonably rapid calculation of the multiple diffraction integral for cases where accurate results are required and where the necessary data on the building positions and heights is available [Saunders, 94]. Such data is usually too expensive for general use in macrocells, although satellite imagery is reducing the data cost considerably nowadays. Two simplified solutions with reduced data and computational requirements are therefore examined here.

8.4.4 The Flat Edge Model

In this model [Saunders, 91], the situation is simplified by assuming all of the buildings to be of equal height and spacing. The values used can be average values for the area under consideration, or can be calculated individually for each direction from the base station if the degree of urbanisation varies significantly. The geometry is shown in Figure 8.8, illustrating the following parameters additional to the definitions in Section 8.2: distance r_1 from the base station to the first building [m] and elevation angle α of the base station antenna from the top of the final building [rad]. In Figure 8.8, buildings are arranged normal to the great circle path.

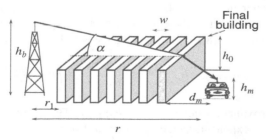

Figure 8.8: Geometry for the flat edge model

As this will not normally be the case in practice, the value of w used should be an effective one to account for the longer paths between the buildings for oblique incidence.

The excess path loss is then expressed as

$$L_{ex} = L_{n-1}(t)L_{ke} \tag{8.15}$$

where L_{ke} accounts for single-edge diffraction over the final building and L_{n-1} accounts for multiple diffraction over the remaining $(n-1)$ buildings. It turns out, provided $r_1 \gg nw$ (i.e. the base station is relatively distant from the first building), that the multiple diffraction integral (6.24) can be completely solved in this special case. The result is that L_{n-1} is a function of a parameter t only, where t is given by

$$t = -\alpha\sqrt{\frac{\pi w}{\lambda}} \tag{8.16}$$

It is given by the following formula

$$L_n(t) = \frac{1}{n}\sum_{m-0}^{n-1} L_m(t)F_s(-jt\sqrt{n-m}) \quad \text{for } n \geq 1 \quad L_0(t) = 1 \tag{8.17}$$

where

$$F_s(jx) = \frac{e^{-jx^2}}{\sqrt{2j}}\left\{\left[S\left(x\sqrt{\frac{2}{\pi}}\right) + \frac{1}{2}\right] + j\left[C\left(x\sqrt{\frac{2}{\pi}}\right) + \frac{1}{2}\right]\right\} \tag{8.18}$$

and $S(.)$ and $C(.)$ are the standard Fresnel sine and cosine integrals defined in Appendix B. This formulation is extremely quick and simple to compute and it applies for any values of α, even when the base station antenna height is below the rooftop level. The number of buildings can be increased to extremely high values with no difficulties.

The flat edge model may be calculated either directly from (8.17), or the results may be estimated from the prediction curves in Figure 8.9, which show the cases where $h_b \geq h_0$. An alternative approach is to use the approximate formula

$$\begin{aligned}L_n(t) &= -20\log A_n(t) \\ &= -(c_1 + c_2\log n)\log(-t) - (c_3 + c_4\log n) \text{ [dB]}\end{aligned} \tag{8.19}$$

where $c_1 = 3.29, c_2 = 9.90, c_3 = 0.77, c_4 = 0.26$. This approximates the value of (8.17) with an accuracy better than ±1.5 dB for $1 \leq n \leq 100$ and $-1 \leq t < 0$. It also enables us to investigate the behaviour of the effective path loss exponent for the flat edge model, as for fixed n, we can rewrite (8.19) with L being the path loss as a power ratio as:

$$L \propto (-t)^{-(c_2/10)\log n} = \left(\alpha\sqrt{\frac{\pi w}{\lambda}}\right)^{-(c_2/10)\log n} \approx \left(\frac{h_b - h_0}{r}\sqrt{\frac{\pi w}{\lambda}}\right)^{-(c_2/10)\log n} \tag{8.20}$$

where the approximation holds if $(h_b - h_0) \ll r$. This is the excess field strength, so the overall path loss exponent, including an extra 2 from the free space part of the loss, is as follows:

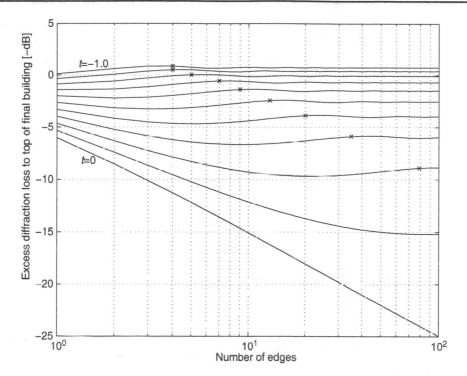

Figure 8.9: Flat edge model prediction curves for elevated base antennas: curves relate to t varying from 0 to -1 in steps of 0.1. The crosses indicate the number of edges required for a settled field according to (8.24)

$$\text{Path loss exponent} = 2 + (c_2/10) \log n \tag{8.21}$$

This expression is shown in Figure 8.10, where it is apparent that, for reasonably large numbers of buildings, the path loss exponent for the flat edge model is close to 4, just as observed in practical measurements. More generally, we can state

Multiple building diffraction accounts for the variation of path loss with range which is observed in macrocell measurements.

Figure 8.9 shows that, for $(h_b > h_0)$ (i.e. $t < 0$), the field at the top of the final building eventually settles to a constant value as the number of edges increases. This number, n_s, corresponds to the number required to fill the first Fresnel zone around the ray from the base station to the final building. The first Fresnel zone radius r_1 is given approximately by

$$r_1 \approx \sqrt{\lambda s} \tag{8.22}$$

where s is the distance along the ray from the field point, provided $s \ll r$. Hence, for small α,

$$\alpha = \tan^{-1} \frac{r_1}{n_s w} \approx \frac{\sqrt{\lambda n_s w}}{n_s w} \tag{8.23}$$

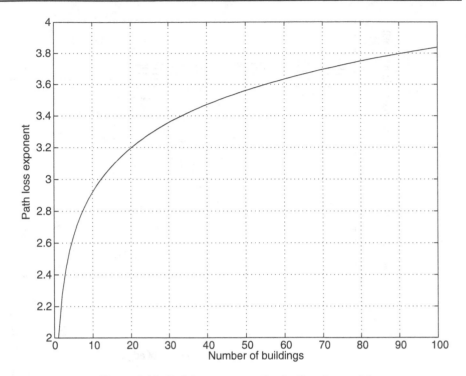

Figure 8.10: Path loss exponent for the flat edge model

So

$$n_s \approx \frac{\lambda}{\alpha^2 w} = \frac{\pi}{t^2} \qquad (8.24)$$

This is marked in Figure 8.9. Note that the number of edges required for settling rises very rapidly with decreasing α. Whenever $\alpha \leq 0$ the field does not settle at all, but decreases monotonically for all n.

The flat edge model is completed by modelling the final building diffraction loss and the reflections from the buildings across the street using the Ikegami model from Section 8.4.2. Thus the total path loss is given by

$$L_T = L_n(t) + L_F + L_E \qquad (8.25)$$

where $L_n(t)$ can be found from (8.17), Figure 8.9 or (8.19), L_F is the free space loss (5.20) and L_E is given in (8.14).

8.4.5 The Walfisch–Bertoni Model

This model can be considered as the limiting case of the flat edge model when the number of buildings is sufficient for the field to settle, i.e. $n \geq n_s$. The multiple diffraction process was investigated in [Walfisch, 88] using a numerical evaluation of the Kirchhoff–Huygens integral and a power law formula is fitted to the results for the settled field. The Walfisch–Bertoni model was the first to actually demonstrate that multiple building diffraction accounts for the variation of distance with range which is observed in measurements.

The settled field approximation is as follows:

$$A_{\text{settled}}(t) \approx 0.1 \left(\frac{\alpha}{0.03} \sqrt{\frac{w}{\lambda}} \right)^{0.9} = 0.1 \left(\frac{-t}{0.03} \right)^{0.9} \tag{8.26}$$

This is valid only for $0.03 \leq t \leq 0.4$. For large ranges, we can again put

$$t \approx -\alpha \sqrt{\frac{\pi w}{\lambda}} \approx -\frac{h_b - h_m}{r} \sqrt{\frac{\pi w}{\lambda}} \tag{8.27}$$

Hence $L_{\text{settled}} \propto r^{-1.8}$. The free space loss is proportional to r^{-2}, so this model predicts that total propagation loss is proportional to $r^{-3.8}$, which is close to the r^{-4} law which is commonly assumed in empirical models and found in measurements. A single knife-edge approximation with a reflection from the building opposite is again used, just as in the Ikegami model, to account for the diffraction from the final building. The complete model is expressed as

$$L_{ex} = 57.1 + L_A + \log f_c + 18 \log R - 18 \log(h_b - h_0)$$
$$- 18 \log \left[1 - \frac{R^2}{17(h_b - h_0)} \right] \tag{8.28}$$

where

$$L_A = 5 \log \left[\left(\frac{w}{2} \right)^2 + (h_0 - h_m)^2 \right] - 9 \log w$$
$$+ 20 \log \left\{ \tan^{-1} \left[\frac{2(h_0 - h_m)}{w} \right] \right\} \tag{8.29}$$

The use of the settled field approximation requires that large numbers of buildings are present, particularly when α is small. Despite this limitation, the Walfisch–Bertoni model is the first to have accounted for observed path loss variation using realistic physical assumptions rather than relying upon forcing agreement using propagation models of entirely different situations, such as the use of the plane earth model in clutter factor models.

A later paper by some of the same authors [Maciel, 93] provides an alternate means of computing the settled rooftop field which is valid over a wider range of the parameter t.

$$L_n(t) = -10 \log(G_2 Q^2(t)) \tag{8.30}$$

where

$$Q(t) = -3.502t - 3.327t^2 - 0.962t^3 \tag{8.31}$$

The antenna gain G_2 is that in the direction to the highest building edge visible from the base station antenna when it is below the surrounding rooftops, or the gain in the horizontal plane when the base station antenna is above the surrounding rooftops.

The authors state that this expression predicts L_n with an accuracy greater than 0.5 dB over the range $-1 \leq t \leq -0.01$, again provided that the number of buildings in the path is large enough for the field to settle. This expression was found to describe measurements made in Denmark for small r with good accuracy [Eggers, 90].

8.4.6 COST 231/Walfisch–Ikegami Model

The Walfisch–Bertoni model for the settled field has been combined with the Ikegami model for diffraction down to street level plus some empirical correction factors to improve agreement with measurements in a single integrated model by the COST 231 project [COST 231, 99].

For non-line-of-sight conditions the total loss is given by

$$L = L_F + L_{msd} + L_{sd} \tag{8.32}$$

where L_F is the free space loss, L_{msd} accounts for multiple knife-edge diffraction to the top of the final building and L_{sd} accounts for the single diffraction and scattering process down to street level. L is given a minimum value of L_F in case the other terms become negative. The individual terms are

$$L_{sd} = -16.9 + 10 \log f_c + 10 \log \frac{(h_0 - h_m)^2}{w_m} + L(\phi) \tag{8.33}$$

where w_m is the distance between the building faces on either side of the street containing the mobile (typically $w_m = w/2$), and the final term accounts for street orientation at an angle ϕ to the great circle path:

$$L(\phi) = \begin{cases} -10 + 0.354\phi & \text{for } 0° < \phi < 35° \\ 2.5 + 0.075\,(\phi - 35°) & \text{for } 35° \le \phi < 55° \\ 4.0 - 0.114\,(\phi - 55°) & \text{for } 55° \le \phi \le 90° \end{cases} \tag{8.34}$$

Finally, the rooftop diffraction term is given by

$$L_{msd} = L_{bsh} + k_a + k_d \quad \log R + k_f \quad \log f_c - 9 \log w \tag{8.35}$$

where

$$L_{bsh} = \begin{cases} -18 \log[1 + (h_b - h_0)] & \text{for } h_b > h_0 \\ 0 & \text{for } h_b \le h_0 \end{cases} \tag{8.36}$$

$$k_a = \begin{cases} 54 & \text{for } h_b > h_0 \\ 54 - 0.8\,(h_b - h_0) & \text{for } R \ge 0.5 \text{ km and } h_b \le h_0 \\ 54 - 0.8\dfrac{(h_b - h_0)R}{0.5} & \text{for } R < 0.5 \text{ km and } h_b \le h_0 \end{cases} \tag{8.37}$$

$$k_d = \begin{cases} 18 & \text{for } h_b > h_0 \\ 18 - 15\dfrac{(h_b - h_0)}{h_0} & \text{for } h_b \le h_0 \end{cases} \tag{8.38}$$

$$k_f = -4 + 0.7\left(\frac{f_c}{925} - 1\right) \quad \text{for medium-sized city and}$$

$$\text{suburban areas with medium tree density} \qquad (8.39)$$

$$k_f = -4 + 1.5\left(\frac{f_c}{925} - 1\right) \quad \text{for metropolitan centres}$$

For approximate work, the following parameter values can be used:

$$h_0 = \begin{cases} 3n_{\text{floors}} & \text{for flat roofs} \\ 3n_{\text{floors}} + 3 & \text{for pitched roofs} \end{cases} \qquad (8.40)$$

$$w = 20 - 50\,\text{m}, \quad d_m = w/2, \quad \phi = 90°$$

where n_{floors} is the number of floors in the building. The model is applicable for $800\,\text{MHz} \leq f_c \leq 2000\,\text{MHz}$, $4\,\text{m} \leq h_b \leq 50\,\text{m}$, $1\,\text{m} \leq h_m \leq 3\,\text{m}$ and $0.02\,\text{km} \leq R \leq 5\,\text{km}$.

An alternative approach is to replace the L_{msd} term by $L_n(t)$ from the flat edge model. This would enable the path loss exponent to vary according to the number of buildings and to be uniformly valid for $h_b \leq h_0$. Note, however, that for very low base station antennas other propagation mechanisms, such as diffraction around vertical building edges and multiple reflections from building walls, are likely to be significant. Models intended for microcells are then more likely to be appropriate, as described in Chapter 12.

8.5 ITU-R MODELS

It is often difficult to select the best model for a given application from the many described in this chapter. Recommendations produced by the International Telecommunications Union are a good reference source in this situation, as they summarise in simple form some recommended procedures. They may not always represent the most accurate model for a given case, but they have the benefit of being widely accepted and used for coordination and comparison purposes. Two recommendations in particular are relevant to this chapter and are briefly summarised below. The full text of the recommendations may be obtained online from http://www.itu.int/ITU-R/

8.5.1 ITU-R Recommendation P.1411

This recommendation [ITU, 1411] contains a model relevant to macrocell applications which is applicable to non-line-of-sight systems operating from 20 m to 5 km. The model is essentially a version of the COST 231 Walfisch–Ikegami model with simplified and generalised calculation procedures. The validity range is quoted as follows:

$$h_b : 4 - 50\,\text{m},$$
$$h_m : 1 - 3\,\text{m},$$
$$f_c : 800 - 2000\,\text{MHz for } h_b \leq h_o$$
$$f_c : 800 - 5000\,\text{MHz for } h_b > h_o.$$

Other models from this recommendation are relevant to microcell situations and will be examined in Chapter 12.

8.5.2 ITU-R Recommendation P.1546

For longer ranges up to 1000 km this recommendation [ITU, 1546] contains useful models. The model is based on series of curves (or tables), originating from measurements, allowing predictions for wide area macrocells and for broadcasting and fixed wireless access applications. As well as the basic prediction curves there are a set of correction methods to account for factors including terrain variations, effective base station antenna height, paths which cross mixtures of land and sea and predictions for different time percentage availability. The method gives similar results to the Okumura–Hata model for distances up to 10 km. Overall the method is valid for $f_c = 30$ MHz to 3 GHz and distances from 1 to 1000 km. As such it is applicable to a wide range of system types, including wide area macrocells, Private Mobile Radio systems and terrestrial video and audio broadcasting networks.

8.6 COMPARISON OF MODELS

The path loss predictions of most of the models described in this chapter are compared in Table 8.3. It shows the exponents of path loss variation predicted by each model. Thus a –2 in the h_b column means the model predicts that path loss is inversely proportional to the square of the base station antenna height. In some cases it is difficult to express the variation in this form, but otherwise it is useful as a means of comparison.

Table 8.3 Comparison of macrocell propagation models

Model	Path loss exponent	h_b	h_m	f_c
Free space (Chapter 5)	2	0	0	2
Plane earth (Chapter 5)	4	−2	−2	0
Egli 8.3.1	4	−2	$-1(h_m < 10)$ $-2(h_m > 10)$	2
Okumura–Hata (Section 8.3.2) COST231–Hata (Section 8.3.3)	a	b	See (8.5)	≈ 2.6
Lee (Section 8.3.4)	2–4.3	−2	−1	NS
Ibrahim (Section 8.3.5)	4	−2	−2	$10^{f_c/400}$
Allsebrook (Section 8.4.1)	4	−2	$(h_0 - h_m)^2$	≈ 2
Ikegami (Section 8.4.2)	1	0	$(h_0 - h_m)^2$	2
Flat edge (Section 8.4.4)	2–4	≈ -2	$(h_0 - h_m)^2$	≈ 2.1
Walfisch–Bertoni (Section 8.4.5)	3.8	≈ -1.8	≈ -1	2.1
COST-231 Walfisch–Ikegami (Section 8.4.6) and [ITU, 1411] NLOS model (Section 8.5.1)	3.8	≈ -1.8	≈ -1	See (8.32)

NS = not specified, $a = 4.5 - 0.66 \log h_b$, $b = -1.38 - 0.66 \log r$

8.7 COMPUTERISED PLANNING TOOLS

The methods described in this chapter are most often implemented for practical planning within computer software. The development of such software has been motivated and enabled by a number of factors:

- The enormous increase in the need to plan cellular systems accurately and quickly.
- The development of fast, affordable computing resources.
- The development of geographical information systems, which index data on terrain, clutter and land usage in an easily accessible and manipulable form.

Such techniques have been implemented in a wide range of commercially available and company-specific planning tools. Some of them are listed at http://www.simonsaunders.com/apbook. Although most are based on combined empirical and simple physical models, it is anticipated there will be progressive evolution in the future towards more physical or physical-statistical methods as computing resources continue to cheapen, clutter data improves in resolution and cost and as research develops into numerically efficient path loss prediction algorithms.

8.8 CONCLUSION

Propagation path loss modelling is the fundamental method of predicting the range of a mobile radio system. The accuracy of the path loss predictions is crucial in determining whether a particular system design will be viable. In macrocells, empirical models have been used with great success, but deterministic physical models are being increasingly applied as a means of improving accuracy, based on the use of multiple rooftop diffraction as the key propagation mechanism. This accuracy comes at the expense of increased input data requirements and computational complexity. Another generation of models is expected to appear which combine sound physical principles with statistical parameters, which can economically be obtained in order to provide the optimum balance between accuracy and complexity.

The path loss may be taken, very roughly, to be given by

$$\frac{P_R}{P_T} = \frac{1}{L} = k\frac{h_m h_b^2}{r^4 f_c^2} \qquad (8.41)$$

where k is some constant appropriate to the environment. It should be emphasised that this expression, and all of the models described in this chapter, account only for the effects of typical clutter on flat or gently rolling terrain. When the terrain variations are sufficient to cause extra obstruction loss, then the models must be supplemented by calculations of terrain loss in a manner similar to Chapter 6.

REFERENCES

[Allsebrook, 77] K. Allsebrook and J. D. Parsons, Mobile radio propagation in British cities at frequencies in the VHF and UHF bands, *IEEE Transactions on Vehicular Technology*, 26 (4), 95–102, 1977.

[Clarke, 68] R. H. Clarke, A statistical theory of mobile radio reception, *Bell System Technical Journal*, 47 (6), 957–1000, 1968.

[COST-207, 89] Final report of the COST-207 management committee, *Digital land mobile radio communications*, Commission of the European Communities, L-2920, Luxembourg, 1989.

[COST 231, 99] COST 231 Final report, *Digital Mobile Radio: COST 231 View on the Evolution Towards 3rd Generation Systems*, Commission of the European Communities and COST Telecommunications, Brussels, 1999.

[Delisle, 85] G. Y. Delisle, J. Lefevre, M. Lecours and J. Chouinard, Propagation loss prediction: a comparative study with application to the mobile radio channel, *IEEE Transactions on Vehicular Tech*nology, 26 (4), 295–308, 1985.

[Edwards, 69] R. Edwards and J. Durkin, Computer prediction of service areas for VHF mobile radio networks, *Proceedings of the IEE*, 116 (9), 1493–1500, 1969.

[Eggers, 90] P. Eggers and P. Barry, Comparison of a diffraction-based radiowave propagation model with measurements, *Electron Letters*, 26 (8), 530–531, 1990.

[Egli, 57] J. J. Egli, Radio Propagation above 40MC over irregular terrain, *Proceedings of the IRE*, 1383–91, 1957.

[Hata, 80] M. Hata, Empirical formula for propagation loss in land mobile radio services, *IEEE Transactions on Vehicular Technology*, 29, 317–25, 1980.

[Huish, 88] P. W. Huish and E. Gurdenli, Radio channel measurements and prediction for future mobile radio systems, *British Telecom Technology Journal*, 6 (1), 43–53, 1988.

[Ibrahim, 83] M. F. Ibrahim and J. D. Parsons, Signal strength prediction in built-up areas, *Proceedings of the IEE*, 130F (5), 377–84, 1983.

[Ikegami, 91] F. Ikegami, T. Takeuchi and S. Yoshida, Theoretical prediction of mean field strength for urban mobile radio, *IEEE Transactions on Antennas and Propagation*, 39 (3), 299–302, 1991.

[ITU, 529] International Telecommunication Union, *ITU-R Recommendation P.529-2: Prediction methods for the terrestrial land mobile service in the VHF and UHF bands*, Geneva, 1997.

[ITU, 1411] International Telecommunication Union, *ITU-R Recommendation P.1411-3: Propagation data and prediction methods for the planning of short-range outdoor radiocommunication systems and radio local area networks in the frequency range 300 MHz to 100 GHz*, Geneva, 2005a.

[ITU, 1546] International Telecommunication Union, *ITU-R Recommendation P.1546-2: Method for point-to-area predictions for terrestrial services in the frequency range 30 MHz to 3 000 MHz*, Geneva, 2005b.

[Kozono, 77] S. Kozono and K. Watanabe, Influence of environmental buildings on UHF land mobile radio propagation, *IEEE Transactions on Commununications*, 25 (10), 1133–43, 1977.

[Lee, 82] W. C. Y. Lee, *Mobile Communications Engineering*, McGraw-Hill, New York, 1982, ISBN 0-070-37039-7.

[Lee, 93] W. C. Y. Lee, *Mobile Design Fundamentals*, John Wiley, New York, 1993, ISBN 0-471-57446-5.

[Maciel, 93] L. R. Maciel, H. L. Bertoni and H. Xia, Propagation over buildings for ranges of base station antenna height, *IEEE Transactions on Vehicular Technology*, 42 (1), 41–45, 1993.

[Okumura, 68] Y. Okumura, E. Ohmori, T. Kawano and K. Fukuda, Field strength and its variability in VHF and UHF land mobile radio service, *Review of the Electrical Communications Laboratories*, 16, 825–73, 1968.

[Parsons, 92] J. D. Parsons, *The Mobile Radio Propagation Channel*, John Wiley & Sons, Ltd, Chichester, ISBN 0-471-96415-8 1992.

[Saunders, 91] S. R. Saunders and F. R. Bonar, Explicit multiple building diffraction attenuation function for mobile radio wave propagation, *Electronics Letters*, 27 (14), 1276–77, 1991.

[Saunders, 94] S. R. Saunders and F. R. Bonar, Prediction of mobile radio wave propagation over buildings of irregular heights and spacings, *IEEE Transactions on Antennas and Propagation*, 42 (2), 137–44, 1994.

[Walfisch, 88] J. Walfisch, and H. L. Bertoni, A theoretical model of UHF propagation in urban environments, *IEEE Transactions on Antennas and Propagation*, 36 (12), 1788–96, 1988.

PROBLEMS

8.1 Use the Okumura–Hata model to compare the maximum cell radius available from systems having maximum acceptable path loss of 130 dB in urban and suburban areas at 900 MHz and 1800 MHz.

8.2 Use the flat edge model to predict the range corresponding to a path loss of 110 dB at 900 MHz. Assume $h_b = 15\,\text{m}, h_0 = 10\,\text{m}, L_E = 15\,\text{dB}, w = 40\,\text{m}$ and $h_m = 1.5\,\text{m}$.

8.3 Repeat the previous example using the Okumura–Hata suburban model.

8.4 If two co-channel base stations with omnidirectional antennas are separated by 5 km, calculate the minimum value of signal-to-interference ratio encountered, assuming a cell radius of 1 km and negligible directivity in the elevation plane.

8.5 Repeat Problem 8.4, assuming that the antennas are 15 m above level ground, with the elevation pattern of a vertical half-wave dipole.

8.6 Discuss the limitations of the Okumura–Hata model, assuming that the frequency range at which has been validated is appropriate.

8.7 Estimate the range of a communication system operating in the GSM 900 MHz band using the Lee macrocell model for a suburban area, if the base station is assumed to have an effective height of 25 m. Make appropriate assumptions for any parameters not provided.

8.8 Prove Eq. (8.22).

8.9 Repeat Problem 8.7 using the flat edge model, making use of the approximate formulation (8.19).

9 Shadowing

'Beware lest you lose the substance by grasping at the shadow'.
Aesop, Greek slave and fable author

9.1 INTRODUCTION

The models of macrocellular path loss described in Chapter 8 assume that path loss is a function only of parameters such as antenna heights, environment and distance. The predicted path loss for a system operated in a particular environment will therefore be constant for a given base-to-mobile distance. In practice, however, the particular clutter (buildings, trees) along a path at a given distance will be different for every path, causing variations with respect to the nominal value given by the path loss models, as shown by the large scatter evident in the measurements in Figure 8.2. Some paths will suffer increased loss, whereas others will be less obstructed and have an increased signal strength, as illustrated in Figure 9.1. This phenomenon is called *shadowing* or *slow fading*. It is crucial to account for this in order to predict the reliability of coverage provided by any mobile cellular system.

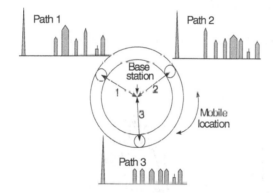

Figure 9.1: Variation of path profiles encountered at a fixed range from a base station

9.2 STATISTICAL CHARACTERISATION

If a mobile is driven around a base station (BS) at a constant distance, then the local mean signal level will typically appear similar to Figure 9.2, after subtracting the median (50%) level in decibels. If the probability density function of the signal is then plotted, a typical result is Figure 9.3. The distribution of the underlying signal powers is *log-normal*; that is, the signal measured in decibels has a normal distribution. The process by which this distribution comes about is known as *shadowing* or *slow fading*. The variation occurs over distances comparable to the widths of buildings and hills in the region of the mobile, usually tens or hundreds of metres.

Antennas and Propagation for Wireless Communication Systems Second Edition   Simon R. Saunders and
Alejandro Aragón-Zavala
© 2007 John Wiley & Sons, Ltd

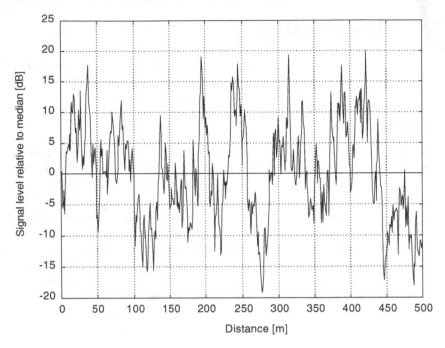

Figure 9.2: Typical variation of shadowing with mobile position at fixed BS distance

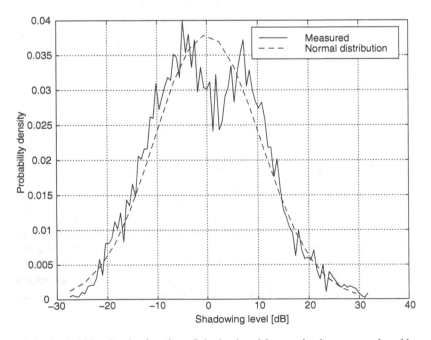

Figure 9.3: Probability density function of shadowing. Measured values are produced by subtracting the empirical model shown in Figure 8.2 from the total path loss measurements. Theoretical values come from the log-normal distribution

The standard deviation of the shadowing distribution (in decibels) is known as the *location variability*, σ_L. The location variability varies with frequency, antenna heights and the environment; it is greatest in suburban areas and smallest in open areas. It is usually in the range 5–12 dB (Section 9.5); the value in Figures 9.2 and 9.3 is 8 dB.

9.3 PHYSICAL BASIS FOR SHADOWING

The application of a log-normal distribution for shadowing models can be justified as follows. If contributions to the signal attenuation along the propagation path are considered to act independently, then the total attenuation A, as a power ratio, due to N individual contributions $A_1, \ldots, A_N$ will be simply the product of the contributions:

$$A = A_1 \times A_2 \times \ldots \times A_N \qquad (9.1)$$

If this is expressed in decibels, the result is the sum of the individual losses in decibels:

$$L = L_1 + L_2 + \ldots + L_N \qquad (9.2)$$

If all of the L_i contributions are taken as random variables, then the central limit theorem holds (Appendix A) and L is a Gaussian random variable. Hence A must be log-normal.

In practice, not all of the losses will contribute equally, with those nearest the mobile end being most likely to have an effect in macrocells. Moreover, as shown in Chapter 8, the contributions of individual diffracting obstacles cannot simply be added, so the assumption of independence is not strictly valid. Nevertheless, when the different building heights, spacings and construction methods are taken into account, along with the attenuation due to trees, the resultant distribution function is indeed very close to log-normal [Chrysanthou, 90] [Saunders, 91].

9.4 IMPACT ON COVERAGE

9.4.1 Edge of Cell

When shadowing is included, the total path loss becomes a random variable, given by

$$L = L_{50} + L_s \qquad (9.3)$$

where L_{50} is the level not exceeded at 50% of locations at a given distance, as predicted by any standard path loss model (the *local median* path loss) described in Chapter 8. L_s is the shadowing component, a zero-mean Gaussian random variable with standard deviation σ_L. The probability density function of L_s is therefore given by the standard Gaussian formula (Appendix A equation (A.16)):

$$p(L_S) = \frac{1}{\sigma_L \sqrt{2\pi}} \exp\left[-\frac{L_S^2}{2\sigma_L^2} \right] \qquad (9.4)$$

In order to provide reliable communications at a given distance, therefore, an extra fade margin has to be added into the link budget according to the reliability required from the system. In Figure 9.4, the cell range would be around 9.5 km if shadowing were neglected, then only 50% of locations at the edge of the cell would be properly covered. By adding the fade margin, the cell radius is reduced to around 5.5 km but the reliability is greatly increased, as a much smaller proportion of points exceed the maximum acceptable path loss.

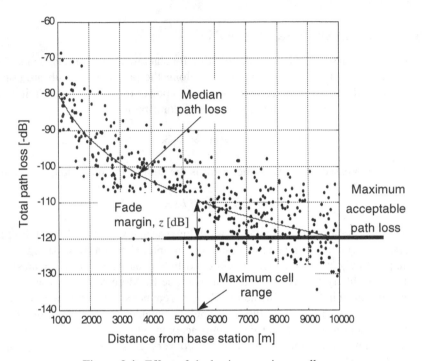

Figure 9.4: Effect of shadowing margin on cell range

The probability that the shadowing increases the median path loss by at least z [dB] is then given by

$$\Pr[L_S > z] = \int\limits_{L_S=z}^{\infty} p(L_S)dL_S = \int\limits_{L_S=z}^{\infty} \frac{1}{\sigma_L\sqrt{2\pi}}\exp\left[-\frac{L_S^2}{2\sigma_L^2}\right]dL_S \qquad (9.5)$$

It is then convenient to normalise the variable z by the location variability:

$$\Pr[L_S > z] = \int\limits_{x=z/\sigma_L}^{\infty} \frac{1}{\sqrt{2\pi}}\exp\left[-\frac{x^2}{2}\right]dx = Q\left(\frac{z}{\sigma_L}\right) \qquad (9.6)$$

where the $Q(.)$ function is the *complementary cumulative normal distribution*. Values for Q are tabulated in Appendix B, or they can be calculated from erfc(.), the standard cumulative error function, using

$$Q(t) = \frac{1}{\sqrt{2\pi}} \int\limits_{x=t}^{\infty} \exp\left(-\frac{x^2}{2}\right)dx = \frac{1}{2}\mathrm{erfc}\left(\frac{t}{\sqrt{2}}\right) \qquad (9.7)$$

$Q(t)$ is plotted in Figure 9.5 and can be used to evaluate the shadowing margin needed for any location variability in accordance with Eq. (9.7) by putting $t = z/\sigma_L$, as described in Example 9.1.

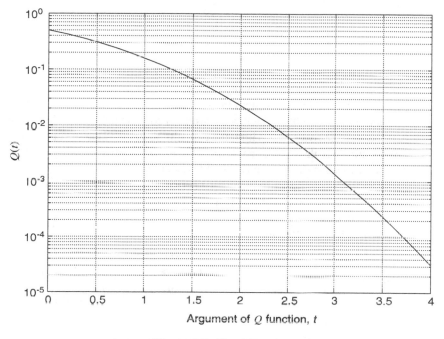

Figure 9.5: The Q function

Example 9.1

A mobile communications system is to provide 90% successful communications at the fringe of coverage. The system operates in an environment where propagation can be described by a plane earth model plus a 20 dB clutter factor, with shadowing of location variability 6 dB. The maximum acceptable path loss for the system is 140 dB. Antenna heights for the system are $h_m = 1.5$ m and $h_b = 30$ m. Determine the range of the system. How is this range modified if the location variability increases to 8 dB?

Solution

The total path loss is given by the sum of the plane earth loss, the clutter factor and the shadowing loss:

$$L_{total} = L_{PEL} + L_{clutter} + L_S$$
$$= 40 \log r - 20 \log h_m - 20 \log h_b + 20 + L_S$$

To find L_S, we take the value of $t = z/\sigma_L$ for which the path loss is less than the maximum acceptable value for at least 90% of locations, or when $Q(t) = 10\% = 0.1$.

From Figure 9.5 this occurs when $t \approx 1.25$. Multiplying this by the location variability gives

$$L_S = z = t\sigma_L = 1.25 \times 6 = 7.5 \text{ dB}$$

Hence

$$\log r = \frac{140 + 20\log 1.5 + 20\log 30 - 20 - 7.5}{40} = 3.64$$

So the range of the system is $r = 10^{3.64} = 4.4$ km.

If σ_L rises to 8 dB, the shadowing margin $L_s = 10$ dB and $d = 3.8$ km. Thus shadowing has a decisive effect on system range.

In the example above, the system was designed so that 90% of locations at the *edge* of the cell have acceptable coverage. Within the cell, although the value of shadowing exceeded for 90% of locations is the same, the value of the total path loss will be less, so a greater percentage of locations will have acceptable coverage.

The calculation in the example may be rearranged to illustrate this as follows. The probability of outage, i.e. the probability that $L_T > 140$ dB is

$$
\begin{aligned}
\text{Outage probability } p_{\text{out}} &= \Pr(L_T > 140) \\
&= \Pr(L_{\text{PEL}} + L_{\text{clutter}} + L_S > 140) \\
&= \Pr(L_S > 140 - L_{\text{PEL}} - L_{\text{clutter}}) \\
&= Q\left(\frac{140 - L_{\text{PEL}} - L_{\text{clutter}}}{\sigma_L}\right) = p_{\text{out}}
\end{aligned}
\tag{9.8}
$$

Consequently, the fraction of locations covered at a range r is simply

$$\text{Coverage fraction} = p_e(r) = (1 - p_{\text{out}}) \tag{9.9}$$

Note that the outage calculated here is purely due to inadequate signal level. Outage may also be caused by inadequate signal-to-interference ratio, and this is considered in Section 9.6.2. In general terms, Eq. (9.9) can be expressed as

$$p_e(r) = \left[1 - Q\left(\frac{L_m - L(r)}{\sigma_L}\right)\right] = \left[1 - Q\left(\frac{M}{\sigma_L}\right)\right] \tag{9.10}$$

where L_m is the maximum acceptable path loss and $L(r)$ is the median path loss model, evaluated at a distance r. $M = (L_m - L(r))$ is the fade margin chosen for the system.

This variation is shown in Figure 9.6, using the same values as Example 9.1. The shadowing clearly has a significant effect on reducing the cell radius from the value predicted using the median path loss alone, which would be around 6.7 km. It is also important to have a good knowledge of the location variability; this is examined in Section 9.5.

9.4.2 Whole Cell

Figure 9.6 shows that, although locations at the edge of the cell may only have a 90% chance of successful communication, most mobiles will be closer to the base station than this, and

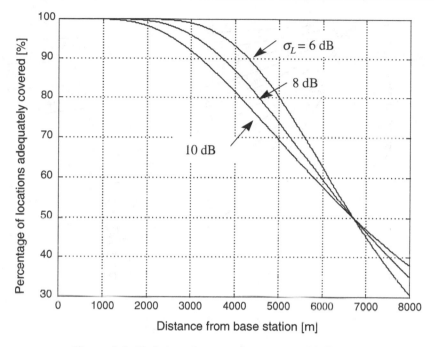

Figure 9.6: Variation of coverage percentage with distance

they will therefore experience considerably better coverage. It is perhaps more appropriate to design the system in terms of the coverage probability experienced over the whole cell. The following analysis is similar to that in [Jakes, 94].

Figure 9.7 shows a cell of radius r_{max}, with a representative ring of radius r, small width Δr, within which the coverage probability is $p_e(r)$. The area covered by the ring is $(2\pi r)\Delta r$. The coverage probability for the whole cell, p_{cell}, is then the sum of the area associated with all such rings from radius 0 to r_{max}, multiplied by the corresponding coverage percentages and

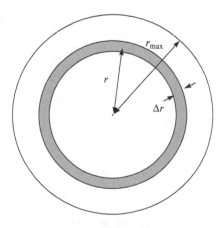

Figure 9.7: Overall cell coverage area by summing contributions at all distances

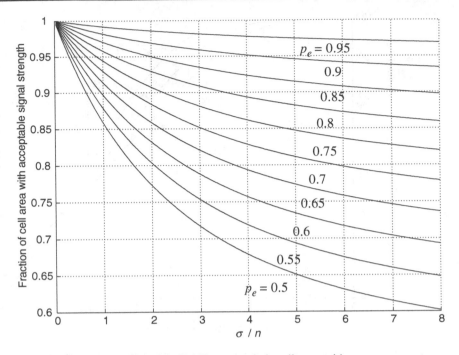

Figure 9.8: Probability of availability over whole cell area, with p_e as a parameter

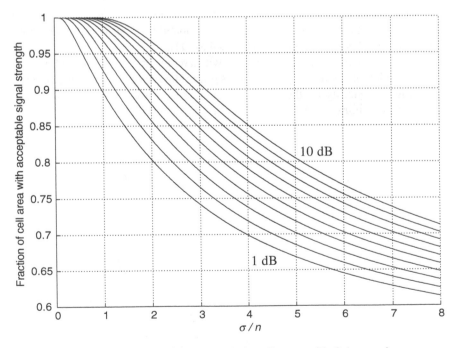

Figure 9.9: Probability of availability over whole cell area, with fade margin as a parameter, varying from 1–10 dB in steps of 1 dB

divided by the area of the whole cell, (πr_{max}^2). As the radius of the rings is reduced, the summation becomes an integral in the limit $\Delta r \to 0$, and we have

$$p_{cell} = \frac{1}{\pi r_{max}^2} \int_{r=0}^{r_{max}} p_e(r) \times 2\pi r \, dr = \frac{2}{r_{max}^2} \int_{r=0}^{r_{max}} r p_e(r) dr \qquad (9.11)$$

After substituting using (9.10) and (9.7), this yields

$$p_{cell} = \frac{1}{2} + \frac{1}{r_{max}^2} \int_{D=0}^{r_{max}} r \, \mathrm{erf} \left(\frac{L_m - L(r)}{\sigma_L \sqrt{2}} \right) dr \qquad (9.12)$$

where $\mathrm{erf}(x) = 1 - \mathrm{erfc}(x)$.

This may be solved numerically for any desired path loss model $L(r)$. In the special case of a power law path loss model, the result may be obtained analytically. If the path loss model is expressed as (8.2)

$$L(r) = L(r_{ref}) + 10n \log \frac{r}{r_{ref}} \qquad (9.13)$$

where n is the path loss exponent, then the eventual result is

$$p_{cell} - p_e(r_{max}) + \frac{1}{2}\exp(A) \times (1 - \mathrm{erf}\,B) \qquad (9.14)$$

where

$$A = \left(\frac{\sigma_L \sqrt{2}}{10n \log e} \right)^2 + \frac{2M}{10n \log e} \qquad B = \frac{\sigma_L \sqrt{2}}{10n \log e} + \frac{M}{\sigma_L \sqrt{2}} \qquad (9.15)$$

Note the direct dependence of p_{cell} on $p_e(r_{max})$, the cell edge availability. Results from (9.14) are illustrated in Figures 9.8 and 9.9.

9.5 LOCATION VARIABILITY

Figure 9.10 shows the variation of the location variability σ_L with frequency, as measured by several studies. It is clear there is a tendency for σ_L to increase with frequency and that it depends upon the environment. Suburban cases tend to provide the largest variability, due to the large variation in the characteristics of local clutter. Urban situations have rather lower variability, although the overall path loss would be higher. No consistent variation with range has been reported; the variations in the [Ibrahim, 83] measurements at 2–9 km are due to differences in the local environment. Note also that it may be difficult to compare values from the literature as shadowing should properly exclude the effects of multipath fading, which requires careful data averaging over an appropriate distance. See Chapters 10 and 19 for discussion of this point.

Figure 9.10 also includes plots of an empirical relationship fitted to the [Okumura, 68] curves and chosen to vary smoothly up to 20 GHz. This is given by

$$\sigma_L = 0.65(\log f_c)^2 - 1.3 \log f_c + A \qquad (9.16)$$

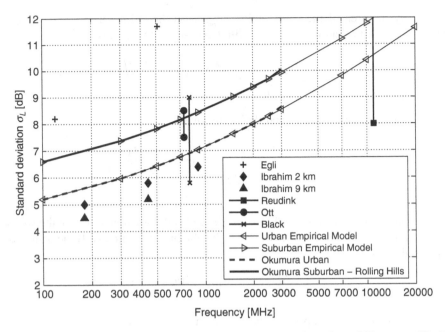

Figure 9.10: Location variability versus frequency. Measured values from [Okumura, 68], [Egli, 57], [Reudink, 72], [Ott, 74], [Black, 72] and [Ibrahim, 83]. [After Jakes, 94]

where $A = 5.2$ in the urban case and 6.6 in the suburban case. Note that these values apply only to macrocells; the levels of shadowing in other cell types will be described in Chapters 12–14.

9.6 CORRELATED SHADOWING

So far in this chapter, the shadowing on each propagation path from base station to mobile has been considered independently. In this section we consider the way in which the shadowing experienced on nearby paths is related. Consider the situation illustrated in Figure 9.11. Two mobiles are separated by a small distance r_m and each can receive signals from two base

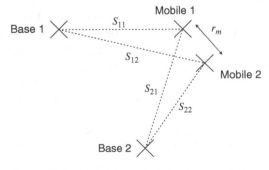

Figure 9.11: Definitions of shadowing correlations

stations. Alternatively, the two mobile locations may represent two positions of a single mobile, separated by some time interval. Each of the paths between the base and mobile locations is marked with the value of the shadowing associated with that path. Each of the four shadowing paths can be assumed log-normal, so the shadowing values S_{11}, S_{12}, S_{21} and S_{22} are zero-mean Gaussian random variables when expressed in decibels. However, they are not independent of each other, as the four paths may include many of the same obstructions in the path profiles. There are two types of correlations to distinguish:

- Correlations between two mobile locations, receiving signals from a single base station, such as between S_{11} and S_{12} or between S_{21} and S_{22}. These are *serial correlations*, or simply the *autocorrelation* of the shadowing experienced by a single mobile as it moves.
- Correlations between two base station locations as received at a single mobile location, such as between S_{11} and S_{21} or between S_{12} and S_{22}. These are *site-to-site correlations* or simply *cross-correlations*.

These two types are now examined individually in terms of their effects on system performance and their statistical characterisation.

9.6.1 Serial Correlation

The serial correlation is defined by Eq. (9.17):

$$\rho_s(r_m) = \frac{E[S_{11}S_{12}]}{\sigma_1 \sigma_2} \tag{9.17}$$

where σ_1 and σ_2 are the location variabilities corresponding to the two paths. It is reasonable to assume here that the two location variabilities are equal as the two mobile locations will typically be sufficiently close together that they encounter the same general category of environment, although the particular details of the environment close to the mobile may be significantly different. Equation (9.18) may therefore be applied:

$$\rho_s(r_m) = \frac{E[S_{11}S_{12}]}{\sigma_L^2} \tag{9.18}$$

The serial correlation affects the rate at which the total path loss experienced by a mobile varies in time as it moves around. This has a particularly significant effect on power control processes, where the base station typically instructs the mobile to adjust its transmit power so as to keep the power received by the base station within prescribed limits. This process has to be particularly accurate in CDMA systems, where all mobiles must be received by the base station at essentially the same power in order to maximise system capacity. If the shadowing autocorrelation reduces very rapidly in time, the estimate of the received power which the base station makes will be very inaccurate by the time the mobile acts on the command, so the result will be unacceptable. If, on the contrary, too many power control commands are issued, the signalling overhead imposed on the system will be excessive.

Measurements of the shadowing autocorrelation process suggest that a simple, first-order, exponential model of the process is appropriate [Marsan, 90]; [Gudmundson, 91], characterised by the *shadowing correlation distance* r_c, the distance taken for the normalised autocorrelation to fall to 0.37 (e^{-1}), as shown in Figure 9.12. This distance is typically a few

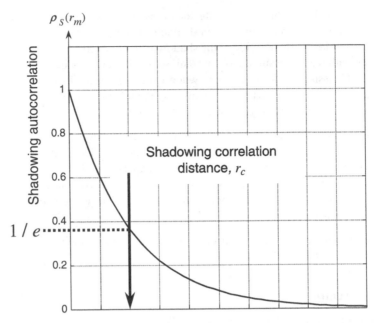

Figure 9.12: Shadowing autocorrelation function

tens or hundreds of metres, with some evidence that the decorrelation distance is greatest at long distances (e.g. $r_c = 44$ m at 1.6 km range, $r_c = 112$ m at 4.8 km range [Marsan, 90]). This corresponds to the widths of the buildings and other obstructions which are found closest to the mobile. The path profile changes most rapidly close to such obstructions as the mobile moves around the base station.

Such a model allows a simple structure to be used when simulating the shadowing process; Figure 9.13 shows an appropriate method. Independent Gaussian samples with zero mean and unity standard deviation are generated at a rate T, the simulation sampling interval. Individual samples are then delayed by T, multiplied by the coefficient a and then summed with the new samples. Finally, the filtered samples are multiplied by $\sigma_L\sqrt{1-a^2}$, so that they have a standard deviation of σ_L as desired.

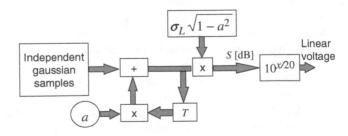

Figure 9.13: Method for generating correlated shadowing process

The result is a process with the correlation function shown in Figure 9.12. Any desired correlation distance can obtained by setting a in accordance with

$$a = e^{-vT/r_c} \tag{9.19}$$

where v is the mobile speed in [m s^{-1}]. A typical output waveform is shown in Figure 9.14.

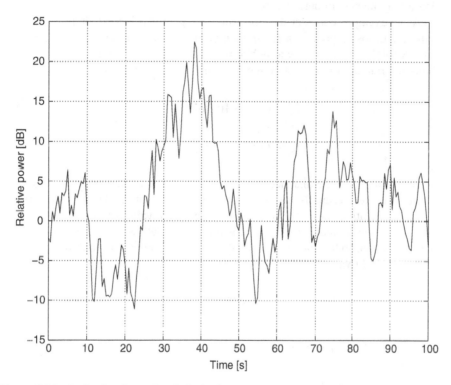

Figure 9.14: A simulated correlated shadowing process, generated using the approach shown in Figure 9.13. Parameters are vehicle speed $= 50 \, \mathrm{km \, h}^{-1}$, correlation distance $= 100 \, \mathrm{m}$, location variability $= 8 \, \mathrm{dB}$

9.6.2 Site-to-Site Correlation

The site-to-site cross-correlation is defined as follows:

$$\rho_c = \frac{E[S_{11}S_{21}]}{\sigma_1 \sigma_2} \tag{9.20}$$

In this case the two paths may be very widely separated and different in length. Although they may also involve rather different environments, the location variability associated with the paths may also be different.

The two base stations involved in the process may be on the same channel, in which case the mobile will experience some level of interference from the base station to which it is not currently connected. The system is usually designed to avoid this by providing sufficient separation between the base stations so that the interfering base station is considerably further away than the desired one, resulting in a relatively large signal-to-interference ratio (C/I). If the shadowing processes on the two links are closely correlated, the C/I will be maintained and the system quality and capacity is high. If, by contrast, low correlation is produced, the interferer may frequently increase in signal level while the desired signal falls, significantly degrading the system performance.

As an example, consider the case where the path loss is modelled by a power law model, with a path loss exponent n. It can then be shown that the downlink carrier-to-interference ratio R [dB] experienced by a mobile receiving only two significant base stations is itself a Gaussian random variable, with mean μ_R and variance σ_R^2, given by

$$\mu_R = E[R] = 10n \log\left(\frac{r_2}{r_1}\right) \tag{9.21}$$

$$\sigma_R^2 = E[R^2] - (E[R])^2 = \sigma_1^2 + \sigma_2^2 - 2\rho_c\sigma_1\sigma_2 \tag{9.22}$$

where r_1 and r_2 are the distances between the mobile and base stations 1 and 2, respectively. Clearly the mean is unaffected by the shadowing correlation, whereas the variance decreases as the correlation increases, reaching a minimum value of 0 when $\rho_c = 1$. Just as the probabilities associated with shadowing on a single path were calculated in Section 9.4 using the Q function, so this can be applied to this case, where the probability of R being less than some threshold value R_T is

$$\Pr[R < R_T] = 1 - Q\left(\frac{R_T - \mu_R}{\sigma_R}\right) \tag{9.23}$$

In the case where $\sigma_1 = \sigma_2 = \sigma_L$ (i.e. the location variability is equal for all paths), Eq. (9.23) becomes

$$\Pr[R < R_T] = 1 - Q\left(\frac{R_T - \mu_R}{\sigma_L\sqrt{2(1 - \rho_c)}}\right) \tag{9.24}$$

This is plotted in Figure 9.15 for $\sigma_L = 8$ dB with various values of ρ_c. Figure 9.15 effectively represents the outage probability for a cellular system in which the interference is dominated by a single interferer. The difference between $\rho_c = 0.8$ and $\rho_c = 0$ is around 7 dB for an outage probability of 10%. With a path loss exponent $n = 4$, the reuse distance r_2 would then have to be increased by 50% to obtain the same outage probability. This represents a very significant decrease in the system capacity compared to the case when the correlation is properly considered. Further discussion of the effects of the correlation on cellular system reuse is given in [Safak, 91], including the effects of multiple interferers, where the distribution of the total power is no longer log-normal, but can be estimated using methods described in [Safak, 93]. It is therefore clear that the shadowing cross-correlation has a decisive effect upon the system capacity and that the use of realistic values is essential to allow accurate system simulations and hence economical and reliable cellular system design.

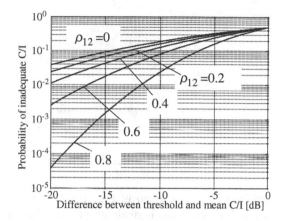

Figure 9.15: Effect of shadowing correlation on interference outage statistics

Here are some other system design issues which may be affected by the shadowing cross-correlation:

- Optimum choice of antenna beamwidths for sectorisation.
- Performance of soft handover and site diversity, including simulcast and quasi-synchro-nous operation, where multiple base sites may be involved in communication with a single mobile. Such schemes give maximum gain when the correlation is low, in contrast to the conventional interference situation described earlier.
- Design and performance of handover algorithms. In these algorithms, a decision to hand over to a new base station is usually made on the basis of the relative power levels of the current and the candidate base stations. In order to avoid 'chatter', where a large number of handovers occur within a short time, appropriate averaging of the power levels must be used. Proper optimisation of this averaging window and of the handover process in general requires knowledge of the dynamics of both serial and site-to-site correlations, particularly for fast-moving mobiles.
- Optimum frequency planning for minimised interference and hence maximised capacity.
- Adaptive antenna performance calculation (Chapter 18).

Unfortunately, there is currently no well-agreed model for predicting the correlation. Here an approximate model is proposed which has some physical basis, but which requires further testing against measurements. It includes two key variables:

- The angle between the two paths between the base stations and the mobile. If this angle is small, the two path profiles share many common elements and are expected to have high correlation. Hence the correlation should decrease with increasing angle-of-arrival difference.
- The relative values of the two path lengths. If the angle-of-arrival difference is zero, the correlation is expected to be one when the path lengths are equal. As one of the path lengths is increased, it incorporates elements which are not common to the shorter path, so the correlation decreases.

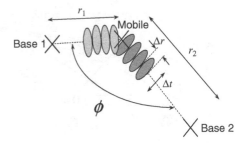

Figure 9.16: Physical model for shadowing cross-correlation

An illustration of these points is given in Figure 9.16. The angle-of-arrival difference is denoted by ϕ and each of the paths is made up of a number of individual elements which contribute to the shadowing process with sizes Δr along the path and Δt transverse to it. If all the elements are assumed independent and equal in their contribution to the overall scattering process, then the following simple model for the cross-correlation may be deduced:

$$\rho_c = \begin{cases} \sqrt{\frac{r_1}{r_2}} & \text{for } 0 \leq \phi < \phi_\text{T} \\ \frac{\phi_\text{T}}{\phi}\sqrt{\frac{r_1}{r_2}} & \text{for } \phi_\text{T} \leq \phi \leq \pi \end{cases} \qquad (9.25)$$

where r_1 is taken to be the smaller of the two path lengths.

The threshold angle ϕ_T between the two regimes in Eq. (9.25) is related to the transverse element size Δt as it occurs when the elements along the two paths no longer overlap fully. This in turn must depend upon the serial correlation distance r_c described in Section 9.6.1. Simple geometry then suggests that

$$\phi_T = 2\sin^{-1}\frac{r_c}{2r_1} \qquad (9.26)$$

Although the correlation distance is typically very much less than the path length, ϕ_T will be very small so the correlation will usually be dominated by the second part of (9.25).

In practice the assumption that each of the shadowing elements is entirely uncorrelated is likely to be invalid, as both terrain and buildings have a definite structure associated with them, which will lead to some significant correlation even when the angle-of-arrival difference is close to $180°$. This structural correlation is examined further in the context of mobile satellite systems in Chapter 14. In order to account for this in a simple manner, the variation with ϕ is permitted to alter at a different rate, parameterised by an exponent γ. This value will vary in practice according to the size and heights of terrain and clutter, and according to the heights of the base station antennas relative to them. The final model is then

$$\rho_c = \begin{cases} \sqrt{\frac{D_1}{D_2}} & \text{for } 0 \leq \phi < \phi_\text{T} \\ \left(\frac{\phi_\text{T}}{\phi}\right)^\gamma \sqrt{\frac{D_1}{D_2}} & \text{for } \phi_\text{T} \leq \phi \leq \pi \end{cases} \qquad (9.27)$$

where ϕ_T is given by (9.26).

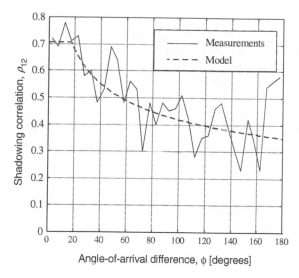

Figure 9.17: Comparison of measurements [Graziano, 78] and model (9.27) for shadowing cross-correlation. The model parameters were $\gamma = 0.3$, $D_1 = 1$ km, $D_2 = 2$ km and $r_c = 300$ m

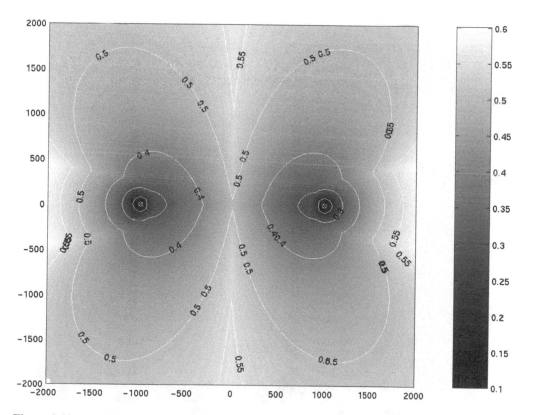

Figure 9.18: Example calculation of shadowing cross-correlation using Eq. (9.27) with parameters as in Figure 9.17. The two base stations are located at ($\pm1000,0$ meters)

Few measurements of ρ_c exist, but Figure 9.17 compares (9.27) with measurements from [Graziano, 78]. These measurements represent an average over many different path lengths, so some representative values were used in the model calculations. The model is also used in Figure 9.18 to calculate the shadowing cross-correlation around a pair of base stations. Figure 9.19 shows the resulting mean carrier-to-interference ratio, if the base stations are transmitting on the same channel and the mobile is connected to the nearest base station at all times, assuming a path loss exponent of 4. This effectively represents the C/I which would be experienced in the absence of shadowing. Finally, Figure 9.20 shows the outage probability for a threshold C/I of 9 dB, calculated using Eq. (9.23). Notice how small the areas are for 10% outage, compared with the 50% values, where the 50% values would result from predictions which neglected the shadowing correlation. This figure emphasises the importance and usefulness of sectorisation in effectively avoiding the areas where the outage probability is high, and it shows that correlation must be properly accounted for in order to predict these effects.

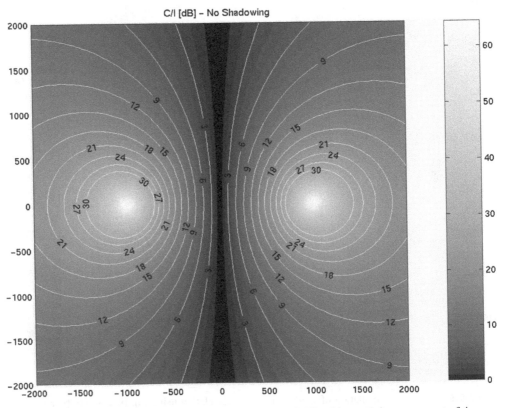

Figure 9.19: Mean C/I ratio [dB], calculated using (9.21) with a path loss exponent of 4

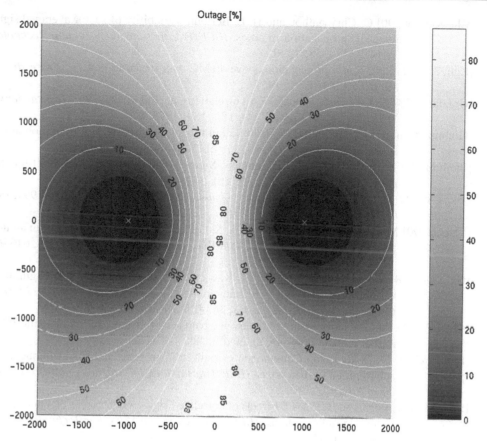

Figure 9.20: Outage probability for a threshold C/I of 9 dB, calculated using (9.23)

9.7 CONCLUSION

The inclusion of shadowing into propagation models transforms the coverage radius of a cell from a fixed, predictable value into a statistical quantity. The properties of this quantity affect the coverage and capacity of a system in ways which can be predicted using the techniques introduced in this chapter. In particular, the shadowing affects the dynamics of signal variation at the mobile, the percentage of locations which receive sufficient power and the percentage which receive sufficient signal-to-interference ratio.

REFERENCES

[Black, 72] D. M. Black and D. O. Reudink, Some characteristics of mobile radio propagation at 836 MHz in the Philadelphia area, *IEEE Transactions on Vehicular Technology*, 21 (2), 45–51, 1972.

[Chrysanthou, 90] C. Chrysanthou and H. L. Bertoni, Variability of sector averaged signals for UHF propagation in cities, *IEEE Transactions on Vehicular Technology*, 39 (4), 352–358, 1990.

[Egli, 57] J. J. Egli, Radio propagation above 40 MC over irregular terrain, *Proceedings of the IRE*, 1383–1391, 1957.

[Graziano, 78] V. Graziano, Propagation correlations at 900 MHz, *IEEE Transactions on Vehicular Technology*, 27 (4), 182–189, 1978.

[Gudmundson, 91] M. Gudmundson, Correlation model for shadow fading in mobile radio systems, *Electronics Letters*, 27, 2145–2146, 1991.

[Ibrahim, 83] M. F. Ibrahim and J. D. Parsons, Signal strength prediction in built-up areas, *Proceedings of the. IEE*, 130F (5), 377–384, 1983.

[Jakes, 94] W. C. Jakes, *Microwave mobile communications*, IEEE, New York, 1994, ISBN 0-78031-069-1

[Marsan, 90] M. J. Marsan, G. C. Hess and S. S. Gilbert, Shadowing variability in an urban land mobile environment at 900 MHz, *Electronics Letters*, 26 (10), 646–648, 1990.

[Okumura, 68] Y. Okumura, E. Ohmori, T. Kawano and K. Fukuda, Field strength and its variability in VHF and UHF land mobile radio service, *Review of the Electrical Communications Laboratories*, 16, 825–873, 1968.

[Ott, 74] G. D. Ott, Data processing summary and path loss statistics for Philadelphia HCNTS measurements program, Quoted in Jakes (1994).

[Reudink, 72] D. O. Reudink, Comparison of radio transmission at X-band frequencies in suburban and urban areas, *IEEE Transactions on Antennas and Propagation*, 20, 470, 1972.

[Safak, 93] A. Safak and R. Prasad, Effects of correlated shadowing signals on channel reuse in mobile radio systems, *IEEE Transactions on Vehicular Technology*, 40 (4), 708–713, 1991.

[Safak, 93] A. Safak, On the analysis of the power sum of multiple correlated log-normal components, *IEEE Transactions on Vehicular Technology*, 42 (1), 58–61, 1993.

[Saunders, 91] S. R. Saunders and F. R. Bonar, Mobile radio propagation in built-up areas: A numerical model of slow fading, In *Proceedings of the 41st IEEE Vehicular Technology Society Conference*, 295–300, 1991.

PROBLEMS

9.1 A mobile system is to provide 95% successful communication at the fringe of coverage, with location variability 8 dB. What fade margin is required? What is the average availability over the whole cell, assuming a path loss exponent of 4?

9.2 Repeat problem 9.1 for 90% successful communication at the fringe of the coverage, for the same location variability. How is the fade margin affected? Explain the impact of this on the design and implementation of the network.

9.3 Section 9.3 states that loss contributions closest to the mobile have the most significant impact on shadowing in macrocells. Why?

9.4 Section 9.6.1 cites evidence that the decorrelation distance is largest at large base-to-mobile distances. Suggest some plausible physical mechanism for this effect.

9.5 Assuming a shadowing decorrelation distance of 50 m, how long does shadowing take to decorrelate for a mobile travelling at 50 km h^{-1}.

9.6 Prove Eqs. (9.21) and (9.22).

9.7 Two co-channel base stations produce shadowing with a cross-correlation of 0.5 in a certain region. If the location variability is 8 dB, what is the standard deviation of the carrier-to-interference ratio experienced by the mobile?

9.8 Discuss the effects in radiowave propagation and link budget calculations if shadowing is neglected.

10 Narrowband Fast Fading

'Enter through the narrow gate; for the gate is wide and the road is easy that leads to destruction, and there are many who take it'.
Bible: New Testament, Matthew 7:13

10.1 INTRODUCTION

After path loss and shadowing have been carefully predicted for particular locations in a mobile system using the methods of Chapters 8 and 9, there is still significant variation in the received signal as the mobile moves over distances which are small compared with the shadowing correlation distance. This phenomenon is *fast fading* and the signal variation is so rapid that it can only usefully be predicted by statistical means. This chapter explains the causes and effects of the fast fading and shows how a baseband signal representation allows convenient analysis of these phenomena. It is assumed here that these effects do not vary with frequency; this restriction will be removed in Chapter 11.

10.2 BASEBAND CHANNEL REPRESENTATION

Signals transmitted over mobile radio channels invariably contain a set of frequencies within a bandwidth which is narrow compared with the centre frequency of the channel. A 3 kHz voice signal, amplitude modulated onto a carrier at 900 MHz has a fractional bandwidth of 6 kHz/900 MHz — 0.0007%. Such signals are *bandpass signals* and can be expressed in the form

$$s(t) = a(t) \cos[2\pi f_c t + \theta(t)] \qquad (10.1)$$

where $a(t)$ is the envelope of $s(t)$, $\theta(t)$ denotes the phase and f_c is the carrier frequency. Since all the information in the signal is contained within the phase and envelope variations, it is usual to analyse the signal in an alternative form which emphasises the role of these components:

$$u(t) = a(t)e^{j\theta(t)} \qquad (10.2)$$

Equation (10.2) expresses the *complex baseband* representation of $s(t)$, namely $u(t)$. It is entirely equivalent to $s(t)$, in that the real signal can always be recovered from the complex baseband simply by multiplying by the time factor involving the carrier frequency and finding the real part:

$$s(t) = \text{Re}[u(t)e^{j2\pi f_c t}] \qquad (10.3)$$

Antennas and Propagation for Wireless Communication Systems Second Edition   Simon R. Saunders and Alejandro Aragón-Zavala
© 2007 John Wiley & Sons, Ltd

This process can be visualised via Figure 10.1, which shows how the complex baseband can be viewed as a frequency shifted version of the bandpass signal, with the same spectral shape. The baseband signal has its frequency content centred around zero frequency (DC).

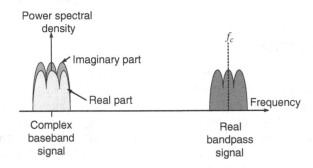

Figure 10.1: Complex baseband representation of signal spectrum

The mean power in such a signal is given by the expected value of its magnitude squared divided by 2,

$$P_s = \frac{E[\ |u(t)|^2]}{2} = \frac{E[u(t)u^*(t)]}{2} \tag{10.4}$$

This gives the same result as finding the mean-square value of the real signal. Linear processes, such as filtering, can easily be represented as operations on complex baseband signals by applying a complex baseband representation of these processes, too.

Complex baseband notation will be used throughout this chapter and extensively in later chapters.

10.3 THE AWGN CHANNEL

The simplest practical case of a mobile radio channel is an additive white Gaussian noise (AWGN) channel. When the signal is transmitted over such channels, the signal arriving at the demodulator is perturbed only by the addition of some noise and by some fixed, multiplicative path loss (including shadowing). This channel applies in a mobile system, where the mobile and surrounding objects are not in motion. It also assumes the signal bandwidth is small enough for the channel to be considered *narrowband*, so there is no variation in the path loss over the signal bandwidth (Chapter 11). The noise is white (it has a constant power spectral density) and Gaussian (it has a normal distribution). Most of this noise is created within the receiver itself, but there may also be contributions from interferers, whose characteristics may often be assumed equivalent to white noise. The resultant system can be modelled using the block diagram in Figure 10.2.

The received signal at time t, $y(t)$, is then given simply by

$$y(t) = Au(t) + n(t) \tag{10.5}$$

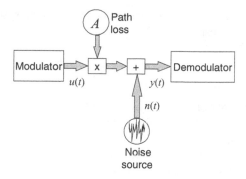

Figure 10.2: The AWGN channel

where $n(t)$ is the noise waveform, $u(t)$ is the modulated signal and A is the overall path loss, assumed not to vary with time.

Assuming complex baseband representation for all signals, the noise is composed of real and imaginary components $x_n(t)$ and $y_n(t)$, respectively,

$$n(t) = x_n(t) + jy_n(t) \tag{10.6}$$

Both $x_n(t)$ and $y_n(t)$ are zero mean, independent, real Gaussian processes, each with a standard deviation of σ_n.

The mean noise power is then given by

$$
\begin{aligned}
P_n &= \frac{E[n(t)n^*(t)]}{2} = \frac{E[(x_n(t) + jy_n(t))(x_n(t) + jy_n(t))^*]}{2} \\
&= \frac{E[x^2(t)] + E[y^2(t)]}{2} = \frac{\sigma_n^2 + \sigma_n^2}{2} = \sigma_n^2
\end{aligned}
\tag{10.7}
$$

Note that expectations of terms involving products of $x_n(t)$ and $y_n(t)$ together are zero because of the assumption of zero-mean, independent processes.

The signal-to-noise ratio (SNR) at the input of the demodulator is then

$$\gamma = \frac{\text{Signal power}}{\text{Noise power}} = \frac{E[A^2 u^2(t)]}{2P_n} = \frac{A^2 E[u^2(t)]}{2P_n} = \frac{A^2}{2P_n} \tag{10.8}$$

where the last step assumes that the variance of the modulator output signal is 1.

Another way of expressing the SNR is more appropriate for signals, such as digital signals, which consist of symbols with a finite duration T. If each symbol has energy E_s, then $E_s = A^2 T / 2$. Similarly, if the noise is contained within a bandwidth $B = 1/T$, and has power spectral density N_0, then $P_n = BN_0 = N_0/T = \sigma_n^2$. Note that the signal may not necessarily be contained within this bandwidth, so Eq. (10.9) may not necessarily hold. Nevertheless, γ and Es/N_0 will always be proportional. We write

$$\gamma = \frac{E_s}{N_0} \tag{10.9}$$

It is usual to express the error rate performance of a digital system in terms of this parameter or in terms of the corresponding SNR per bit,

$$\gamma_b = \frac{\gamma}{m} = \frac{E_b}{N_0} \qquad (10.10)$$

where m is the number of bits per symbol. The SNR is the key parameter in calculating the system performance in the AWGN channel. Here we calculate the bit error rate (BER) performance of binary phase shift keying (BPSK) in the AWGN channel.

It can be shown [Proakis, 89] that the error rate performance of any modulation scheme in AWGN with power spectral density N_0 depends on the *Euclidean distance d* between the transmitted waveforms corresponding to different transmitted bits according to

$$P_e = Q\left(\sqrt{\frac{A^2 d^2}{2N_0}} \right) \qquad (10.11)$$

In the case of BPSK, the two signals corresponding to a binary 1 and 0 can be represented in complex baseband notation by

$$u_1 = \sqrt{\frac{2E_s}{T}} \qquad u_0 = -\sqrt{\frac{2E_s}{T}} \qquad (10.12)$$

where the duration of each symbol is T, the energy in the symbol is E_s and $A = 1$. Thus the signals consist of segments of carrier wave of duration T and a phase difference of 180°. These signals can be represented as points in signal space (which plots the imaginary part of the complex baseband signal versus the real part) as shown in Figure 10.3.

It is clear from the figure that in this case $d = 2\sqrt{E_s}$. Hence

$$P_e = Q\left(\sqrt{\frac{A^2 d^2}{2N_0}} \right) = Q\left(\sqrt{\frac{4E_s}{2N_0}} \right) = Q(\sqrt{2\gamma}) \qquad (10.13)$$

This result is illustrated in Figure 10.4.

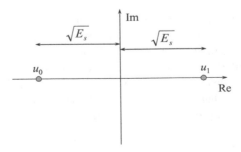

Figure 10.3: Signal space representation of BPSK

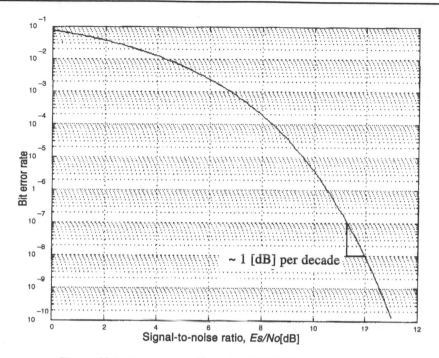

Figure 10.4: Error rate performance of BPSK in AWGN channel

The rapid decrease in error rate as SNR increases (Figure 10.4) is characteristic of an AWGN channel. This decrease is the fastest which could take place for uncoded modulation in mobile radio channels, so the AWGN channel is a 'best case' channel. For high signal-to-noise ratios, the bit error rate decreases by a factor of approximately 10 (i.e. a decade) for every 1 dB increase in SNR.

10.4 THE NARROWBAND FADING CHANNEL

For the most part, mobile radio performance will not be as good as the pure AWGN case. The detailed characteristics of the propagation environment result in fading, which shows itself as a multiplicative, time-variant process applied to the channel (Figure 10.5). The channel is a narrowband one because the fading affects all frequencies in the modulated signal equally, so it can be modelled as a single multiplicative process.

Since the fading varies with time, the SNR at the demodulator input also varies with time. It is thus necessary, in contrast to the AWGN case, to distinguish between the instantaneous SNR, $\gamma(t)$ and the mean SNR, Γ.

The received signal at time t is now given simply by

$$y(t) = A\alpha(t)u(t) + n(t) \tag{10.14}$$

where $\alpha(t)$ is the complex fading coefficient at time t. If the fading is assumed constant over the transmitted symbol duration, then $\gamma(t)$ is also constant over a symbol, and is given by

$$\gamma(t) = \frac{\text{Signal power}}{\text{Noise power}} = \frac{A^2|\alpha(t)|^2 E[\,|u(t)|^2]}{2P_n} = \frac{A^2|\alpha(t)|^2}{2P_n} \tag{10.15}$$

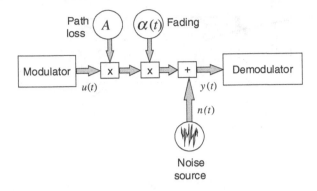

Figure 10.5: The narrowband fading channel

and

$$\Gamma = E[\gamma(t)] \tag{10.16}$$

It is usual to take the fading as having unit variance and lump any change in mean signal power into the path loss, so

$$\Gamma = E[\gamma(t)] = \frac{A^2}{2P_n} \tag{10.17}$$

Two things are therefore needed in order to find the performance of a system in the narrowband fading channel: the mean SNR and a description of the way the fading causes the instantaneous SNR to vary relative to this mean. In order to achieve this, we examine the physical causes of fading.

10.5 WHEN DOES FADING OCCUR IN PRACTICE?

Figure 10.6 shows how fading might arise in practice. A transmitter and receiver are surrounded by objects which reflect and scatter the transmitted energy, causing several waves

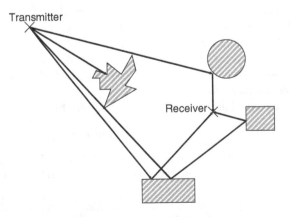

Figure 10.6: Non-line-of-sight multipath propagation

to arrive at the receiver via different routes. This is *multipath propagation*. Since the direct wave from the transmitter to the receiver is blocked, this situation is called non-line-of-sight (NLOS) propagation. Each of the waves has a different phase and this phase can be considered as an independent uniform distribution, with the phase associated with each wave being equally likely to take on any value.

By contrast, Figure 10.7 shows the line-of-sight (LOS) case, where a single strong path is received along with multipath energy from local scatterers.

No attempt is usually made to predict the exact value of the signal strength arising from multipath fading, as this would require a very exact knowledge of the positions and electromagnetic characteristics of all scatterers. Instead a statistical description is used, which will be very different for the LOS and NLOS cases.

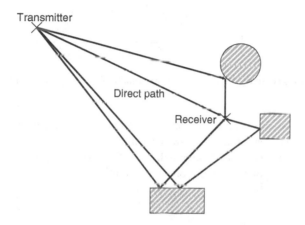

Figure 10.7: Line-of-sight multipath propagation

10.6 THE RAYLEIGH DISTRIBUTION

The central limit theorem shows that, under certain conditions, a sum of enough independent random variables approaches very closely to a normal distribution (Appendix A). In the NLOS case, the real and imaginary parts of the multipath components fulfil these conditions since they are composed of the sum of a large number of waves. Some simple simulations are used here to illustrate how this affects the distribution of the resulting fading amplitude.

First 1000 samples from a normal distribution with zero mean and unit standard deviation are produced, labelled $x(1)$ to $x(1000)$; their probability density function is plotted in Figure 10.8. They can be considered to represent samples of the total signal received by a mobile at 1000 locations taken within an area small enough to experience the same total path loss, from a transmitter radiating a continuous wave (CW) signal.

Note the distinctive bell-shaped curve of the normal distribution. Now another 1000 samples are generated, independent of the first set and labelled $y(1)$ to $y(1000)$. These two sets of samples are the real and imaginary *in-phase* and *quadrature* or I and Q components of a complex baseband signal

$$\alpha = x + jy \tag{10.18}$$

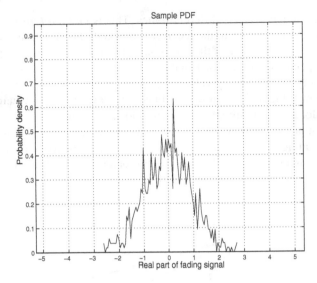

Figure 10.8: Probability density function of real part of NLOS fading signal

These points are plotted on an Argand diagram (imaginary versus real) as shown in Figure 10.9. The variable α is now a *complex Gaussian random variable*, just as was the additive noise in the AWGN channel. Now the probability density function (p.d.f.) of the distance of each point from the origin is examined. This is the distribution of r, where

$$r = |\alpha| = \sqrt{x^2 + y^2} \tag{10.19}$$

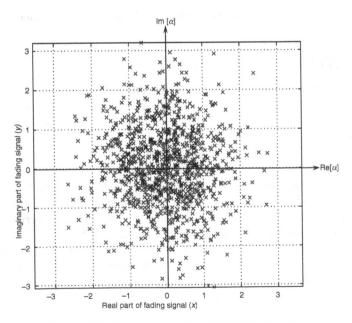

Figure 10.9: Complex samples of NLOS fading signal

The p.d.f. of r is plotted in Figure 10.10 as the fluctuating curve. The resulting distribution is highly asymmetrical. The theoretical distribution shown in Figure 10.10 as a smooth line is the *Rayleigh distribution*, given theoretically by

$$p_R(r) = (r/\sigma^2)e^{-r^2/2\sigma^2} \tag{10.20}$$

Here σ is the standard deviation of either the real or imaginary parts of x, which was 1 in the example. The total power in α is

$$\text{Power} = \frac{E\lfloor|\alpha|^2\rfloor}{2} = \frac{E[x^2] + E[y^2]}{2} = \frac{\sigma^2 + \sigma^2}{2} = \sigma^2 \tag{10.21}$$

The good agreement between the two curves in Figure 10.10 shows that we have verified the following important result:

The magnitude of a complex Gaussian random variable is
a Rayleigh-distributed random variable.

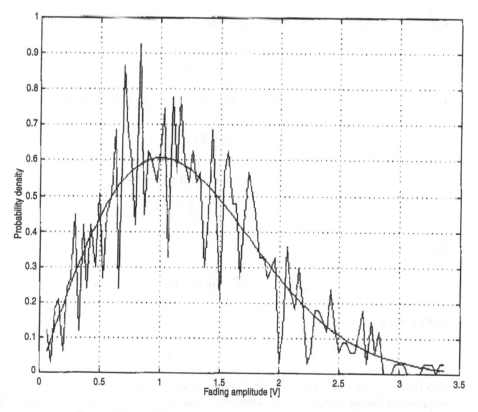

Figure 10.10: Theoretical and experimental Rayleigh distributions

The Rayleigh distribution is an excellent approximation to measured fading amplitude statistics for mobile fading channels in NLOS situations. Such channels are *Rayleigh-fading channels* or simply *Rayleigh channels.*

Some other useful properties of the Rayleigh distribution are given below:

$$
\begin{aligned}
Mean(r) &= \sigma\sqrt{\frac{\pi}{2}} \\
Median(r) &= \sigma\sqrt{\ln(4)} \\
Mode(r) &= \sigma \\
Variance(r) &= \frac{4-\pi}{2}\sigma^2
\end{aligned}
\tag{10.22}
$$

10.7 DISTRIBUTION OF THE SNR FOR A RAYLEIGH CHANNEL

It is useful to know the distribution of the instantaneous signal-to-noise ratio γ which arises when the input signal at a receiver is Rayleigh:

$$
\gamma = \frac{\text{Signal power}}{\text{Noise power}} = \frac{(A^2 r^2)/2}{P_N} = \frac{A^2 r^2}{2P_N}
\tag{10.23}
$$

The average SNR for the channel Γ is the mean of γ,

$$
\Gamma = \frac{A^2 E[r^2]}{2P_N} = \frac{2A^2\sigma^2}{2P_N} = \frac{A^2\sigma^2}{P_N}
\tag{10.24}
$$

To find the distribution of γ given the distribution of r, we use the identity

$$
p_\gamma(\gamma) = p_R(r)\frac{dr}{d\gamma}
\tag{10.25}
$$

Combining this with the Rayleigh distribution from Eq. (10.20),

$$
p_\gamma(\gamma) = (r/\sigma^2)e^{-r^2/2\sigma^2} \times P_N/(A^2 r)
\tag{10.26}
$$

The final result for the p.d.f. is

$$
p_\gamma(\gamma) = \frac{1}{\Gamma}e^{-\gamma/\Gamma} \quad \text{for } \gamma > 0; \quad 0 \text{ otherwise}
\tag{10.27}
$$

and for the c.d.f. we have

$$
\Pr(\gamma < \gamma_s) = 1 - e^{-\gamma_s/\Gamma}
\tag{10.28}
$$

The result is illustrated in Figure 10.11. This result can be used to calculate the mean SNR required to obtain an SNR above some threshold for an acceptable percentage of the time. This is illustrated in Figure 10.12 and in Example 10.1.

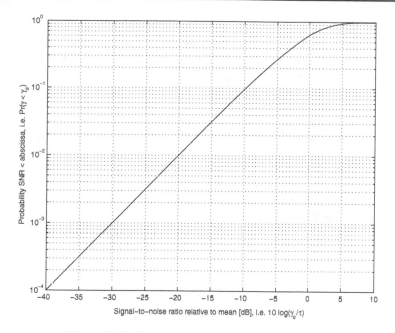

Figure 10.11: SNR distribution for a Rayleigh channel

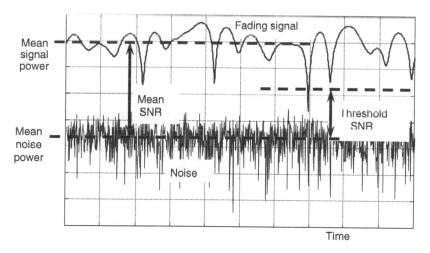

Figure 10.12: Variation of instantaneous SNR relative to mean value

Example 10.1

A mobile receiver is known to produce acceptable bit error rates when the instantaneous SNR is at or above a threshold value of 9 dB. What mean SNR is required in a Rayleigh channel for acceptable error rates to be obtained 99.9% of the time?

Solution

For 99.9% success rate, the probability of the SNR being less than 9 dB is $(100 - 99.9)/100 = 1 \times 10^{-3}$. From Figure 10.11, or using Eq. (10.28), this is achieved for an SNR of 30 dB below the mean. Hence the mean SNR should be $30 + 9 = 39$ dB.

The SNR distribution in Eq. (10.27) can also be used to predict the error rate performance of digital modulation schemes in a Rayleigh channel, provided it can be assumed that the SNR is constant over one symbol duration. In this case, the Rayleigh error rate performance can be predicted directly from the AWGN case. For BPSK the AWGN error rate performance was given by the $Q(\sqrt{2\gamma})$ function in Eq. (10.13). The average error rate performance in the Rayleigh channel can be taken as being the average AWGN performance, appropriately weighted by the Rayleigh statistics of the instantaneous SNR, γ, as illustrated in Figure 10.13.

Hence the average bit error probability P_e is given by

$$\begin{aligned} P_e &= E[P_e(\gamma)] \\ &= \int_0^\infty P_e(\gamma) p_\gamma(\gamma) \, d\gamma \end{aligned} \tag{10.29}$$

Substituting Eqs. (10.13) and (10.27) lead to

$$P_e = \int_0^\infty Q(\sqrt{2\gamma}) \frac{1}{\Gamma} e^{-\gamma/\Gamma} d\gamma = \frac{1}{2}\left[1 - \sqrt{\frac{\Gamma}{1+\Gamma}}\right] \tag{10.30}$$

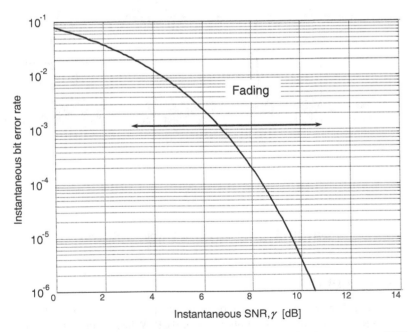

Figure 10.13: BER for BPSK in Rayleigh fading varies with instantaneous SNR

This is plotted in Figure 10.14 along with the AWGN curve. Note that the performance obtained is significantly worse in the Rayleigh case. Also plotted is an approximation which holds for large Γ, $P_e \approx 1/(4\Gamma)$. This inverse proportionality is characteristic of uncoded modulation in a Rayleigh channel, leading to a reduction of BER by one decade for every 10 dB increase in SNR. This contrasts sharply with the $\sim$1 dB per decade variation in the AWGN channel. The performance of other modulation schemes in Rayleigh channels may be analysed in a similar way; see [Proakis, 89].

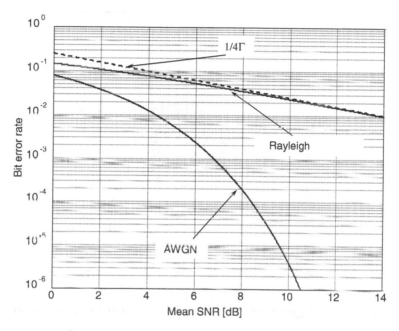

Figure 10.14: BPSK performance in Rayleigh channel

10.8 THE RICE DISTRIBUTION

In the LOS situation, the received signal is composed of a random multipath component, whose amplitude is described by the Rayleigh distribution, plus a coherent line-of-sight component which has essentially constant power (within the bounds set by path loss and shadowing). The power of this component will usually be greater than the total multipath power before it needs to be considered as affecting the Rayleigh distribution significantly.

Simulations are again used in this section to illustrate the fading distribution produced in the LOS case. First a real constant (2 in this case) is added to the complex Gaussian variable α of Section 10.7 to represent the voltage of the line-of-sight signal. The new p.d.f. of $|\alpha|$ is shown as the fluctuating curve in Figure 10.15.

Compare this with Figure 10.10. Obviously the variable still cannot be less than zero, but the distribution is now much closer to being symmetrical. The theoretical distribution which applies in this case, plotted as the smooth curve in Figure 10.15, is the Rice distribution

[Rice, 48] given theoretically by

$$p_R(r) = \frac{r}{\sigma^2} e^{-(r^2+s^2)/2\sigma^2} \, I_0\left(\frac{rs}{\sigma^2}\right) \tag{10.31}$$

where σ^2 is the variance of either of the real or imaginary components of the multipath part alone and s is the magnitude of the LOS component. Note that the Rice distribution is sometimes also referred to as the Nakagami-Rice distribution or the Nakagami-n distribution.

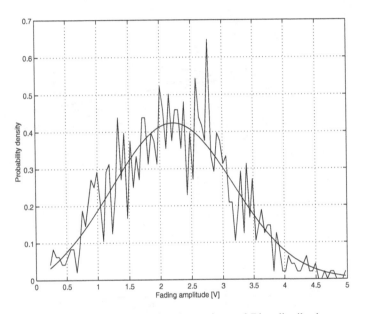

Figure 10.15: Theoretical and experimental Rice distributions

Notice the similarity to the Rayleigh formulation (10.20). Indeed, if s is set to zero, the two distributions are identical. The function I_0 is the modified Bessel function of the first kind and zeroth order (Figure 10.16).

The Rice p.d.f. is often expressed in terms of another parameter, k, usually known as the *Rice factor* and defined as

$$k = \frac{\text{Power in constant part}}{\text{Power in random part}} = \frac{s^2/2}{\sigma^2} = \frac{s^2}{2\sigma^2} \tag{10.32}$$

Then the Rice p.d.f. can be written in either of these forms:

$$p_R(r) = \frac{2kr}{s^2} e^{-kr^2/s^2} e^{-k} I_0\left(\frac{2kr}{s}\right) = \frac{r}{\sigma^2} e^{-r^2/(2\sigma^2)} e^{-k} I_0\left(\frac{r\sqrt{2k}}{\sigma}\right) \tag{10.33}$$

This is plotted in Figure 10.17 for several values of k, keeping the total signal power constant. For very large values of k, the line-of-sight component dominates completely, very little fading is encountered, and the channel reverts to AWGN behaviour.

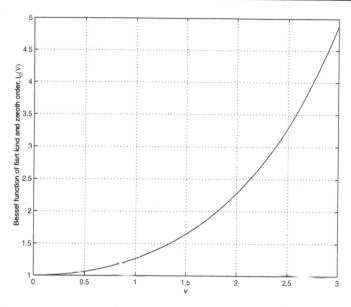

Figure 10.16: Modified Bessel function of the first kind and zeroth order

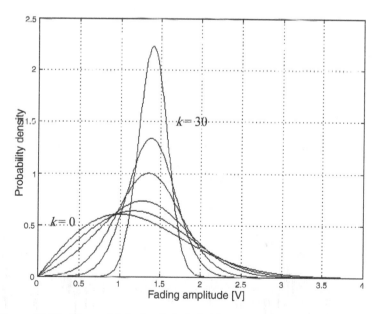

Figure 10.17: Rice p.d.f. for $k = 0$, 1, 2, 5, 10 and 30 with constant total power

The cumulative Rice distribution, found by numerically integrating the results of Eq. (10.33), is shown in Figure 10.18 for several values of k, keeping the total power constant. Typical fading signals for the same k values are illustrated in Figure 10.19. As the k factor increases, the probability of encountering a deep fade reduces, so the mean error rate will

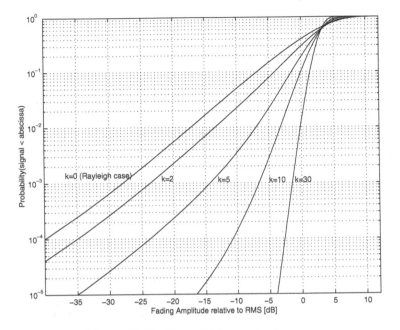

Figure 10.18: Rice c.d.f. for varying k values

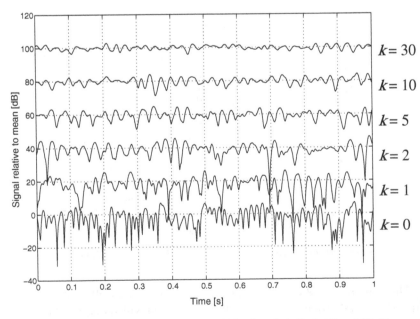

Figure 10.19: Time series of Rice-fading signals for $k = 0$, 1, 2, 5, 10 and 30. Curves are offset upwards by 20 dB for each increasing k value for clarity

decrease. The Rice channel is therefore a more 'friendly' channel than the Rayleigh case, which represents in some senses a 'worst-case' mobile channel. For deep fades, however, the Rice distribution always tends towards a slope of 10 dB per decade of probability, just as in the Rayleigh distribution. This limit is only encountered at very deep fade levels in the case of large values of k.

Note that the Rice distribution applies whenever one path is much stronger than the other multipath. This may occur even in NLOS cases where the power scattered from one object is particularly strong.

The bit error rate for a slow-fading Rice channel is intermediate between the AWGN and Rayleigh cases; the precise value depends on the k value. For example, Figure 10.20 shows the variation of BER with k factor for BPSK, from which it is clear that the Rice channel behaves like SNR in the limit as $k \to \infty$.

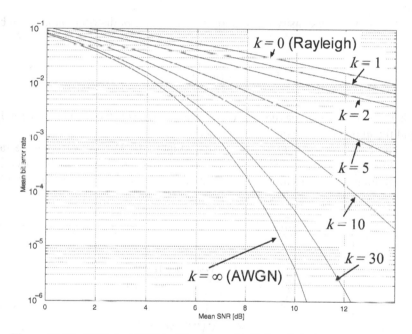

Figure 10.20: BER for BPSK in Rice channel with $k = 0, 1, 2, 5, 10, 30$ and ∞

Example 10.2

If the mobile receiver of Example 10.1 is operated in a Rice channel with $k = 10$, what mean SNR is required?

Solution

At a fading probability of 10^{-3} and $k = 10$, the SNR is approximately 7 dB below the mean, using Figure 10.18. Hence the mean SNR required is $7 + 9 = 16$ dB. Thus the transmitted power can be 23 dB less than in the Rayleigh channel for the same BER performance.

10.9 THE NAKAGAMI- *m* DISTRIBUTION

For most purposes, the Rayleigh and Rice cases are sufficient to characterise the performance of systems in mobile channels. Some channels, however, are neither Rice nor Rayleigh. If, for example, two paths are of comparable power, and stronger than all the others, the signal statistics will not be well approximated by a Rice distribution.

An alternative distribution, for such cases was originally proposed in [Nakagami, 60] and is known as the Nakagami-*m* distribution. This assumes that the received signal is a sum of vectors with random magnitude and random phases, leading to more flexibility and potentially more accuracy in matching experimental data than the use of Rayleigh or Rician distributions [Charash, 79]. The Nakagami-*m* distribution has been used extensively in the literature to model complicated fading channels (e.g. [Zhang, 03]). It is sometimes less attractive than a Rayleigh or Rice distribution, in that it has a less direct physical interpretation.

The probability density function of the Nakagami-*m* distribution is given by

$$p_R(r) = \frac{2}{\Gamma(m)} \left(\frac{m}{\Omega}\right)^m r^{2m-1} e^{-\frac{m}{\Omega}r^2}, \quad r \geq 0 \tag{10.34}$$

where $\Gamma(\cdot)$ is the gamma function (see Appendix B); Ω is the second moment, i.e., $\Omega = E(r^2)$ and the *m* parameter defines the fade depth.

The Nakagami-*m* p.d.f is illustrated in Figure 10.21. It covers a wide range of fading conditions; it is a one-sided Gaussian distribution when $m = 0.5$, and it becomes a Rayleigh distribution if $m = 1$ (with $\Omega = 2\,\sigma^2$). The corresponding cumulative distribution is shown in

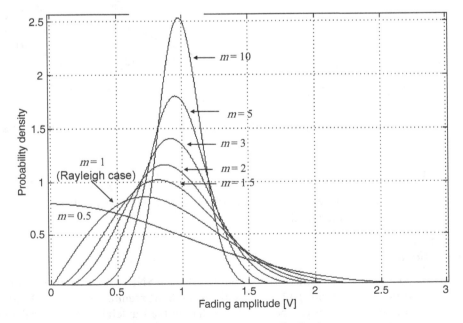

Figure 10.21: Nakagami-*m* distribution

Figure 10.22. In contrast to the Rice distribution, the probability slope for deep fades can exceed 10 dB per decade.

In order to use the Nakagami-*m* distribution to model a given set of empirical data, the figure *m* from the data must be estimated, and some reported methods in the literature have given satisfactory results [Cheng, 01].

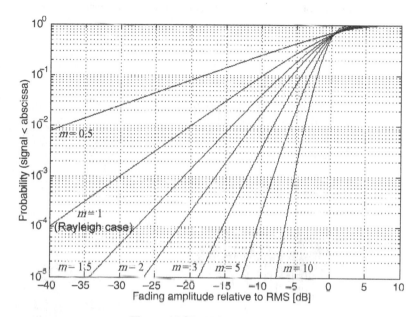

Figure 10.22: Nakagami-*m* c.d.f

10.10 OTHER FADING DISTRIBUTIONS

In addition to the Rice, Rayleigh and Nakagami-*m* distributions introduced here, many other distributions can be used to model fading for particular applications. These include the Gaussian, gamma, exponential and Pearson χ^2 distributions. It is also possible to create mixed distributions, such as one which combines lognormal and Rayleigh statistics so that the overall statistics of shadowing and fast fading can be considered in a single distribution. See [ITU, 1057] for further details.

10.11 SECOND-ORDER FAST-FADING STATISTICS

The fading statistics considered so far in this chapter give the probability that the signal is above or below a certain level. This says nothing, however, about how rapidly the signal level changes between different levels. This is important, since the rates of change of the signal will determine, for example, the ability of a digital system to correct for transmission errors, or the subjective impact of fast fading on voice quality in an analogue system. These dynamic effects are specified by the *second-order fading statistics*.

The effect of the second-order statistics is illustrated in Figure 10.23. These fading records both consist of the same individual data points. However, the lower pair show uncorrelated Rayleigh fading, where each sample is independent of the value of the previous one, whereas the upper pair is filtered, producing fading with second-order statistics similar to those typically observed in the real mobile radio channel. The filtered data clearly has a degree of smoothness and structure which is not present in the other case, and this structure must be understood in terms of its cause and impact on the system performance.

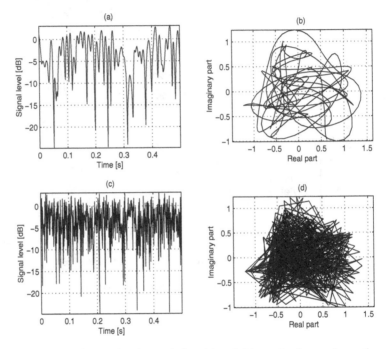

Figure 10.23: Effect of second-order statistics: (a) and (b) are the time series and complex plane plots of a Rayleigh signal with memory; (c) and (d) are the equivalent plots for a Rayleigh signal with no memory between samples

Second-order statistics are concerned with the distribution of the signal's *rate of change*, rather than with the signal itself. This distribution is most commonly specified by the spectrum of the signal. The uncorrelated fading displayed in Figure 10.23 has a flat spectrum, whereas the correlated fading has a very specific spectral shape. In order to understand this shape and how it comes about, we examine its physical causes, starting with the Doppler effect.

10.11.1 The Doppler Effect

Figure 10.24 shows a mobile moving in a straight line with a speed v. Its direction of motion makes an angle α to the arrival direction of an incoming horizontal wave at frequency f, where the source of the wave is assumed stationary. The Doppler effect results in a change of the apparent frequency of the arriving wave, as observed by the mobile, by a factor proportional to the component of the mobile speed in the direction of the wave. If the mobile is moving towards the source of the wave, the apparent frequency is increased; the apparent frequency decreases for motion away from the source. This occurs because the mobile crosses the wavefronts of the incoming wave at a different rate from when it is stationary.

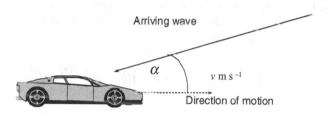

Figure 10.24: The Doppler effect

More precisely, the apparent change in frequency, the *Doppler shift* f_d, is given by the rate at which the mobile crosses wavefronts of the arriving signal,

$$f_d = \frac{v}{\lambda} \cos \alpha = f_c \frac{v}{c} \cos \alpha = f_m \cos \alpha \qquad (10.35)$$

where f_m is the maximum Doppler shift, given by

$$f_m = f_c \frac{v}{c} \qquad (10.36)$$

which occurs for $\alpha = 0$. Thus the Doppler shift associated with an incoming wave can have any apparent frequency in the range $(f_c - f_m) \leq f \leq (f_c + f_m)$.

When multipath propagation occurs, waves arrive with several directions, each of which then has its own associated Doppler frequency. The bandwidth of the received signal is therefore spread relative to its transmitted bandwidth (Figure 10.25). This is the phenomenon of *Doppler spread*. The exact shape of the resulting spectrum depends on the relative amplitudes and directions of each of the incoming waves, but the overall spectral width is called the *Doppler bandwidth*.

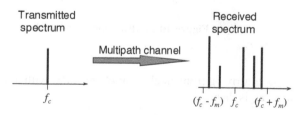

Figure 10.25: Effect of Doppler spread on signal spectrum

Example 10.3

A mobile system is operated at 900 MHz. What is the greatest Doppler shift which could be experienced by a mobile travelling at 100 km h^{-1}? What is the greatest possible Doppler bandwidth for the system?

Solution

The maximum positive Doppler shift is experienced when the mobile moves directly towards the source of the incoming waves, so that $\cos \alpha = 1$.
The Doppler shift is then

$$f_d = f_c \frac{v}{c} \cos \alpha = 900 \times 10^6 \times \frac{100 \times 10^3}{60 \times 60 \times 3 \times 10^8} = 83.3 \, \text{Hz}$$

The maximum negative Doppler shift is the negative of this value, experienced when the mobile is moving away from the incoming wave. The overall maximum Doppler spread is therefore $2 \times 83.3 = 166.6$ Hz.

Note how this example assumes the source of the wave is stationary. If the source is actually a moving scatterer, such as a vehicle, then the maximum Doppler frequency may be even greater (Problem 10.3).

10.11.2 The Classical Doppler Spectrum

The shape and extent of the Doppler spectrum has a significant effect on the second-order fading statistics of the mobile signal. In order to analyse these effects, it is necessary to assume some reasonable model of the statistics of the arrival angle of the multipaths. In the most commonly used model for mobile Doppler spread, the arriving waves at the mobile are assumed to be equally likely to come from any horizontal angle [Clarke, 68]; [Gans, 72]. The p.d.f. $p(\alpha)$ for the arrival angle, α, is as shown in Figure 10.26.

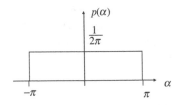

Figure 10.26: P.d.f. for arrival angle α

The mean power arriving from an element of angle $d\alpha$, with a mobile antenna gain in the direction α of $G(\alpha)$, is then

$$P(\alpha) = G(\alpha)p(\alpha) \, d\alpha \tag{10.37}$$

The power spectrum of the received signal, $S(f)$, is then found by equating the power in an element of α to the power in an element of spectrum, thus

$$P(f) = S(f)df \tag{10.38}$$

Note that the same Doppler shift occurs for two values of α, one the negation of the other. The positive and negative parts of the p.d.f. are, therefore, handled separately

$$S(f)df = G(\alpha)p(\alpha)d\alpha + G(-\alpha)p(-\alpha)\,d\alpha \tag{10.39}$$

Hence

$$|S(f)| = \frac{G(\alpha)p(\alpha) + G(-\alpha)p(-\alpha)}{|df/d\alpha|} \tag{10.40}$$

From Eq. (10.35) we obtain

$$f = \frac{f_c v}{c}\cos\alpha = f_m \cos\alpha$$
$$\left|\frac{df}{d\alpha}\right| = f_m|-\sin\alpha| \tag{10.41}$$

Hence, taking $G(\alpha) = 1.5$, which corresponds to a short (Hertzian) dipole,

$$|S(f)| = \frac{G(\alpha)p(\alpha) + G(-\alpha)p(-\alpha)}{|df/d\alpha|} = \frac{1.5/2\pi + 1.5/2\pi}{f_m|-\sin\alpha|} \tag{10.42}$$

Now substitute for $\sin\alpha = \sqrt{1 - \cos^2\alpha} = \sqrt{1 - (f/f_m)^2}$ to give

$$S(f) = \frac{1.5}{\pi f_m\sqrt{1 - (f/f_m)^2}} \quad \text{for} |f| < f_m \tag{10.43}$$

and $S(f) = 0$ for $|f| \geq f_m$.

This result, illustrated in Figure 10.27, is called the *classical Doppler spectrum*; it is used as the basis of many simulators of mobile radio channels. The assumption of uniform angle-of-arrival probability may be questionable over short distances where propagation is dominated by the effect of particular local scatterers, but nevertheless it represents a good reference model for the long-term average Doppler spectrum. The large peaks at $\pm f_m$ lead to a typical separation between signal nulls of around one half-wavelength, although there is no strict regularity to the signal variations. This wide spread of arrival angles at the mobile also alters the effective gain so an angle-averaged value – the *mean effective gain* – must be used in link budget calculations (Section 15.2.7). Other models, e.g. [Aulin, 79], have included the angle-of-arrival spread in the vertical direction. This tends to shift some of the energy away from the peaks in Figure 10.27 and towards lower Doppler frequencies, thereby reducing the average fading rate.

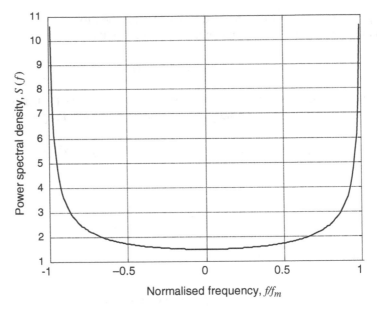

Figure 10.27: The classical Doppler spectrum

Example 10.4

The orientation of a certain street with respect to the base station and the buildings on each side of the street results in the angular spread of waves being restricted to arrival angles $\pm\beta/2$ with respect to the perpendicular to the direction of motion. Within this range, all arrival angles are equally likely. Calculate the Doppler spectrum for the case $\beta = \pi/4$. Will the rate of fading be more or less than in the standard classical model?

Solution

The situation is illustrated in Figure 10.28. Since the waves parallel to the direction of motion are now excluded, we expect the spectrum to contain less power at the largest Doppler shifts. The p.d.f. for α is illustrated in Figure 10.29. Since the total area under any p.d.f. must be unity, the maximum value of $p(\alpha)$ is $1/(2\beta)$.

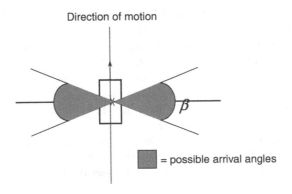

Figure 10.28: Effect of restricted arrival angles on Doppler spread

Since $p(\alpha)$ only exists for angles in the range $\pi/2 - \beta/2 \le |\alpha| \le \pi/2 + \beta/2$, $S(f)$ correspondingly has components only in the range $f_m \cos(\pi/2 + \beta/2) \le f \le f_m \cos(\pi/2 - \beta/2)$. Within this range, the analysis for $S(f)$ proceeds exactly as for the case of the classical spectrum, with $p(\alpha)$ replaced by $1/(2\beta)$. The final result is therefore $S(f) = 1.5/(\beta f_m \sqrt{1 - (f/f_m)^2})$ for β in the above range, and 0 elsewhere.

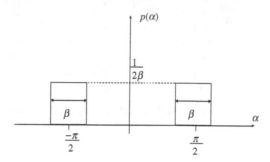

Figure 10.29: P.d.f. of azimuth angle for the case shown in Figure 10.28

The result is illustrated in Figure 10.30 for $\beta - \pi/2$. Since the spectrum is now restricted to $-f_m/\sqrt{2} \le f \le f_m/\sqrt{2}$, it contains less high-frequency power than in the classical spectrum. The fade rate is therefore constrained to be less than in the classical case.

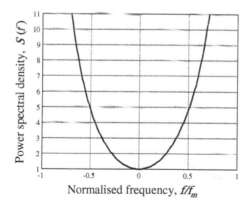

Figure 10.30: Doppler spectrum for restricted arrival angles

It is difficult to measure or directly observe the Doppler spectra calculated above because their fractional bandwidth is so small. Other parameters which may be measured more directly, such as the *level-crossing rate* (l.c.r.) and the *average fade duration* (a.f.d.), come as direct consequences of a given Doppler spectrum. They are illustrated in Figure 10.31: the l.c.r. is the number of positive-going crossings of a reference level r in unit time, while the a.f.d. is the average time between negative and positive level-crossings.

For the classical spectrum, the level-crossing rate is as follows [Jakes, 94]:

$$N_R = \sqrt{2\pi} f_m \bar{r} \exp(-\bar{r}^2) \tag{10.44}$$

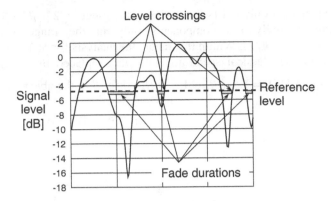

Figure 10.31: Level-crossings and fade durations

where

$$\bar{r} = \frac{r}{\sigma\sqrt{2}} = \frac{r}{r_{RMS}} \tag{10.45}$$

The normalised level-crossing rate is illustrated in Figure 10.31.

The average fade duration $\tau_{\bar{r}}$ (s) for a signal level of $\bar{r}$ is given by Eq. (10.46) [Jakes, 94] and is illustrated in Figure 10.32,

$$\tau_{\bar{r}} = \frac{e^{\bar{r}^2} - 1}{\bar{r}f_m\sqrt{2\pi}} \tag{10.46}$$

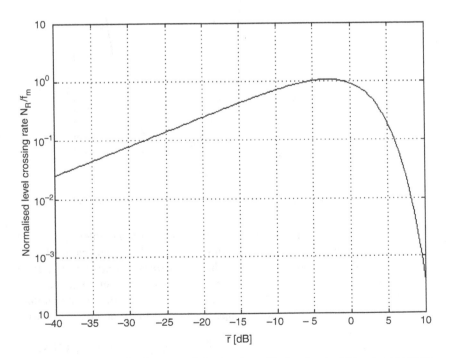

Figure 10.32 Normalised average fade duration with the classical Doppler spectrum

It is clear from these two figures that the signal spends most of its time crossing signal levels just below the RMS value, and that deep fades below this value have far shorter duration than enhancements above it. This means that, although outages may be frequent, they will not last for very long. Correspondingly, it is clear that a stationary mobile antenna has only a very small chance of being in a very deep fade.

Note that both the average fade duration (10.44) and level-crossing rate (10.46) expressions are direct consequences of the classical Doppler spectrum. Nevertheless, this is not the only spectrum which yields these results, as it may be shown that any Doppler spectrum having the same variance will share the same l.c.r. and a.f.d.

Example 10.5

For a mobile system operating at 900 MHz and a speed of 100 km h^{-1}, how many times would you expect the signal to drop more than 20 dB below its RMS value in 1 min? How many times for 30 dB fades? How would these values change if the carrier frequency were increased to 1800 MHz and the speed reduced to 50 km h^{-1}?

Solution

From Example 10.3 the maximum Doppler frequency in this case is 83.3 Hz. From Figure 10.33 or Eq. (10.44) with $\rho = -20$ dB $= 0.1$, the normalised level-crossing rate $N_R/f_m = 0.25$, hence $N_R = 83.3 \times 0.25 = 21$ s^{-1}. Hence the expected number of level-crossings in 1 m is $21 \times 60 = 1260$.

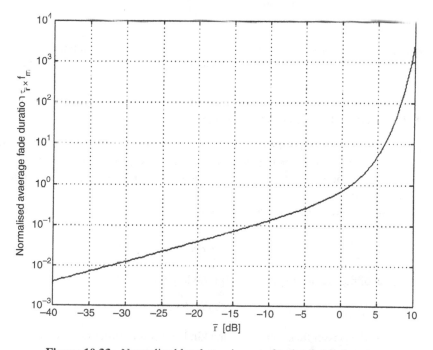

Figure 10.33: Normalised level-crossing rate for the classical spectrum

For $\rho = -30\,\text{dB}$, $N_R/f_m \approx 0.079$, so the expected number of level-crossings in 1 m becomes $0.079 \times 83.3 \times 60 = 395$.

By doubling the frequency and halving the speed, the maximum Doppler frequency is unchanged, so the expected number of level-crossings is maintained at the same value. Figure 10.34 indicates the way in which the BER varies with time for a classical Rayleigh channel for BPSK modulation. It is clear that errors occur in a very bursty manner, being concentrated at the times when deep fades occur. In practical systems the order of bits transmitted through the channel is varied so that these bursts are given a more even distribution with time, which allows error correction and detection codes to operate more efficiently. This process is known as *interleaving* and must be carefully optimised for best performance, accounting for both the characteristics of the code used and the second-order statistics of the channel.

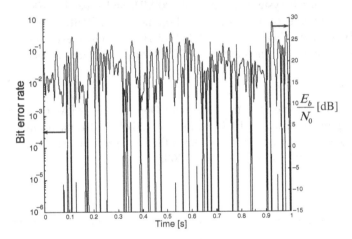

Figure 10.34: Variation of instantaneous BER with time for a classical Rayleigh channel, with mean $E_b/N_0 = 20\,\text{dB}$. Right axis indicates instantaneous E_b/N_0 [dB], left axis is instantaneous BER

10.12 AUTOCORRELATION FUNCTION

Another way to view the effect of the Doppler spread is in the time domain. For a random process, the inverse Fourier transform of the power spectral density is the autocorrelation function. This function expresses the correlation between a signal at a given time and its value at some time delay, τ later.

The normalised autocorrelation function for the received complex fading signal $\alpha(t)$ is defined by

$$\rho(\tau) = E\big[\alpha(t)\alpha^*(t+\tau)\big]/E\big[|\alpha|^2\big] \tag{10.47}$$

In the case of the classical spectrum with Rayleigh fading, the result is as follows [Jakes, 94]:

$$\rho(\tau) = J_0(2\pi f_m \tau) \tag{10.48}$$

where J_0 is the Bessel function of the first kind and zeroth order (see Appendix B, Section 5). This is shown in Figure 10.35.

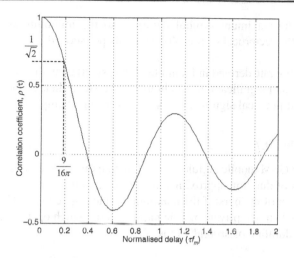

Figure 10.35: Autocorrelation function for the classical spectrum

This process leads to a definition of the *coherence time* T_c for a channel as the time over which the channel can be assumed constant. This is assured if the normalised autocorrelation function remains close to unity for the duration under consideration. The coherence time is therefore inversely proportional to the Doppler spread of the channel, i.e.

$$T_C \propto \frac{1}{f_m} \tag{10.49}$$

To determine the proportionality constant, a threshold level of correlation for the complex envelope has to be chosen, usually $1/\sqrt{2}$. According to [Steele, 98] a useful approximation to the coherence time for the classical channel is then

$$T_C \approx \frac{9}{16\pi f_m} \tag{10.50}$$

The significance of this is that signals which have less than the coherence time of the channel are received approximately undistorted by the effects of Doppler spread (and hence mobile speed).

Example 10.6

What is the minimum symbol rate to avoid the effects of Doppler spread in a mobile system operating at 900 MHz with a maximum speed of $100\ \text{km h}^{-1}$?

Solution

As in previous examples, the maximum Doppler frequency for this system is $f_m = 83.3\ \text{Hz}$. The coherence time is therefore

$$T_C \approx \frac{9}{16\pi f_m} = \frac{9}{16\pi \times 83.3} \approx 2.1\ \text{ms}$$

This is the maximum symbol duration, so the minimum symbol rate for undistorted symbols is the reciprocal of this, 465 symbols per second.

The symbol rate derived in Example 10.6 is small enough that most practical systems will rarely encounter significant effects in this form. Nevertheless, the coherence time does have an important practical significance as the following example shows.

Example 10.7

In the GSM mobile cellular system, which operates at around 900 MHz, data is sent in bursts of duration approximately 0.5 ms. Can the channel be assumed constant for the duration of the burst if the maximum vehicle speed is 100 km h^{-1}? How about for the TETRA digital private mobile radio system, which operates at around 400 MHz with a burst duration of around 14 ms?

Solution

In the GSM case, the burst duration is considerably less than the channel coherence time calculated in the previous example, so the channel is substantially constant for the whole burst. In the case of TETRA, the coherence time is

$$T_C \approx \frac{9}{16\pi f_m} = \frac{9}{16\pi(v_m f_c/c)}$$
$$= \frac{9}{16\pi \times (100 \times 10^3 \times 400 \times 10^6)/(60 \times 60 \times 3 \times 10^8)} \approx 4.8\,\text{ms}$$

This is considerably less than the burst duration, so the TETRA channel cannot be assumed constant over the burst.

10.13 NARROWBAND MOBILE RADIO CHANNEL SIMULATIONS

It is often necessary to simulate or emulate mobile channels for development and testing of mobile transceivers. Any simulation must be consistent with the known first-and second-order statistics of the mobile channel. One approach to doing this is shown in Figure 10.36. A complex white Gaussian noise generator is used to represent the in-phase and quadrature signal

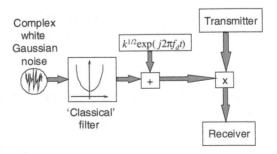

Figure 10.36: Baseband simulation of the Rice-fading channel

components, with unit power. These are passed to a filter, carefully designed to produce a close approximation to the classical Doppler spectrum at its output. The exact shape of the filter is not critical, since the a.f.d. and l.c.r. will be correct provided the variance of the noise spectrum at the filter output matches the variance of the desired classical spectrum. Other approaches to creating the signal at this stage are also available, e.g. a sum-of-sinusoids simulator [Jakes, 94]. A phasor of constant amplitude $\sqrt{k}$, where k is the desired Rice factor, is then added, representing the dominant coherent part of the channel. The phasor is usually given some non-zero frequency shift f_d, representing the Doppler shift associated with the line-of-sight path. The final result can then be used to multiply the signal from any transmitter, either in a computer simulation (such as the waveforms shown in Figure 10.19) or by creating a real-time implementation of the simulator in hardware, permitting real mobile radio equipment to be tested in laboratory conditions which repeatably emulate the practical mobile environment.

10.14 CONCLUSION

This chapter has shown how multipath propagation in mobile channels leads to rapid fading of the received signal level. This fading is described statistically by Rayleigh or Rice distributions with good accuracy, depending on whether non-line-of-sight or line-of-sight conditions prevail. Both cases degrade the signal quality relative to the static case, where the channel can be described by simpler additive white Gaussian noise statistics. The rate of variation of the fading signal, due to the phenomenon of Doppler spread, is controlled by the carrier frequency, the speed of the mobile and the angle of-arrival distribution of waves at the mobile. The Doppler spread leads to characteristic fading behaviour which can be simulated using simple structures to provide comparison of mobile equipment in realistic conditions.

REFERENCES

[Aulin, 79] T. Aulin, A modified model for the fading signal at a mobile radio channel, *IEEE Transactions on Vehicular Technology*, 28 (3), 182–203, 1979.

[Charash, 79] U. Charash, Reception through Nakagami fading multipath channels with random delays, *IEEE Transactions on Communications*, 27 (4), 657–670, 1979.

[Cheng, 01] J. Cheng, N. Beaulieu, Maximum-likelihood based estimation of the Nakagami-*m* parameter, *IEEE Communication Letters*, 5(3), 101–106, 2001.

[Clarke, 68] R. H. Clarke, A statistical theory of mobile radio reception, *Bell System Technical Journal*, 47, 957–1000, 1968.

[Gans, 72] M. J. Gans, A power-spectral theory of propagation in the mobile radio environment, *IEEE Transactions on Vehicular Technology*, 21 (1), 27–38, 1972.

[ITU, 1057] International Telecommunication Union, ITU-R Recommendation P.1057-1, *Probability Distributions Relevant to Radiowave Propagation Modelling*, Geneva 2001.

[Jakes, 94] W. C. Jakes (ed.), *Microwave Mobile Communications*, IEEE Press, New York, ISBN 0-7803-1069-1, 1974.

[Nakagami, 60] M. Nakagami, The m-distribution – a general formula of intensity distribution of rapid fading, reprint from *Statistical Methods of Radiowave Propagation*, Pergamon Press, 1960.

[Proakis, 89] J. G. Proakis, *Digital Communications*, 2nd edn, McGraw-Hill, New York, ISBN 0-07-100269-3, 1989.

[Rice, 48] S. O. Rice, Mathematical analysis of a sine wave plus random noise, *Bell System Technical Journal*, 27 (1), 109–157, 1948.

[Steele, 98] R. Steele and L. Hanzo (eds.), *Mobile Radio Communications*, John Wiley & Sons, Ltd, Chichester, ISBN 0471-97806-X, 1998.

[Zhang, 03] Q. T. Zhang, A generic correlated Nakagami fading model for wireless communications, *IEEE Transactions on Communications*, 51 (11), 1745–1748, 2003.

PROBLEMS

10.1 Prove that the power in a complex baseband signal, given by Eq. (10.1), is the same as in the equivalent real bandpass signal.

10.2 If a mobile moves at a speed $50 \, \mathrm{m \, s^{-1}}$ towards another mobile, also at speed $50 \, \mathrm{m \, s^{-1}}$, where the first mobile transmits at a frequency of 900 MHz to the second, what Doppler shift does the second mobile experience?

10.3 In Problem 10.2, what is the maximum Doppler shift which the second mobile can experience if it receives signals from a stationary base station, again transmitting at 900 MHz, indirectly via reflections from the first mobile?

10.4 In calculating the Rayleigh error performance of BPSK, it was assumed that the fading can be assumed constant over one bit duration. Up to what vehicle speed is this assumption valid for the GSM system, given that the bit rate in GSM is $\sim$270 kbps and the carrier frequency is $\sim$900 MHz?

10.5 Prove Eqs. (10.27) and (10.28).

10.6 A mobile communication system uses binary phase shift keying modulation and a carrier frequency f_c. The resulting transmitted waveform can be expressed as $s(t) = A \cos(2\pi f_c t + n\pi)$, where $n = 0$ represents a binary 0 and $n = 1$ represents a binary 1.

(a) Express the transmitted signal in complex baseband representation.

(b) Explain briefly why complex baseband representation is usually used in describing and analysing mobile radio signals.

(c) Calculate the mean signal power when $A = 10 \, \mu\mathrm{V}$.

(d) When the mobile receiver is stationary, a bit error rate (BER) of 10^{-6} is produced when the noise power at the receiver input is 5 pW. Estimate the signal-to-noise ratio (SNR) which would be required to produce a BER of 10^{-7}.

(e) Draw a graph showing the SNR in dB along the x-axis and the log of BER along the y axis. Sketch the variation of BER versus SNR when the mobile receiver is stationary, using the values from part (d).

(f) When the mobile receiver moves through a built-up area with no line of sight present, a signal-to-noise ratio of 14 dB is required to produce a BER of 10^{-2}. Add a sketch of the variation of BER versus SNR for this case to your graph from part (e). Estimate the SNR required to produce a BER of 10^{-3} in this case.

(g) What type of channel does the mobile receiver encounter in part (f)? Briefly describe the physical causes of this channel and its first-order statistical properties. Explain how the characteristics of this channel would differ if a line-of-sight signal was also received.

(h) What extra parameters would be needed in order to calculate the level-crossing rate of the received signal in part (f)? What assumptions would be necessary?

11 Wideband Fast Fading

> *'A wide screen just makes a bad film twice as bad'.*
> Samuel Goldwyn

11.1 INTRODUCTION

Mobile radio systems for voice and low bit rate data applications can consider the channel as having purely narrowband characteristics, but the wideband mobile radio channel has assumed increasing importance in recent years as mobile data rates have increased to support multimedia services. In non-mobile applications, such as television and fixed links, wideband channel characteristics have been important for a considerable period.

In the narrowband channel, multipath fading comes about as a result of small path length differences between rays coming from scatterers in the near vicinity of the mobile. These differences, on the order of a few wavelengths, lead to significant phase differences. Nevertheless, the rays all arrive at essentially the same time, so all frequencies within a wide bandwidth are affected in the same way.

By contrast, if strong scatterers exist well-off of the great circle path between the base and mobile, the time differences may be significant. If the relative delays are large compared to the basic unit of information transmitted on the channel (usually a symbol or a bit), the signal will then experience significant distortion, which varies across the channel bandwidth. The channel is then a wideband channel, and any models need to account for these effects.

Figure 11.1 shows a situation in which the signal arriving at a mobile is composed of two 'beams' of considerably different delays. Each beam is composed of several individual waves due to the rough scattering characteristics of the obstacles and may thus be subject to fading as

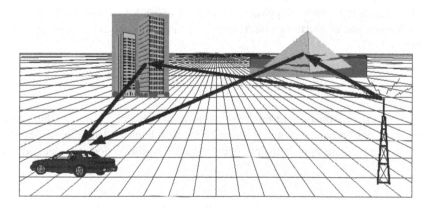

Figure 11.1: Physical causes of wideband fading

Antennas and Propagation for Wireless Communication Systems Second Edition   Simon R. Saunders and Alejandro Aragón-Zavala
© 2007 John Wiley & Sons, Ltd

the mobile moves. If the relative delay of the beams is τ, then wideband fading will be experienced whenever the system transmits signals which are comparable in duration to τ or shorter. Thus the definition of a wideband channel includes the characteristics of both the signal and the channel.

11.2 EFFECT OF WIDEBAND FADING

The wideband channel is composed of energy which reaches the mobile from the direct (not necessarily line-of-sight) path and from scatterers which lie off the path. If single scattering alone is considered, all scatterers lying on an ellipse with the base and mobile at its foci will contribute energy with the same delay (Figure 11.2), given in terms of the distances between the terminals and the scatterer as

$$\tau = \frac{r_1 + r_2}{c} \tag{11.1}$$

where c is the speed of light.

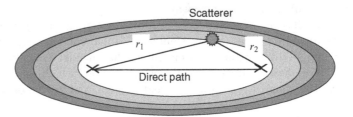

Figure 11.2: Equal-delay ellipses focused at the transmitter and receiver locations

The signal received at the mobile, y, will be composed of a sum of waves from all scatterers, whose phase θ and amplitude a depend on the reflection coefficient and scattering characteristics of the scatterer, and whose time delay τ is given by (11.1). Thus

$$y = a_1 e^{j(\omega\tau_1 + \theta_1)} + a_2 e^{j(\omega\tau_2 + \theta_2)} + \cdots \tag{11.2}$$

Each of the components in this expression constitutes an 'echo' of the transmitted signal. In the narrowband channel described in Chapter 10, the time delays of the arriving waves are approximately equal, so the amplitude does not depend upon the carrier frequency:

$$y \approx e^{j\omega\tau}(a_1 e^{j\theta_1} + a_2 e^{j\theta_2} + \cdots) \tag{11.3}$$

All frequencies in the received signal are therefore affected in the same way by the channel, and the channel can be represented by a single multiplicative component. If the relative time delays for the arriving waves are significantly different, then the channel varies with frequency and the spectrum of the received signal will be distorted. Figure 11.3 shows the

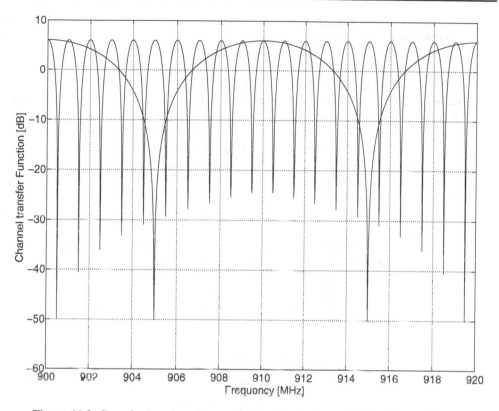

Figure 11.3: Transfer function of two-path channel: relative path delays 1 μs and 0.1 μs

power in the received signal as a function of frequency (the *channel transfer function*) when two paths are received with equal amplitude, *a*. From (11.2) for this case

$$
\begin{aligned}
y &= ae^{j(\omega\tau_1+\theta_1)}\left[1 + e^{j(\omega(\tau_2-\tau_1)+(\theta_2-\theta_1))}\right] \\
&= a\sqrt{[1 + \cos(\omega\Delta\tau + \Delta\theta)]^2 + \sin^2(\omega\Delta\tau + \Delta\theta)} \\
&= a\sqrt{2(1 + \cos(\omega\Delta\tau + \Delta\theta))}
\end{aligned}
\tag{11.4}
$$

When the path delay difference $\Delta\tau = (\tau_2 - \tau_1)$ is 0.1μs, the two paths therefore cancel at frequencies corresponding to multiples of 10 MHz. A transmitted signal with a bandwidth of, say, 1 MHz would experience essentially constant attenuation and the channel could be regarded as narrowband. But when the delay difference is increased to 1μs, the channel amplitude varies significantly across the signal bandwidth and the channel must be regarded as wideband.

Although large-delay ellipses involve longer paths between the scatterer and the terminals, there is a tendency for the power in a given delay range to reduce with increasing delay. This is counterbalanced somewhat by the increasing area of the large delay ellipses and hence the increased number of scatterers which can contribute to a given delay range. The power at a given delay is also typically composed of energy from several scatterers, each of which may contribute energy from several small wavelets having virtually equal delay but different phase due to the roughness of the scattering surface. Each wave is subject to the narrowband fading processes described in Chapter 10.

Thus, the wideband channel is a combination of several paths subject to narrowband fading, combined together with appropriate delays. The precise form of the channel will depend on the environment within which the system is operated. In particular, if the transmitter and receiver are close together (i.e. small cells), then only scatterers with small delay can contribute significant energy. Delay spread is therefore a function of cell size.

To a first approximation, the relative influence of scatterers at different delays is independent of frequency, so delay spread varies only slowly with frequency. This assumption breaks down if the wavelength approaches the size of significant scattering features, thereby changing the scattering coefficient via processes such as the Mie scattering described in Chapter 7.

It is easiest to visualise the impact of delay spread in the time domain. A transmitted symbol is delayed in the channel. This delay would be unimportant if all of the energy arrived at the receiver with the same delay. However, the energy actually becomes spread in time, and the symbol arrives at the receiver with a duration equal to the transmitted duration plus the *delay range* of the channel (Figure 11.4). The symbol is therefore still arriving at the receiver when the

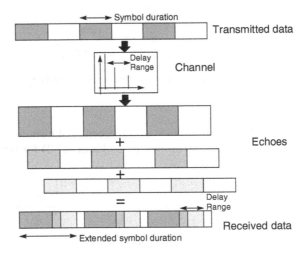

Figure 11.4: ISI due to echoes in the wideband channel

initial energy of the next symbol arrives, and this energy creates ambiguity in the demodulation of the new symbol. This process is known as *intersymbol interference* (ISI). One manifestation of delay spread in analogue modulation systems is the 'ghosting' observed when a television signal is received both directly from the transmitter and indirectly via a strong scatterer, such as a building.

When the error rate performance of a digital system in a wideband channel is examined, it tends to level off at high signal-to-noise ratios (Figure 11.5), in contrast to the *flat* or narrowband fading case, where error rates decrease without limit (Figure 10.14). This arises

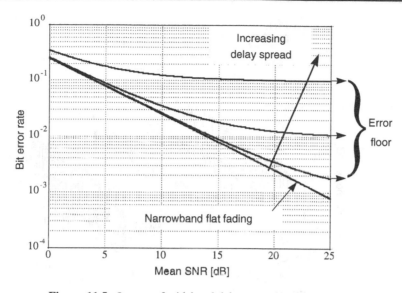

Figure 11.5: Impact of wideband delay spread on bit error rate

because the ISI, rather than the random AWGN noise, is the dominant source of errors. As the signal level is increased, the ISI increases proportionately, so demodulation performance stays the same. The final value of the error rate is known as an error floor, or sometimes the irreducible error rate, although the 'irreducible' is a misnomer because error floors may be reduced using equalisation and other techniques (Chapter 17).

11.3 WIDEBAND CHANNEL MODEL

The standard form of the channel model for wideband mobile channels is shown in Figure 11.6. The effects of scatterers in discrete delay ranges are lumped together into individual 'taps' with the same delay; each tap represents a single beam. The taps each have a gain which varies in time according to the standard narrowband channel statistics of Chapter 10. The taps are usually assumed to be uncorrelated from each other, as each arises from scatterers which are physically distinct and separated by many wavelengths. The channel is therefore a linear filter with a time-variant finite impulse response. It may be implemented for simulation purposes in digital or analogue form. We next look at the parameters which characterise such models.

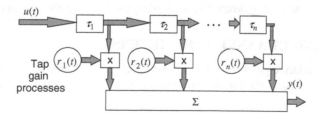

Figure 11.6: Wideband channel model

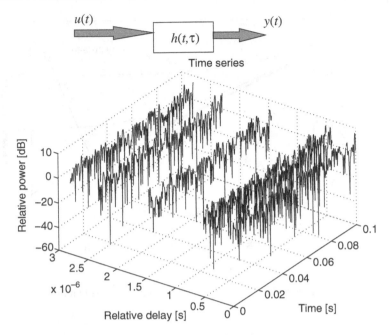

Figure 11.7: Time-variant impulse response

The basic function which characterises the wideband channel is its time-variant impulse response. The output y at a time t can be found from the input u by convolving the input time series $u(t)$ with the impulse response $h(t, \tau)$ of the channel as it appears at time t, so

$$y(t) = u(t)^* h(t, \tau) = \int_{-\infty}^{\infty} h(t, \tau)\, u(t - \tau)\, d\tau \qquad (11.5)$$

where * denotes convolution and τ is the delay variable.

The time-variant impulse response may also be known as the *input delay spread function*. Figure 11.7 shows a simulated example. Along the delay axis, the individual taps appear as delta functions. In practice the impulse response would always be experienced via a receiver filter of finite bandwidth, which would smear the time response. This means we can take scattering processes with similar delays and gather them into discrete bins, as the system resolution will be unable to distinguish between the discrete and continuous case.

Along the time axis, the amplitude of each tap varies just as in the narrowband case, with a fading rate proportional to the vehicle speed and carrier frequency, and with first-order statistics which are usually approximated by the Rayleigh or Rice distributions as described in Chapter 10.

11.4 WIDEBAND CHANNEL PARAMETERS

The mean relative powers of the taps are specified by the *power delay profile* (PDP) for the channel, defined as the variation of mean power in the channel with delay, thus

$$P(\tau) = \frac{E\lfloor |h(t, \tau)|^2 \rfloor}{2} \qquad (11.6)$$

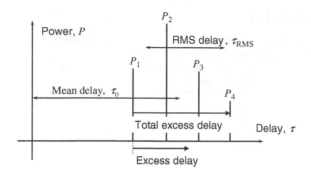

Figure 11.8: Power delay profile

Usually, the PDP is discretised in the delay dimension to yield n individual *taps* of power $P_1, \ldots, P_n$ (Figure 11.8). Each tap-gain process may be Rice or Rayleigh distributed.

The PDP may be characterised by various parameters:

- *Excess delay*: the delay of any tap relative to the first arriving tap.
- *Total excess delay*: the difference between the delay of the first and last arriving tap; this is the amount by which the duration of a transmitted symbol is extended by the channel.
- *Mean delay*: the delay corresponding to the 'centre of gravity' of the profile; defined by

$$\tau_0 = \frac{1}{P_T} \sum_{i=1}^{n} P_i \tau_i \tag{11.7}$$

where the total power in the channel is $P_T = \sum_{i=1}^{n} P_i$.

- *RMS delay spread*: the second moment, or spread, of the taps; this takes into account the relative powers of the taps as well as their delays, making it a better indicator of system performance than the other parameters; defined by

$$\tau_{\text{RMS}} = \sqrt{\frac{1}{P_T} \sum_{i=1}^{n} P_i \tau_i^2 - \tau_0^2} \tag{11.8}$$

It is independent of the mean delay and hence of the actual path length, being defined only by the relative path delays. The RMS delay spread is a good indicator of the system error rate performance for moderate delay spreads (within one symbol duration). If the RMS delay spread is very much less than the symbol duration, no significant ISI is encountered and the channel may be assumed narrowband. Note that the effect of delayed taps in (11.8) is weighted by the square of the delay. This tends to overestimate the effect of taps with large delays but very small power, so remember that RMS delay spread is not an unambiguous performance indicator. It nevertheless serves as a convenient way of comparing different wideband channels.

Example 11.1

Calculate the total excess delay, mean delay and RMS delay spread for a channel whose PDP is specified in the following table

Relative delay [μs]	Average relative power [dB]
0.0	−3.0
0.2	0.0
0.5	−2.0
1.6	−6.0
2.3	−8.0
5.0	−10.0

Would the channel be regarded as a wideband channel for a binary modulation scheme with a data rate of 25 kbps?

Solution

The total excess delay is simply the difference between the shortest and the longest delays, i.e. 5 μs. To calculate the other parameters, first the relative powers are converted to linear power values and then normalised by the total, as shown in the following table.

The mean delay is then found by multiplying the powers by the delays and summing:

$$\tau_0 = (0.19 \times 0) + (0.38 \times 0.2) + (0.24 \times 0.5) + (0.09 \times 1.6)$$
$$+ (0.06 \times 2.3) + (0.04 \times 5.0) = 0.678\,\mu s$$

The same values are used to calculate the RMS delay spread:

$$\tau_{RMS}^2 = (0.19 \times 0^2) + (0.38 \times 0.2^2) + (0.24 \times 0.5^2)$$
$$+ (0.09 \times 1.6^2) + (0.06 \times 2.3^2) + (0.04 \times 5.0^2)$$
$$- 0.678^2 = 1.163\,\mu s$$

So $\tau_{RMS} = 1.08\,\mu s$

A binary system with a data rate of 25 kbps has a symbol period of 40 μs. This is very much larger than τ_{RMS}, so the channel can be regarded as narrowband.

Average relative power [dB]	Average relative power [W]	Average relative power (normalised)
−3.0	0.50	0.19
0.0	1.00	0.38
−2.0	0.63	0.24
−6.0	0.25	0.09
8.0	0.16	0.06
−10.0	0.10	0.04

Table 11.1: Typical RMS delay spreads

Environment	Approximate RMS delay spread [μs]
Indoor cells	0.01–0.05
Mobile satellite	0.04–0.05
Open area	<0.2
Suburban macrocell	<1
Urban macrocell	1–3
Hilly area macrocell	3–10

Typical values for the RMS delay spread are given in Table 11.1 for various environments. Large cells ordinarily experience greater delay spreads, although this is contradicted in mobile satellite systems due to the enormous elevation angle compared with terrestrial systems (Chapter 14). The values experienced at particular locations may exceed the typical values by a large factor.

The delay spread values do indicate that data rates in indoor cells can be greater than in hilly area macrocells by a factor of several hundred unless special techniques such as equalisation are applied to reduce the effect of ISI (Chapter 17). Future systems will account for the variation between environments by allowing mobiles to adapt to the characteristics of the environment in which they find themselves.

Some particular examples of macrocell PDPs are given in Figure 11.9. These profiles were extracted as typical cases from a huge database of measurements in European countries [Failli, 89], and have been adopted by the GSM system as standard test cases for assessing the performance of equipment at 900 and 1800 MHz. Figure 11.9 also shows the duration of one GSM symbol for comparison. The terminology adopted to describe these models is NAMEx, where NAME is an abbreviation referring to the environment, and x is the mobile speed in [km h^{-1}]. The GSM standards require the mobiles that are tested in RA250, TU50 and HT100 channels, representing rural areas, typical urban cases and hilly terrain, respectively. All taps in these channels have multipath energy whose second-order statistics are described by the classical Doppler spectrum (Chapter 10). In the rural area case, the first tap gain process is assumed Rician with a k factor of around 3 dB, as the first tap usually comes about from a line-of-sight path. The delay spread is small compared to the GSM symbol duration, so the rural area model is essentially narrowband. In the TU channel, delay spread is experienced over more than one symbol, so significant ISI is experienced. The hilly terrain channel is a very demanding one, with delay spread occurring over four or five symbols. The raw bit error rate in such a channel is unacceptable, so the use of an equaliser of some form is essentially mandatory in GSM.

Two more standardised PDPs are given in Table 11.2. These parameters were used in evaluation of candidate macrocell technologies for the third-generation UMTS European system [ETSD, 97]. Channel A is a low delay spread case, occurring approximately 40% of the time, and channel B has a much higher delay spread and occurs around 55% of the time. Channel parameters for cell types other than macrocells are given in Chapters 12–14.

More sophisticated wideband channel models also allow the delays of the taps to vary with time, simulating the effect of the changing mobile position with respect to the scatterers over

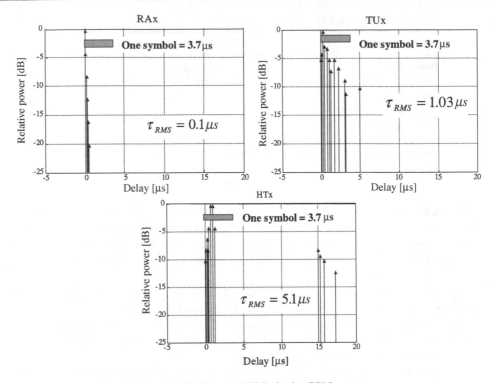

Figure 11.9: Standard PDPs in the GSM system

the long term. Such models can also include the dynamics of shadowing, as described in Chapter 9.

For small urban macrocells (1–3 km), an empirical model for delay spread is available [ITU, 1411] for use in both line-of-sight and non-line-of-sight cases for systems operating around 2 GHz using omnidirectional antennas. The median RMS delay spread S_u in this environment is given by

$$S_u = \exp(A \cdot L + B)[\text{ns}] \tag{11.9}$$

where $A = 0.038, B = 2.3$ and L is total path loss incorporating shadowing effects [dB].

Table 11.2: Evaluation channel for UMTS

Tap number	Channel A		Channel B	
	Relative delay [ns]	Relative mean power [dB]	Relative delay [ns]	Relative mean power [dB]
1	0	0.0	0	−2.5
2	310	−1.0	300	0
3	710	−9.0	8 900	−12.8
4	1090	−10.0	12 900	−10.0
5	1730	−15.0	17 100	−25.2
6	2510	−20.0	20 000	−16.0

11.5 FREQUENCY DOMAIN EFFECTS

Although it is usually natural to consider wideband channel effects in the time domain, it is equally valid to consider the channel as a filter with a time-variant frequency response (Figure 11.10). The resulting function is the time-variant transfer function (TVT) $T(f,t)$, defined as the Fourier transform of the input delay-spread function with respect to the delay variable, τ:

$$T(f,t) = \mathsf{F}[h(t,\tau)] = \int_{-\infty}^{\infty} h(t,\tau)e^{-j2\pi f}\,d\tau \qquad (11.10)$$

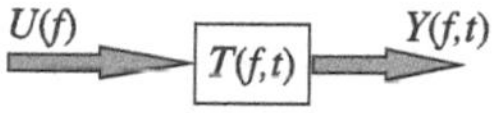

$$U(f) \longrightarrow \boxed{T(f,t)} \longrightarrow Y(f,t)$$

Figure 11.10: Time-variant transfer function

where F denotes the Fourier transform. The spectrum of the output signal at time t, $Y(f,t)$, can then be found by simply multiplying the spectrum of the input signal by the TVT:

$$Y(f,t) = U(f)T(f,t) \qquad (11.11)$$

In practical channels, the TVT is not known in advance and is specified statistically in terms of the correlation between frequency components of the output spectrum separated by a given shift. The correlation between two components of the channel transfer function with a frequency separation Δf and a time separation Δt is defined by

$$\rho(\Delta f, \Delta t) = \frac{E[T(f,t)T^*(f+\Delta f, t+\Delta t)]}{\sqrt{E[|T(f,t)|^2]E[|T(f+\Delta f, t+\Delta t)|^2]}} \qquad (11.12)$$

The correlation is assumed independent of the particular time t and frequency f at which (11.12) is evaluated. This is acceptable if the environment through which the mobile is moving can be taken as homogeneous over the time period used to calculate the correlation. This assumption, together with the previous assumption that the individual scatterers are uncorrelated, is known as the *wide-sense stationary, uncorrelated scattering* (WSSUS) assumption. The WSSUS assumption is very commonly made when analysing mobile channels. The standard channel model shown in Figure 11.6 implicitly makes the WSSUS assumption as the tap gain processes are independently generated and have fixed delays and fading statistics.

If the correlation is examined for signals with $\Delta t = 0$, then the frequency separation Δf for which the correlation equals 0.5, is termed the *coherence bandwidth* of the channel, B_c. The channel is then deemed wideband when the signal bandwidth is large compared with the coherence bandwidth.

Due to the uncertainty relation between Fourier transform pairs [Bracewell, 86], the product of the coherence bandwidth and the RMS delay spread is always greater than a certain value. For a given shape of PDP, the two quantities are inversely proportional:

$$B_c \propto \frac{1}{\tau_{RMS}} \qquad (11.13)$$

More specifically, for a correlation level ρ, the uncertainty relationship is as follows [Fleury, 96]:

$$B_c \geq \frac{1}{2\pi\tau_{RMS}} \cos^{-1}\rho \tag{11.14}$$

The equality holds if and only if the PDP is a simple two-path case with both paths of equal power. Measured values will typically be above this value.

For example, consider an idealised model of a channel in which the PDP is given by the following exponential function:

$$P(\tau) = \frac{1}{2\pi\tau_{RMS}} e^{-\tau/\tau_{RMS}} \ (\tau > 0) \tag{11.15}$$

It can then be shown that, assuming a classical Doppler spectrum for all components [Jakes, 94], the correlation function is given by

$$\rho(\Delta f, \Delta t) = \frac{J_0(2\pi f_m \Delta t)}{\sqrt{1 + (2\pi\Delta f\tau_{RMS})^2}} \tag{11.16}$$

where J_0 is the Bessel function of first kind and zeroth order. Hence, with $\Delta t = 0$, the coherence bandwidth is the value of Δf for which $\rho = 0.5$, i.e.

$$B_c = \frac{\sqrt{3}}{2\pi\tau_{RMS}} \tag{11.17}$$

This is indeed larger than the value specified by (11.14).

11.6 THE BELLO FUNCTIONS

The ultimate in wideband channel characterisation is the set of functions illustrated in Figure 11.11. This is known as the family of *Bello functions* after their originator [Bello, 63]. The input delay spread function $h(t, \tau)$ and the TVT $T(f, t)$ have already been introduced: They are Fourier transform pairs with respect to the delay variable τ. If the input delay spread function is instead Fourier transformed with respect to time, we obtain the

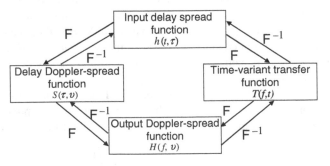

Figure 11.11: The Bello functions

Doppler spectra of each of the taps. The result is the *delay Doppler spread function*. In the case, for example, of the GSM channel models, this would be a set of multiple versions of the classical spectrum, spaced in delay according to the tap spacing. It is useful to check this function when viewing measurement results, as information is then given about the angles of arrival from waves at various time shifts, which can then be correlated with knowledge of the physical environment to see which scatterers are contributing significantly. Finally, transforming either *S* or *T* with respect to the remaining time-based variable yields the *output Doppler spread function*, which shows the Doppler spectrum associated with each frequency component. Figure 11.12 shows a practical example of these functions, taken from measurements for a mobile satellite channel [Parks, 97].

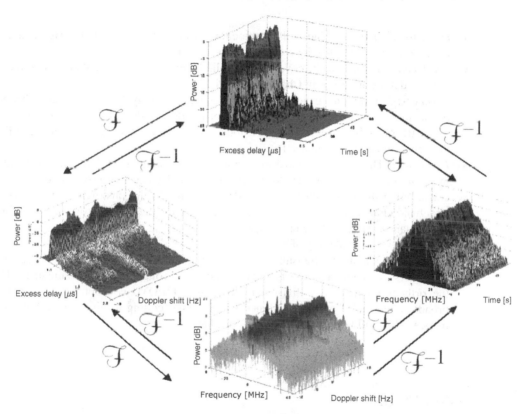

Figure 11.12: Measured Bello functions for a mobile satellite channel at 1.6 GHz [Parks, 97]: (a) input delay spread function; (b) delay Doppler spread function; (c) TVT and (d) output Doppler spread function

11.7 WIDEBAND FADING IN FIXED LINKS

In fixed links, wideband fading occurs due to multipath from either the ground or from tropospheric ducts, as described in Chapter 6. Such channels can be represented by simple 2- or 3-path models similar to Eq. (11.2), with a total excess delay of a few nanoseconds [Rummler, 86]. In contrast to the mobile channel, the individual paths are fairly stable in amplitude and have relative phases and delays which vary only over long periods, of order seconds.

11.8 OVERCOMING WIDEBAND CHANNEL IMPAIRMENTS

It is clear that the wideband nature of the mobile channel can present significant challenges in both modelling the channel and in coping with the distortions it imposes on the received signal. There are several ways this can be mitigated:

- *Directional antennas*: These allow the energy transmitted towards the significant scatterers to be reduced, thereby reducing far-out echoes.
- *Small cells*: By limiting the coverage of a cell, the maximum differential delay is reduced. Further details are given in Chapters 12 and 13.
- *Diversity*: This does not cancel the multipath energy directly, but instead makes better use of the signal energy by reducing the level of the deep fades. In this way the SNR for a given BER can be reduced and error floors are reduced, although not removed. This is described in detail in Chapter 16.
- *Equalisers*: These work to transform the wideband channel back into a narrowband one, by applying an adaptive filter to flatten the channel frequency response or by making constructive use of the energy in the delayed taps. With constructive energy use, the wideband channel performance can actually be better than the narrowband (flat fading) performance. These will be examined in Chapter 17.
- *Data rate*: One simple way of avoiding the effects of delay spread is simply to reduce the modulated data rate. By transmitting the required data simultaneously on a large number of carriers, each with a narrow bandwidth, the data throughput can be maintained. This is the orthogonal frequency division multiplexing (OFDM) concept, as used in digital broadcasting, and it can be combined with channel coding to give very robust performance. This will be examined in depth in Chapter 17.

11.9 CONCLUSION

The wideband channel generalises the concepts of fast fading introduced in Chapter 10 to include the effects of multipath echoes with significant delay. These echoes create signal *dispersion* which manifests itself as frequency-selective distortion of transmitted signals, causing increased bit errors and other degradations. The channel can be characterised using various parameters, of which the RMS delay spread and the coherence bandwidth are the most commonly used. When the signal duration is of the same order as the RMS delay spread or shorter, or equivalently, when the signal bandwidth is of the same order as the coherence bandwidth or greater, then the channel must be modelled as wideband to produce accurate results.

REFERENCES

[Bello, 63] P. A. Bello, Characterisation of randomly time-invariant linear channels, *IEEE Transactions*, CS-11 (4), 360–393, 1963.

[Bracewell, 86] R. N. Bracewell, *The Fourier transform and its applications*, 2nd edn, McGraw-Hill, New York, ISBN 0-07-007015-6, 1986.

[ETSI, 97] European Telecommunication Standards Institute, *ETR/SMG-50402: Selection procedures for the choice of radio transmission technologies of the Universal Mobile Telecommunications System (UMTS)*, 1997.

[ITU, 1411] International Telecommunication Union, ITU-R Recommendation P.1411-3: Propagation data and prediction methods for the planning of short-range outdoor radiocommunication systems and radio local area networks in the frequency range 300 MHz to 100 GHz, Geneva, 2005.

[Failli, 89] M. Failli (chairman), COST 207 final report, *Digital land mobile radio communications*, Commission of the European Communities, Brussels, ISBN 92-825-9946-9 1989.

[Fleury, 96] B. H. Fleury, An uncertainty relation for WSS processes and its application to WSSUS systems, *IEEE Transactions on Communications*, 44 (12), 1632–34, 1996.

[Jakes, 94] W. C. Jakes (ed.), *Microwave Mobile Communications*, IEEE Press, New York, 1974, ISBN 0-7803-1069-1.

[Parks, 97] M. A. N. Parks, S. R. Saunders and B. G. Evans, Wideband characterisation and modelling of the mobile satellite propagation channel at L and S bands, in *Proceedings of the 10th IEE International Conference on Antennas and Propagation*, 2.39 to 2.43, 14–17 April, 1997.

[Rummler, 86] W. D. Rummler, R. P. Coutts and M. Liniger, Multipath fading channel models for microwave digital radio, *IEEE Communications. Magazine*, 24 (11), 30–42, 1986.

PROBLEMS

11.1 Determine the RMS delay spread for a two-tap channel with equal mean tap powers and a difference in arrival time of τ seconds. Find an expression for the spacing between frequencies at which nulls in the spectrum occur.

11.2 Repeat the previous problem for the case where the two taps have powers P_1 and P_2 watts. Under what conditions is the delay spread maximised?

11.3 A non-line-of-sight channel is measured to consist of two taps with $P_1 = -10\,\mathrm{dBW}$, $\tau_1 = 3\,\mu\mathrm{s}$ and $P_2 = -15\,\mathrm{dBW}$, $\tau_2 = 5\,\mu\mathrm{s}$. Find the total excess delay, mean delay and RMS delay spread. What is the impact of introducing a line-of-sight tap with $P_0 = 0\,\mathrm{dBW}$, $\tau_0 = 2\,\mu\mathrm{s}$?

11.4 Discuss how ISI can affect the performance of high data rate applications, and which environments are most susceptible to such effects.

11.5 Calculate the total excess delay, mean delay and RMS delay spread for a channel whose PDP is specified in the following table. Would this channel be regarded as wideband for a quaternary data system with a data rate of 20 Mbps?

Tap number	Relative delay [ns]	Average relative power [dB]
1	0	0
2	4	−2
3	6	−4
4	17	−8
5	18	−10
6	50	−6

11.6 Sketch the Bello functions for a wideband mobile channel with two taps of equal mean power, separated by 1 μs, where the first is Rice distributed with a high k-factor and the second is Rayleigh distributed. The carrier frequency is 2.0 GHz, the mobile velocity is 50 km h^{-1} and the channel bandwidth of interest is 5 MHz. Label the axes of your graphs to show the typical sizes of the main features of interest.

12 Microcells

'In small proportions we just beauties see; And in short measures, life may perfect be'.
Ben Jonson

12.1 INTRODUCTION

The deployment of microcells is motivated by a desire to reduce cell sizes in areas where large numbers of users require access to the system. Serving these users with limited radio spectrum requires frequencies to be reused over very short distances, with each cell containing only a reduced number of users. This could in principle be achieved with base station antennas at the same heights as in macrocells, but this would increase the costs and planning difficulties substantially. As a result, large numbers of microcells have been deployed to serve dense concentrations of cellular mobile users and increasing numbers of microcells are being deployed in cities to provide near-continuous outdoor Wi–Fi operation. In a *microcell* the base station antenna is typically at about the same height as lampposts in a street (3–6 m above ground level), but may often be mounted at a similar height on the side of a building (Figure 12.1). Coverage, typically over a few hundred metres, is then determined mostly by the specific locations and electrical characteristics of the surrounding buildings, with cell shapes being far from circular. Pattern shaping of the base station antenna can yield benefits in controlling interference, but it is not the dominant factor in determining the cell shape. The dominant propagation mechanisms are free space propagation plus multiple reflection and scattering within the cell's desired coverage area, together with diffraction around the corners of buildings and over rooftops, which becomes significant when determining interference between co-channel cells. Microcells thus make increased use of the potential of the environment surrounding the base station to carefully control the coverage area and hence to manage the interference between sites. More general information on microcell systems is available in [Greenstein, 92], [Sarnecki, 93] and [Madfors, 97].

12.2 EMPIRICAL MODELS

In order to model the path loss in microcells, empirical models of the type described in Chapter 8 could be used in principle. However, measurements (e.g. [Green, 90]), indicate that a simple power law path loss model cannot usually fit measurements with good accuracy. This section describes models of a slightly more complex form which can yield reasonable prediction accuracy when their parameters are tuned against measurements.

12.2.1 Dual-Slope Model

A better empirical model than the ones presented in Chapter 8 is a dual-slope model. Two separate path loss exponents are used to characterise the propagation, together with a

Antennas and Propagation for Wireless Communication Systems Second Edition   Simon R. Saunders and Alejandro Aragón-Zavala
© 2007 John Wiley & Sons, Ltd

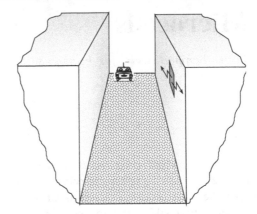

Figure 12.1: A microcell in a built-up area

breakpoint distance of a few hundred metres at which propagation changes from one regime to the other. In this case the path loss is modelled as

$$\frac{1}{L} = \begin{cases} \dfrac{k}{r^{n_1}} & \text{for} \quad r \leq r_b \\[2mm] \dfrac{k}{(r/r_b)^{n_2} r_b^{n_1}} & \text{for} \quad r > r_b \end{cases} \tag{12.1}$$

or, in decibels:

$$L = \begin{cases} 10 n_1 \log\left(\dfrac{r}{r_b}\right) + L_b & \text{for } r \leq r_b \\[2mm] 10 n_2 \log\left(\dfrac{r}{r_b}\right) + L_b & \text{for } r > r_b \end{cases} \tag{12.2}$$

where L_b is the reference path loss at $r = r_b$, r_b is the breakpoint distance, n_1 is the path loss exponent for $r \leq r_b$ and n_2 is the path loss exponent for $r > r_b$. In order to avoid the sharp transition between the two regions of a dual-slope model, it can also be formulated according to an approach suggested by [Harley, 89]:

$$\frac{1}{L} = \frac{k}{r^{n_1} \left(1 + (r/r_b)\right)^{n_2 - n_1}} \tag{12.3}$$

This can be considered in two regions: for $r \ll r_b, (1/L) \approx k r^{-n_1}$, while for $r \gg r_b$, $(1/L) \approx k(r/r_b)^{-n_2}$. Hence the path loss exponent is again n_1 for short distances and n_2 for larger distances. The model is conveniently expressed in decibels as

$$L = L_b + 10 n_1 \log\left(\frac{r}{r_b}\right) + 10(n_2 - n_1) \log\left(1 + \frac{r}{r_b}\right) \tag{12.4}$$

where L_b is the reference path loss close to $r = r_b$. Figure 12.2 compares (12.2) and (12.4). Note that the loss from (12.4) is actually 6 dB less than that from (12.2) at r_b due to the smoothing function.

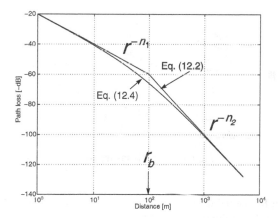

Figure 12.2: Dual-slope empirical loss models. $n_1 = 2, n_2 = 4, r_b = 100$ m and $L_1 = 20$ dB

Typical values for the path loss exponents are found by measurement to be around $n_1 = 2$ and $n_2 = 4$, with breakpoint distances of 200–500 m, but it should be emphasised that these values vary greatly between individual measurements. See for example [Chia, 90], [Green, 90], [Xia, 94] and [Bultitude, 96]. In order to plan the locations of microcells effectively, it is important to ensure that co-channel cells have coverage areas which do not overlap within the breakpoint distance. The rapid reduction of signal level beyond the breakpoint then isolates the cells and produces a large carrier-to-interference ratio, which can be exploited to maximise system capacity.

12.2.2 The Lee Microcell Model

Although the dual slope model of the previous section can account well for the general behaviour of the path loss in a microcell, it takes no particular account of the geometry of the streets, which may have a profound effect on the path loss. A very simple empirical model for dense urban microcells which nevertheless provides some account for path-specific geometry has been proposed by [Lee, 98]. This model assumes flat terrain over cell sizes of less than 1 km and assumes that there is a strong correlation between building size and total path loss along the propagation path. Figure 12.3 shows the geometry for this model. Base station antenna heights are assumed to be below 15 m. The digitised building layout information is used to calculate the proportional length of the direct wave path which passes through or across building blocks. The total path loss is a combination of a line of sight-portion L_{LOS} and an excess attenuation $\alpha(B)$ to account for building blockage:

$$L = L_{LOS} + \alpha(B) \tag{12.5}$$

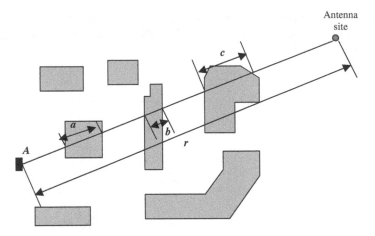

Figure 12.3: Geometry for Lee microcell model

The line-of-sight contribution is determined via a dual-slope model with an empirically determined path loss exponent of 4.3 for distances beyond the breakpoint r_b at 900 MHz:

$$L_{LOS} = \begin{cases} 20\log\left(\dfrac{4\pi r}{\lambda}\right) & \text{if } r \leq r_b \\ 20\log\left(\dfrac{4\pi r_b}{\lambda}\right) + 43\log\left(\dfrac{r}{r_b}\right) & \text{if } r > r_b \end{cases} \tag{12.6}$$

The breakpoint distance is determined as the point at which the first Fresnel zone just touches the ground:

$$r_b = \frac{4h_b h_m}{\lambda} \tag{12.7}$$

The building blockage attenuation α is an empirical function of the length of the straight-line path occupied by buildings, B. Thus for the example in Figure 12.3, B is as follows:

$$B = a + b + c \tag{12.8}$$

The attenuation function α is illustrated in Figure 12.4. The attenuation reaches a nearly constant value of 18 dB for distances beyond 150 m. The authors state that this constant is expected to be variable with the street width, with 18 dB measured for streets in the range 15–18 m. No values are given for other street widths, however. This model is incomplete in that values are only given for 900 MHz. The form of the model is however a useful starting point for tuning parameters for a particular measurement dataset.

12.2.3 The Har–Xia-Bertoni Model

While the Lee model used the building geometry in its calculations, the building loss was calculated in a very simple fashion which is unlikely to account for the full observed variation in path loss. In [Har, 99], the authors present empirical path loss formulas for microcells in low- and high-rise environments, established from measurements conducted in the

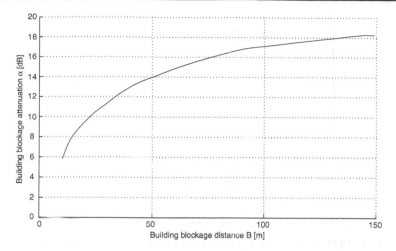

Figure 12.4: Building blockage attenuation for Lee microcell model

San Francisco Bay area in the 900 MHz and 1.9 GHz bands for radial distances up to 3 km for three different transmitter locations at each of the three test routes, for each frequency and at three different base station antenna heights (3.2, 8.7 and 13.4 m). [Xia, 93]; [Xia, 94]). In contrast to the Lee model, the formulas presented vary with the details of the signal path between base station and mobile.

Figure 12.5 shows how the signal paths are categorised into several cases (lateral, staircase, transverse and LOS) depending on the geometry of the shortest unblocked path between the base station and mobile.

The average height of surrounding rooftops h_o is used as a parameter in the path loss prediction through the relative base station antenna height Δh defined as

$$\Delta h = h_b - h_o \tag{12.9}$$

Table 12.1 shows a summary of the path loss formulas derived in [Har, 99] for low- and high-rise environments and various route types, as shown in Figure 12.3. These formulas are valid for $0.05 < R < 3$ km, operation frequencies in the range $900 < f_c < 2000$ MHz, and average building height of $-8 < \Delta h < 6$ m. Δh_m is the height of the last building relative to the mobile m, d_m is the distance of the mobile from the last rooftop m and R_{bk} represents the break point distance in kilometres.

$$R_{bk} = \frac{r_b}{1000} \tag{12.10}$$

Notice the use of the sign function in Table 12.1, which is given by

$$\mathrm{sgn}(x) = \begin{cases} +1 & x \geq 0 \\ -1, & \text{otherwise} \end{cases} \tag{12.11}$$

Although these models are fundamentally empirical, the form and structure of them is derived by consideration of several physical models, giving reasonably good confidence in the extension of the formulas beyond the parameter range of the measurements.

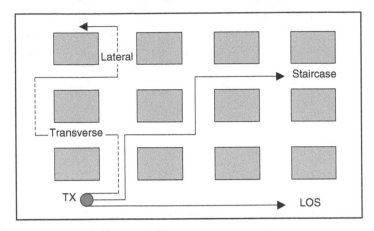

Figure 12.5: Staircase, zig-zag (transverse and lateral) and line-of-sight test routes relative to the street grid for [Har, 99]

Table 12.1: Summary of Path Loss formulas for Har–Xia-Bertoni model

Environment	Route type	Path loss formula
Low-rise	NLOS routes	$L = [11.24 + 42.59 \log f_c] -$ $[4.99 \log f_c] \mathrm{sgn}(\Delta h) \log(1 + |\Delta h|) +$ $[40.67 - 4.57 \mathrm{sgn}(\Delta h) \log(1 + |\Delta h|)] \log R +$ $20 \log(h_m/7.8) + 10 \log(20/d_m)$
High-rise	Lateral route	$L = 97.94 + 12.49 \log f_c - 4.99 \log h_b +$ $(46.84 - 2.34 \log h_b) \log R$
	Combined staircase or transverse routes	$L = 53.99 + 29.74 \log f_c - 0.99 \log h_b +$ $(47.23 + 3.72 \log h_b) \log R$
Low-rise + high-rise	LOS route	$L = -37.06 + 39.40 \log f_c - 0.09 \log h_b +$ $(15.8 - 5.73 \log h_b) \log R \qquad R < R_{bk}$ $L = (48.38 - 32.1 \log R_{bk}) - 137.1 +$ $45.70 \log f_c + (25.34 - 13.90 \log R_{bk}) \log h_b +$ $(32.10 + 13.90 \log h_b) \log R +$ $20 \log(1.6/h_m) \qquad R > R_{bk}$

12.3 PHYSICAL MODELS

Although the empirical models in the previous section can provide reasonable results, it is desirable to understand the physical mechanisms underlying them to gain insight and to create physical models with potentially greater accuracy. In creating physical models for microcell

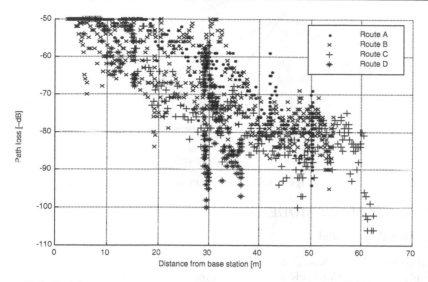

Figure 12.6: Path loss measurements at 900 MHz from a suburban microcell. Routes A, B and C are radial streets, often with a line-of-sight present; route D is a transverse street with most locations obstructed. The route D measurements vary over almost 45 dB, despite the range being almost constant at around 30 m (Reproduced by permission of Red-M Services Ltd.)

propagation, it is useful to distinguish line-of-sight and non-line-of-sight situations. We shall see that it is possible to make some reasonable generalisations about the LOS cases, while the NLOS cases require more site-specific information. Figures 12.6 and 12.7 show a practical measurement, in which it is clear that the obstructed path suffers far greater variability at a given range than the others. Such effects must be accounted for explicitly in the models.

Figure 12.7: Measurement routes corresponding to Figure 12.6 (Reproduced by permission of Red-M Services Ltd.)

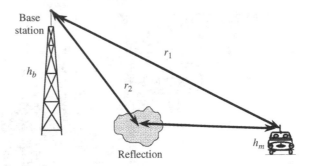

Figure 12.8: Two-ray model of line-of-sight propagation

12.4 LINE-OF-SIGHT MODELS

12.4.1 Two-Ray Model

In a line-of-sight situation, at least one direct ray and one reflected ray will usually exist (Figure 12.8). This leads to a similar approach to the derivation of the plane earth loss in Chapter 5, except it is no longer appropriate to assume the direct and reflected path lengths are necessarily similar, or to assume the reflection coefficient necessarily has a magnitude of unity. The loss is then

$$\frac{1}{L}\left(\frac{\lambda}{4\pi}\right)^2 \left| \frac{e^{-jkr_1}}{r_1} + R\frac{e^{-jkr_2}}{r_2} \right|^2 \tag{12.12}$$

where R is the Fresnel reflection coefficient (Chapter 3) for the relevant polarisation. At long distances, the reflection approaches grazing incidence so the reflection coefficient is very close to -1. The path loss exponent thus tends towards 4 at long distances, just as in the plane earth loss. At short distances, constructive and destructive fading between the two waves occur, with the mean value being close to the free space loss and hence the path loss exponent is essentially 2. Hence the two-ray model produces two regimes of propagation, as desired (see Figure 12.9).

Although the reflection coefficient causes a phase change of $180°$, when $r_2 = (r_1 + \lambda/2)$, there is a further phase inversion, leading to the two waves being perfectly in phase. This is exactly the definition of the first Fresnel zone which was given in Chapter 3. Thus rapid destructive fading can occur only at shorter distances than this. At longer distances the combined signal reduces smoothly towards zero, because at infinity the two rays are of equal length but in antiphase. For high frequencies the distance at which the first Fresnel zone first touches the ground is given approximately by

$$r_b = \frac{4h_b h_m}{\lambda} \tag{12.13}$$

It has been suggested that this forms a physical method for calculating the breakpoint distance for use in empirical models such as (12.2) and (12.4) [Xia, 93].

The two-ray model forms a useful idealisation for microcells operated in fairly open, uncluttered situations, such as on long, straight motorways where a line-of-sight path is always present and little scattering from other clutter occurs.

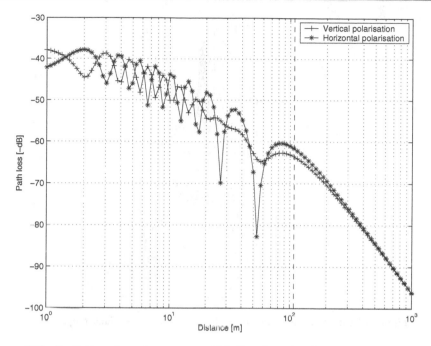

Figure 12.9: Predictions from the two-ray model. Here $h_b = 6\,\mathrm{m}, h_m = 1.5\,\mathrm{m}, f_c = 900\,\mathrm{MHz}$ and the constitutive parameters of the ground are $\epsilon_r = 15$ and $\sigma = 0.005\,\mathrm{S\,m^{-1}}$

12.4.2 Street Canyon Models

Although a line-of-sight path frequently exists within microcells, such cells are most usually situated within built-up areas where the situation is more complicated than the two-ray model would suggest. The buildings around the mobile can all interact with the transmitted signal to modify the simple two-ray regime described in Section 12.4.1. A representative case is illustrated in Figure 12.10. It assumes that the mobile and base stations are both located in a long straight street, lined on both sides by buildings with plane walls. Models which use this canonical geometry are *street canyon* models.

Six possible ray paths are illustrated in Figure 12.10. Many more are possible, but they tend to include reflections from more than two surfaces. These reflections are typically attenuated to a much greater extent, so the main signal contributions are accounted for by those illustrated.

The characteristics of this approach are illustrated by reference to a four-ray model, which considers all three of the singly reflected paths from the walls and the ground. The structure follows Eq. (12.12), but the reflections from the vertical building walls involve the Fresnel reflection coefficients for the opposite polarisation to the ground. A typical result is shown in Figure 12.11. In comparison to Figure 12.9 the multipath fading is more rapid, and the differences between vertical and horizontal polarisation are less. The average path loss exponent is close to 2 over the whole range examined.

Figure 12.12 shows the variation of field predicted by this model as the base station antenna height is varied at a particular range. Neither polarisation shows any definite advantage in increasing the antenna height, and the particular position of the nulls in this figure is strongly

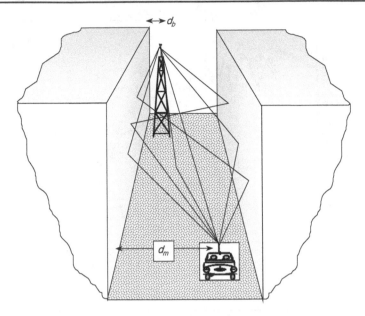

Figure 12.10: Street canyon model of line-of-sight microcellular propagation

dependent on the range. In general, for line-of-sight microcells, the base station height has only a weak effect on the cell range. There is some effect due to obstruction from clutter (in this case vehicles, street furniture and pedestrians), but we will see in later sections that increasing the base station height significantly increases the distance over which interference

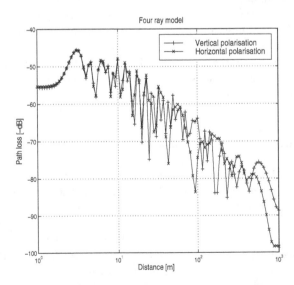

Figure 12.11: Predictions from four-ray model: $h_b = 6\,\text{m}, h_m = 1.5\,\text{m}, w = 20\,\text{m}, \quad d_m = 10\,\text{m}$ and $d_b = 5\,\text{m}, f_c = 900\,\text{MHz}$ and the constitutive parameters of the ground and buildings are $\varepsilon_r = 15$ and $\sigma = 0.005\,\text{S m}^{-1}$

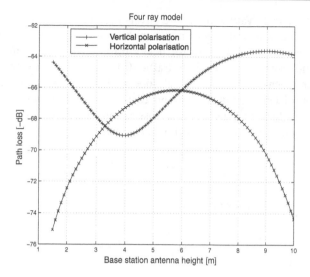

Figure 12.12: Base station antenna height variation according to the four-ray model. Here $r = 50$ m, other parameters as Figure 12.11

is caused. Thus the antenna should be maintained as low as possible, consistent with providing a line of sight to locations to be covered.

12.4.3 ITU-R P.1411 Street Canyon Model

ITU-R recommendation P.1411 [ITU, 1411] provides a model for line-of-sight propagation in street canyons which extends and enhances the basic dual-slope empirical model.

At UHF frequencies (30–300 MHz) the formulation for r_b is the same as previously:

$$r_b = \frac{4h_b h_m}{\lambda} \qquad (12.14)$$

The loss is then given in the same form as (12.2), but with two distinct cases. The lower bound model again sets $n_1 = 2$ and $n_2 = 4$ and provides an analytical formulation for L_b as follows:

$$L_b = 20 \ \log\left(\frac{\lambda^2}{8 \ \pi \ h_b h_m}\right) \qquad (12.15)$$

The upper bound model, however, sets $n_1 = 2.5$ and $n_2 = 4$ and increases the value of L_b by 20 dB.

For SHF frequencies from 3–15 GHz and path lengths up to about 1 km, the model considers the impact of clutter on the road, consisting of road traffic and pedestrians. Road traffic will influence the effective road height and will thus affect both the breakpoint distance and the breakpoint loss. The values of h_b and h_m are replaced by $(h_b - h_s)$ and $(h_m - h_s)$, respectively, in both (12.14) and (12.15). Here h_s is the effective road height due to such objects as vehicles on the road and pedestrians near the roadway. Hence h_s depends on the traffic on the road. Values of h_s are given derived from measurements in a variety of conditions. The values depend on the mobile and base station antenna heights and on frequency, but are around 1.5 m for heavy traffic (10–20% of the roadway covered with vehicles) and around 0.5 m for light traffic (0.1–0.5% of the roadway covered).

However, when $h_m \leq h_s$, no Fresnel-zone related breakpoint exists. Experiments indicated that for short distances, less than around 20 m, the path loss has similar characteristics to the UHF range. At longer ranges, the path loss exponent increases to around 3. An approximate lower bound path loss for $r \geq r_s$ is thus given by:

$$L_{\text{lower}} = L_s + 30 \, \log\left(\frac{r}{r_s}\right) \tag{12.16}$$

and an approximate upper bound for $r \geq r_s$ is:

$$L_{\text{upper}} = L_s + 20 + 30 \log\left(\frac{r}{r_s}\right) \tag{12.17}$$

where $r_s \approx 20$ m. The loss at $r_s L_s$ is defined as:

$$L_s = 20 \, \log\left(\frac{\lambda}{2\pi r_s}\right) \tag{12.18}$$

At frequencies above about 10 GHz, the breakpoint distance is far beyond the expected maximum cell radius of around 500 m. The path loss exponent then approximately follows the free-space loss over the whole cell range with a path-loss exponent of about 2.2. Attenuation by atmospheric gases and by rain must also be added in the same way as described in Chapter 7.

12.4.4 Random Waveguide Model

At long distances, the street-canyon LoS models presented so far all suggest a path loss exponent of 4. Some experimental evidence, however, suggests an increase in the path loss exponent from 4 to as much as 7 at long distances [Blaunstein, 95]. If the street canyon model is modified to include the gaps between the buildings, the path loss exponent does indeed increase. [Blaunstein, 98a] modelled this effect by treating the street as a waveguide, with randomly distributed slits representing the gaps between the building. This model implicitly considers the multiple reflections on building walls, multiple diffraction on their edges and reflections from the road surface.

Figure 12.13 shows the geometry for this model, showing the street in plan view. The distances between the buildings (or slits for the waveguide model) are defined as l_m, where $m = 1, 2, 3, \ldots$ The buildings on the street are replaced by randomly distributed non-transparent screens with lengths L_m, the electrical properties of which are described by the wave impedance Z_{EM} given by

$$Z_{EM} = \frac{1}{\sqrt{\varepsilon}}, \quad \varepsilon = \varepsilon_r - j\frac{4\pi\sigma}{\omega} \tag{12.19}$$

where ε_r is the relative permittivity of the walls and σ is their conductivity [S m^{-1}].

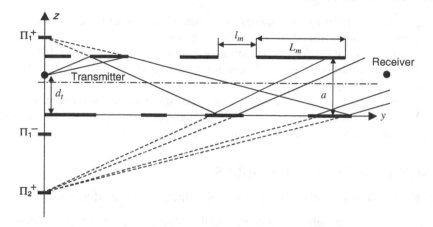

Figure 12.13: Waveguide multislit model geometry proposed in [Blaunstein, 98]

A geometrical theory of diffraction (GTD) calculation is applied for the rays reflected from the walls and diffracted from the building edges. In this approximation, the resulting field can be considered as a sum of the fields arriving at the mobile at a height h_m, from the virtual image sources Π_1^+, Π_1^- and Π_2^+, as presented in Figure 12.13.

The path loss is then approximately [Blaunstein, 98b]

$$L = 32.1 - 20\log|R_n| - 20\log\left[\frac{1 - (M|R_n|)^2}{1 + (M|R_n|)^2}\right] \\ + 17.8\log r + 8.6\left\{-|lnM|R_n||\frac{[(\pi n - \varphi_n)/a]r}{\rho_n^{(0)}a}\right\} \tag{12.20}$$

where it is assumed that $L_m \gg \lambda$ and $l_m \gg \lambda$.

Here M is the parameter of brokenness, defined as

$$M = \frac{L_m}{L_m + l} \tag{12.21}$$

so that $M = 1$ for an unbroken waveguide without slits or separations between buildings. $\rho_n^{(0)}$ is defined as

$$\rho_n^{(0)} = \sqrt{k^2 + (\pi n/a)^2} \tag{12.22}$$

where k is the wave number and n_r represents the number of reflections, $n_r = 0, 1, 2, 3, \ldots R_{nr}$ is the reflection coefficient of each reflecting wall, given by

$$R_{nr} = \frac{(\pi n + j|\ln M|)/a - kZ_{EM}}{(\pi n + j|\ln M|)/a + kZ_{EM}} = |R_{nr}|e^{j\varphi_n} \tag{12.23}$$

The Blaunstein waveguide model has been validated for frequencies in the 902–925 MHz frequency band [Blaunstein, 98b], and Blaunstein demonstrates that the model is in good agreement with the two-ray model of Section 12.4.1 for wide streets whereby the path loss exponent changes from 2 for $r < r_b$ to 4 when $r > r_b$.

Suggested values for the constitutive parameters, are given in [Blaunstein, 98b] for ferroconcrete walls as $R_{nr} = 1$ and for brick building walls as $\varepsilon_r = 15 - 17, \sigma = 0.05 -0.08 \, \mathrm{S \, m^{-1}}, |R_{nr}| = 0.73 - 0.81$.

This model is an example of a *physical-statistical* model, as it combines a sound basis in physical propagation mechanisms with a statistical description of the environment. More physical-statistical models will be described in Section 14.6.

12.5 NON-LINE-OF-SIGHT MODELS

12.5.1 Propagation Mechanisms and Cell Planning Considerations

When the line-of-sight path in a microcell is blocked, signal energy can propagate from the base to the mobile via several alternative mechanisms:

- diffraction over building rooftops
- diffraction around vertical building edges
- reflection and scattering from walls and the ground.

These mechanisms are described further in [Dersch, 94]. At relatively small distances from the base station and low base antenna heights, there is a large angle through which the signal must diffract over rooftops in order to propagate and the diffraction loss is correspondingly large. Then propagation is dominated by the other two mechanisms in the list, where the balance between the diffraction and reflection depends on the specific geometry of the buildings. Figure 12.14 shows the motion of a mobile across the shadow boundary created by a vertical building edge. As this building is in an isolated situation, the only possible source of energy in the shadow region is via diffraction and the energy will drop off very rapidly with increasing distance. This contrasts with the case illustrated in Figure 12.15, where the building is now surrounded by others which act as reflecting surfaces. The reflected ray is then likely to be much stronger than the diffracted ray, so the signal remains strong over much larger distances.

At even larger distances, particularly those involved in interference between co-channel microcells, the rooftop-diffracted signal (Figure 12.16) again begins to dominate due to the large number of diffractions and reflections required for propagation over long distances. Figure 12.17 shows the plan view of buildings arranged in a regular Manhattan grid structure.

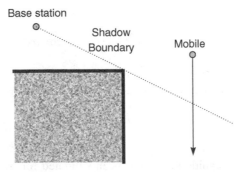

Figure 12.14: Street geometry where diffraction dominates

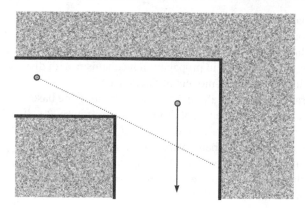

Figure 12.15: Street geometry where reflection dominates

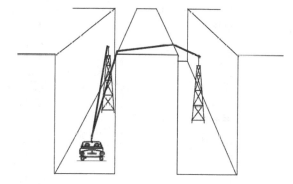

Figure 12.16: Rooftop diffraction as an interference mechanism

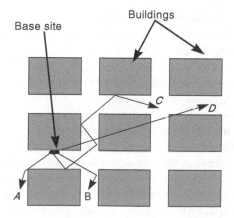

Figure 12.17: Variation of propagation mechanisms with distance for NLOS microcells

The short paths A and B involve only a single reflection or diffraction and are likely to be dominant sources of signal energy. By contrast, the long path C is likely to be very weak as four individual reflection losses are involved, and the rooftop-diffracted path D is then likely to dominate. This variation in propagation mechanism with distance is another source of the two slopes in the empirical models of Section 12.2.

System range is greatest along the street containing the base site. When the mobile turns a corner into a side street, the signal drops rapidly, often by 20–30 dB. The resultant coverage area is therefore broadly diamond-shaped, as illustrated in Figure 12.18, although the precise shape will depend very much upon the building geometry. Confirmed by measurement, the curved segments forming the diamonds in Figure 12.18 have been shown to indicate that the dominant mechanism of propagation into side streets is diffraction rather than reflection [Goldsmith, 93].

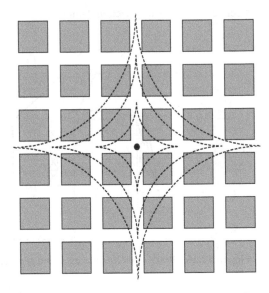

Figure 12.18: Microcellular propagation along and across streets: base site (•), (—) path loss contours

The variation of the microcell shape with base antenna height in a Manhattan grid structure has been investigated in detail using the multiple diffraction integral described in Chapter 6 [Maciel, 94], and it is shown that there is a smooth transition from a diamond shape to nearly circular as the antenna height increases. It has also been shown [Erceg, 94] that the characteristic diamond cell shape is obtained even when considering only the vertical corner diffraction plus reflections from building walls. This work also showed that the distance at which the transition between the various mechanisms occurs depends strongly on the distance between the base station and the nearest street corners.

For low antenna heights, the strong scattering present in microcells prevents the efficient use of cell sectorisation as the free space antenna radiation pattern is destroyed. Efficient frequency reuse can still be provided, however, by taking advantage of the building geometry. In regular street grid structures, co-channel microcells should be separated diagonally across the street directions, and with sufficient spacing to ensure that cells do not overlap within their breakpoint distance, in order to maintain high signal-to-interference levels.

In more typical environments, where the buildings are not regular in size, more advanced planning techniques must be applied, particularly when frequencies are shared between microcell and macrocell layers; see for example [Dehghan, 97] and [Wang, 97].

12.5.2 Recursive Model

This model is intermediate between an empirical model and a physical model [Berg, 95]. It uses the concepts of GTD/UTD, in that effective sources are introduced for non-line-of-sight propagation at the street intersections where diffraction and reflection points are likely to exist. The model breaks down the path between the base station and the mobile into a number of segments interconnected by nodes. The nodes may be placed either just at the street intersections or else at regular intervals along the path, allowing streets which are not linear to be handled as a set of piecewise linear segments. An *illusory* distance for each ray path considered is calculated according to the following recursive expressions:

$$k_j = k_{j-1} + d_{j-1} \times q(\theta_{j-1})$$
$$d_j = k_j \times r_{j-1} + d_{j-1}$$

(12.24)

subject to the initial values

$$k_0 = 1 \quad d_0 = 0$$

(12.25)

where d_n is the illusory distance calculated for the number n of straight line segments along the ray path (e.g. $n = 3$ in Figure 12.19) and r_j is the physical distance [m] for the line segment following the jth node. The result is reciprocal.

The angle through which the path turns at node j is θ_j (degrees). As this angle increases, the illusory distance is increased according to the following function:

$$q(\theta_j) = \left(\frac{\theta_j q_{90}}{90}\right)^\nu$$

(12.26)

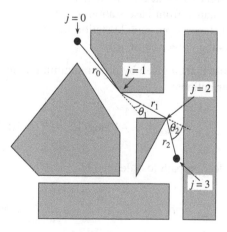

Figure 12.19: Example geometry for recursive microcell model

where q_{90} and are v parameters of the model, with $q_{90} = 0.5$ and $v = 1.5$ suggested in [Berg, 95]. The path loss is then calculated as

$$L = 20 \log \left[\frac{4\pi d_n}{\lambda} D \left(\sum_{j=1}^{n} r_{j-1} \right) \right] \qquad (12.27)$$

where

$$D(r) = \begin{cases} r/r_b & \text{for } r > r_b \\ 1 & \text{for } r \le r_b \end{cases} \qquad (12.28)$$

Equation (12.27) creates a dual-slope behaviour with a path loss exponent of 2 for distances less than the breakpoint r_b and 4 for greater distances. The overall model is simple to apply and accounts for the key microcell propagation effects, namely dual-slope path loss exponents and street-corner attenuation, with an angle dependence which incorporates effects encountered with real street layouts.

12.5.3 ITU-R P.1411 Non-Line-of-Sight Model

For non-line-of-sight situations, this ITU-R recommendation [ITU, 1411] suggests using a multiple knife-edge diffraction loss for the over-rooftop component, similar to the physical models in Chapter 8. For the paths propagated around street corners due to diffracted and reflected from building walls and vertical edges, it defines the geometry shown in Figure 12.20.

The parameters for this model are

w_1:   street width at the base station [m]
w_2:   street width at the mobile station [m]
x_1:   distance from base station to street crossing [m]
x_2:   distance from mobile station to street crossing [m]
α: is the corner angle [rad].

The overall path loss is then given by power summing contributions from the diffracted and reflected path contributions:

$$L_{NLoS2} = -10 \log(10^{-L_r/10} + 10^{-L_d/10}) \qquad (12.29)$$

where L_r is the reflection path loss defined by

$$L_r = 20 \; \log(x_1 + x_2) + x_1 x_2 \frac{f(\alpha)}{w_1 w_2} + 20 \; \log \left(\frac{4\pi}{\lambda} \right) \qquad (12.30)$$

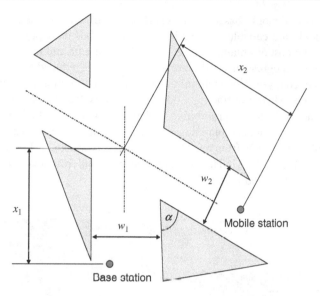

Figure 12.20: Geometry for P.1411 NLOS model

Here

$$f(\alpha) = \frac{3.86}{\alpha^{3.5}} \tag{12.31}$$

where $0.6 < \alpha[\text{rad}] < \pi$. L_d is the diffraction path loss defined by

$$L_d = 10 \ \log[x_1 x_2 (x_1 + x_2)] + 2D_a - 0.1\left(90 - \alpha\frac{180}{\pi}\right) + 20 \ \log\left(\frac{4\pi}{\lambda}\right) \tag{12.32}$$

and D_a is an approximate wedge diffraction coefficient, defined as

$$D_a = \left(\frac{40}{2\pi}\right)\left[\arctan\left(\frac{x_2}{w_2}\right) + \arctan\left(\frac{x_1}{w_1}\right) - \frac{\pi}{2}\right] \tag{12.33}$$

This model strikes a good balance between accounting correctly for the important physical mechanisms and maintaining reasonable computational simplicity.

12.5.4 Site-Specific Ray Models

Prediction of the detailed characteristics of microcells requires a site-specific prediction based on detailed knowledge of the built geometry. Electromagnetic analysis of these situations is commonly based on the GTD and its extensions, as described in Section 3.5.3. These models are capable of very high accuracy, but their practical application has been limited by the cost of obtaining sufficiently detailed data and the required computation time. More recently progress in satellite remote sensing has reduced the cost of the necessary data while advanced ray-tracing techniques and cheap computational resources have advanced the art to the point where ray-tracing models are entirely feasible. Nevertheless, many operators consider the costs to be prohibitive and prefer to deploy their microcells based on the knowledge of experienced planning engineers together with site-specific measurements.

Example predictions from a GTD-based model [Catedra, 98] for a real building geometry are shown in Figure 12.21. This includes contributions from a very large number of multiply reflected and diffracted rays.

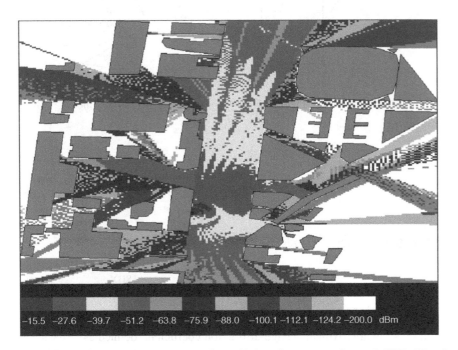

Figure 12.21: Predictions for an urban microcell, based on ray tracing and GTD. The shading indicates the received signal strength [dBm] at a mobile with a quarter-wave monopole antenna and a transmit power of 1 W at 900 MHz (reproduced by permission of Felipe Catedra)

12.6 DISCUSSION

For the practical application of microcell propagation models, there is an important trade-off between the accuracy of the prediction and the speed with which the prediction can be made. Microcells often have to be deployed very quickly, with little engineering effort, often by people who are not necessarily radio experts. Rules of thumb and very rapid statistical planning tools are very important. Also, even with a very high resolution topographic database, propagation may often be dominated by items of street furniture (signs, lampposts, etc.) and by details of the antenna sitting and its interaction with the objects on which it is mounted, which no database could hope to have available. These features may also change rapidly with time, as is certainly the case when dealing with the effects of traffic. For example, when a double-decker bus or a large truck passes close by a microcell antenna, the coverage area of the microcell may change dramatically for a short time. Either these items must be entered by hand or, more likely, systems of the future will have to be capable of adapting their characteristics to suit the environment which they find, by taking measurements from the active network and responding accordingly.

These factors will dramatically change the way in which propagation models are applied, from being processes which are run at the start of a system deployment, and then used to create a fixed set of predictions and recommendations for deployment, to real-time processes which operate within the base station, with assistance from the mobiles, which are optimised on an ongoing basis and are used by the system to assess the likely impact of changes to system parameters such as transmit powers, antenna patterns and channel assignments.

12.7 MICROCELL SHADOWING

The lognormal distribution is applied to predicting shadow fading in microcells, just as for macrocells (Chapter 9). Some measurements have suggested that the location variability increases with range, typically in the range 6–10 dB [Feuerstein, 94]. In order to account for microcell shadowing cross-correlation, the shadowing can be separated into two parts, one of which is caused by obstructions very local to the mobile and is therefore common to all paths, and another which is specific to the transport of energy from the mobile to a particular base station [Arnold, 88].

12.8 NARROWBAND FADING

Although the microcell channel frequently involves the presence of a line-of-sight component, the narrowband fading statistics may be expected to be Rician, and this prediction is confirmed by measurements [Green, 90]. However, the Rice k factor may often be rather small, as the large number of possible multipaths encountered at near-grazing incidence will have amplitudes comparable to, or exceeding, the line-of-sight amplitude. Figure 12.22 predicts the variation of k with range using the four-ray street canyon model from Section 12.4.2, suggesting that a Rayleigh model would be pessimistic for distances less than around 50 m. In practical measurements the k factor is very variable, and it is safest to assume a Rayleigh channel ($k = 0$) as the worst case.

12.9 WIDEBAND EFFECTS

One important motivation for the use of microcells is the reduced delay spread compared with macrocells, permitting the use of higher data rates. Measurements [Bultitude, 89] suggest that RMS delay spreads of order $0.5\,\mu s$ are typical. Simple ray models have been found to substantially underestimate microcell delay spread, as they may not fully account for the influence of rough scattering and scattering from street furniture, particularly on adjacent streets. Although such elements may contribute very little to the total signal power, they may have a significant impact on the delay spread. Certainly the presence or otherwise of a

Table 12.2: Coefficients for ITU P-1411 microcell delay spread model

| | | | | Mean RMS Delay Spread a_s | | Standard Deviation of RMS Delay Spread σ_s | |
	Measurement conditions						
Area	f [GHz]	h_b [m]	h_m [m]	C_a	γ_a	C_σ	γ_σ
	2.5	6.0	3.0	55	0.27	12	0.32
			2.7	23	0.26	5.5	0.35
Urban	3.35–15.75						
		4.0	1.6				
				10	0.51	6.1	0.39
	3.35–8.45		0.5				
	3.35		2.7	2.1	0.53	0.54	0.77
Residential		4.0					
	3.35–15.75		1.6	5.9	0.32	2.0	0.48

line-of-sight component is significant, with measurements of 280 ns with LOS and 520 ns without LOS being typical [Arowojolu, 94].

More recently, delay profile models that attempt to predict multipath propagation characteristics more accurately have been proposed [Ichitsubo, 00). These models have shown good agreement with measured profiles for LOS situations. [ITU, 1411] gives a model for delay spread in street canyon microcells based on measured data at frequencies from 2.5 to 15.75 GHz at distances from 50 to 400 m. The RMS delay spread S at distance of r (m) is predicted to follow a normal distribution with the mean value given by

$$a_s = C_a d^{\gamma_a} \text{ [ns]} \tag{12.34}$$

and the standard deviation given by

$$\sigma_s = C_\sigma d^{\gamma_\sigma} \text{ [ns]} \tag{12.35}$$

where C_a, γ_a, C_σ and γ_σ depend on the antenna height and propagation environment. Some typical values of the coefficients for distances of 50–400 m based on measurements made in urban and residential areas are given in Table 12.2.

12.10 CONCLUSION

Propagation in microcells can be modelled using either empirical or physical models, just like the macrocells in Chapter 8. In either case, however, the clutter surrounding the base station has a significant impact on the cell shape and this must be accounted for to avoid serious prediction errors. In particular, a simple path loss exponent model is inadequate and dual-slope behaviour must be accounted for. This clutter also creates difficulties in deploying antennas, as the clutter disrupts the free space antenna radiation pattern. Nevertheless, the enormous potential offered by microcells in creating high-capacity cellular systems makes them increasingly attractive methods of providing outdoor coverage in areas with high user densities [Dehghan, 97].

REFERENCES

[Arnold, 88] H. W. Arnold, D. C. Cox and R. R. Murray, Macroscopic diversity performance measured in the 800 MHz portable radio communications environment, *IEEE Transactions on Antennas and Propagation*, 36 (2), 277–280, 1988.

[Arowojolu, 94] A. A. Arowojolu, A. M. D. Turkmani and J. D. Parsons, Time dispersion measurements in urban microcellular environments, in *Proceedings IEEE Vehicular Technology Conference*, Vol. 1, 150–154, Stockholm, 1994.

[Berg, 95] J. E. Berg, A recursive method for street microcell path loss calculations, n *Proceedings IEEE International Symposium on Personal, Indoor and Mobile Radio Communications*, (*PIMRC '95*), Vol. 1, 140–143, 1995.

[Blaunstein, 95] N. Blaunstein and M. Levin, Prediction of UHF-wave propagation in suburban and rural environments, in *Proceedings URSI Symposium Comm-Sphere '95.*, 191–200, 1995.

[Blaunstein, 98a] N. Blaunstein, R. Giladi and M. Levin, Characteristics' prediction in urban and suburban environments, *IEEE Transactions.on Vehicular Techology*, 47 (1), 225–234, 1998.

[Blaunstein, 98b] N. Blaunstein, Average field attenuation in the nonregular impedance street waveguide, *IEEE Transactions on Antenna and Propagation*, 46 (12), 1782–1789, 1998.

[Bultitude, 89] R. J. C. Bultitude and G. K. Bedal, Propagation characteristics on micro-cellular urban mobile radio channels at 910 MHz, *IEEE Journal on Selected Areas in Communications*, 7 (1), 31–39, 1989.

[Bultitude, 96] R. J. C. Bultitude and D. A. Hughes, Propagation loss at 1.8 GHz on microcellular mobile radio channels, in *Proceedings IEEE International Symposium on Personal, Indoor and Mobile Radio Communications*, (*PIMRC'96*), 786–90, October 1996.

[Catedra, 98] M. F. Catedra, J. Perez, *et al.*, Efficient ray-tracing techniques for three-dimensional analyses of propagation in mobile communications: application to picocell and microcell scenarios, *IEEE Antennas and Propagation Magazine*, 40 (2), 15–28, 1998.

[Chia, 90] S. T. S. Chia, Radiowave propagation and handover criteria for microcells, *BT Technology Journal*, 8 (4), 50–61, 1990.

[Dehghan, 97] S. Dehghan and R. Steele, Small cell city, *IEEE Communications Magazine*, 35 (8), 52–59, 1997.

[Dersch, 94] U. Dersch and E. Zollinger, Propagation mechanisms in microcell and indoor environments, *IEEE Transactions on Vehicular Technology*, 43 (4), 1058–1066, 1994.

[Ecreg, 94] V. Erceg, A. J. Rustako and P. S. Roman, Diffraction around corners and its effects on the microcell coverage area in urban and suburban environments at 900 MHz, 2 GHz, and 6 GHz, *IEEE Transactions on Vehicular Techology*, 43 (3), 762–766, 1994.

[Feuerstein, 94] M. J. Feuerstein, K. L. Blackard, T. S. Rappaport and S. Y. Seidel, Path loss, delay spread and outage models as functions of antenna height for microcellular system design, *IEEE Transactions on Vehicular Technology*, 43 (3), 487–498, 1994.

[Goldsmith, 93] A. J. Goldsmith and L. J. Goldsmith, A measurement-based model for predicting coverage areas of urban microcells, *IEEE Journal on Selected Areas in Communication*, 11 (7), 1013–1023, 1993.

[Green, 90] E. Green, Radio link design for microcellular systems, *BT Technology Journal*, 8 (1), 85–96, 1990.

[Greenstein, 92] L. J. Greenstein, N. Armitay, T. S. Chu *et al.*, Microcells in personal communication systems, *IEEE Communications Magazine*, 30 (12), 76–88, 1992.

[Har, 99] D. Har, H. Xia and H. Bertoni, Path loss prediction model for microcells, *IEEE Transactions on Vehicular Technology*, 48 (5), 1453–1462, 1999.

[Harley, 89] P. Harley, Short distance attenuation measurements at 900 MHz and 1.8 GHz using low antenna heights for microcells, *IEEE Journal on Selected Areas in Communication*, 7 (1), 5–11, 1989.

[Ichitsubo, 00] S. Ichitsubo, T. Furuno, T. Taga and R. Kawasaki, Multipath propagation model for line-of-sight street microcells in urban area, *IEEE Transactions on Vehicual Technology*, 49 (2), 422–427, 2000.

[ITU, 1411] International Telecommunication Union, ITU-R Recommendation P.1411-3: Propagation data and prediction methods for the planning of short-range outdoor radiocommunication systems and radio local area networks in the frequency range 300 MHz to 100 GHz, Geneva, 2005.

[Lee, 98] W. C. Y. Lee and D. J. Y. Lee, Microcell prediction in dense urban area, *IEEE Transactions on Vehicular Technology*, 47 (1), 246–253, 1998.

[Maciel, 94] L. R. Maciel and H. L. Bertoni, Cell shape for microcellular systems in residential and commercial environments, *IEEE Transactions on Vehicular Technology*, 43 (2), 270–278, 1994.

[Madfors, 97] M. Madfors, K. Wallstedt *et al.*, High capacity with limited spectrum in cellular systems, *IEEE Communication Magazine*, 35 (8), 38–46, 1997.

[Sarnecki, 93] J. Sarnecki, C. Vinodrai, A. Javed, P. O'Kelly and K. Dick, Microcell design principles, *IEEE Communication Magazine*, 31 (4), 76–82, 1993.

[Wang, 97] L. C. Wang, G. L. Stubea and C. T. Lea, Architecture design, frequency planning and performance analysis for a microcell/macrocell overlaying system, *IEEE Transactions on Vehicular Technology*, 46 (4), 836–848, 1997.

[Xia, 93] H. H. Xia, H. L. Bertoni, L. R. Maciel, A. Lindsay-Stewart and R. Rowe, Radio propagation characteristics for line-of-sight microcellular and personal communications, *IEEE Transactions on Antenna and Propagation*, 41 (10), 1439–1447, 1993.

[Xia, 94] H. H. Xia, H. L. Bertoni, L. R. Maciel, A. Lindsay-Stewart and R. Rowe, Microcellular propagation characteristics for personal communications in urban and suburban environments, *IEEE Transactions on Vehicular Technology*, 43 (3), 743–752, 1994.

PROBLEMS

12.1 Calculate the carrier-to-interference ratio for two co-channel microcells separated by 250 m, where the path loss is described by the dual-slope model in Eq. (12.3), with $n_1 = 2, n_2 = 4$ and $r_b = 100$ m.

12.2 Assuming that Eq. (12.13) is valid, calculate the breakpoint distance for a line-of-sight microcell with base station height of 10 m and mobile height of 1.5 m at 2 GHz. Why does this formula not hold for non-line-of-sight cases?

12.3 Investigate how multiple knife-edge rooftop diffraction can affect propagation in microcells. How does it impact on channel reuse schemes?

12.4 A microcellular system us to be used to create cells in an urban area with a range (base-mobile) of 150 m at 2.5 GHz. Predict the necessary maximum acceptable path loss, including shadowing effects, using at least two of the models given in this chapter, stating any necessary assumptions.

12.5 Assuming a quaternary modulation scheme, estimate the maximum data rate available at the edge of the cells described in Problem 12.4 for 95% of locations.

13 Picocells

'Small things have a way of overmastering the great'.
Sonya Levien

13.1 INTRODUCTION

When a base station antenna is located inside a building, a *picocell* is formed (Figure 13.1). Picocells are increasingly used in cellular telephony for high-traffic areas such as railway stations, office buildings and airports. Additionally, the high data rates required by wireless local area networks (WLANs) restrict cell sizes to picocells and impose a further requirement to predict the wideband nature of the picocell environment. Picocell propagation is also relevant in determining propagation into buildings from both macrocellular and microcellular systems, which could either act as a source of interference to the indoor cells or as a means of providing a greater depth of coverage without capacity enhancement.

This chapter describes both empirical and physical models of picocellular propagation, together with some examples of systems and techniques used to improve coverage and capacity for picocell environments. Other enclosed spaces, such as railway and road tunnels and mines, also exhibit similar propagation effects to picocells and are therefore described here.

Figure 13.1: Picocells

13.2 EMPIRICAL MODELS OF PROPAGATION WITHIN BUILDINGS

13.2.1 Wall and Floor Factor Models

Two distinct approaches have been taken to empirical modelling for picocells. The first is to model propagation by a path loss law, just as in macrocellular systems, determining the parameters from measurements. This approach tends to lead to excessively large errors in the

Antennas and Propagation for Wireless Communication Systems Second Edition   Simon R. Saunders and Alejandro Aragón-Zavala
© 2007 John Wiley & Sons, Ltd

indoor case, however, because of the large variability in propagation mechanisms among different building types and among different paths within a single building. The same is true if dual-slope models, similar to those used in microcells, are applied.

A more successful approach [Keenan, 90] is to characterise indoor path loss by a fixed path loss exponent of 2, just as in free space, plus additional loss factors relating to the number of floors n_f and walls n_w intersected by the straight-line distance r between the terminals. Thus

$$L = L_1 + 20 \log r + n_f a_f + n_w a_w \qquad (13.1)$$

where a_f and a_w are the attenuation factors (in decibels) per floor and per wall, respectively. L_1 is the loss at $r = 1\,\text{m}$. No values for these factors were reported in [Keenan, 90]. An example prediction using this model is shown in Figure 13.2 for a series of offices leading off a corridor, with the base station inside one of the offices. Contours are marked with the path loss [−dB].

A similar approach is taken by an ITU-R model [ITU, 1238], except that only the floor loss is accounted for explicitly, and the loss between points on the same floor is included implicitly by changing the path loss exponent. The basic variation with frequency is assumed to be the same as in free space, producing the following total path loss model (in decibels):

$$L_T = 20 \log f_c + 10 n \log r + L_f(n_f) - 28 \qquad (13.2)$$

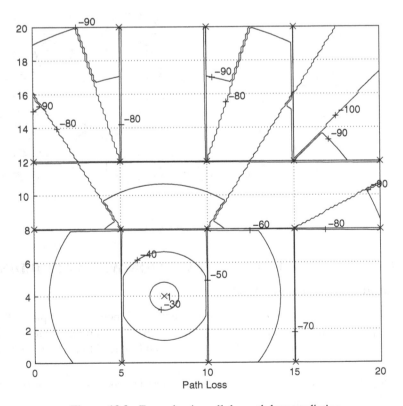

Figure 13.2: Example picocellular path loss prediction

Table 13.1: Path loss exponents n for the ITU-R model $(13.2)^a$

Frequency [GHz]	Environment		
	Residential	Office	Commercial
0.9	—	3.3	2.0
1.2–1.3	—	3.2	2.2
1.8–2.0	2.8	3.0	2.2
4.0	—	2.8	2.2
60.0	—	2.2	1.7

^{a}The 60 GHz figures apply only within a single room for distances less than around 100 m, since no wall transmission loss or gaseous absorption is included.

Table 13.2: Floor penetration factors, $L_f(n_f)$ [dB] for the ITU-R model $(13.2)^a$

Frequency [GHz]	Environment		
	Residential	Office	Commercial
0.9	—	9 (1 floor) 19 (2 floors) 24 (3 floors)	—
1.8–2.0	$4\,n_f$	$15 + 4(n_f - 1)$	$6 + 3(n_f - 1)$

^{a}Note that the penetration loss may be overestimated for large numbers of floors, for reasons described in Section 13.3.3. Values for other frequencies are not given.

where n is the path loss exponent (Table 13.1) and $L_f(n_f)$ is the floor penetration loss, which varies with the number of penetrated floors n_f (Table 13.2).

13.2.2 COST231 Multi-Wall Model

This model of propagation within buildings [COST231, 99] incorporates a linear component of loss, proportional to the number of walls penetrated, plus a more complex term which depends on the number of floors penetrated, producing a loss which increases more slowly as additional floors after the first are added,

$$L_T = L_F + L_c + \sum_{i=1}^{W} L_{wi}n_{wi} + L_f n_f^{((n_f+2)/(n_f+1))-b} \tag{13.3}$$

where L_F is the free space loss for the straight-line (direct) path between the transmitter and receiver, n_{wi} is the number of walls crossed by the direct path of type i, W is the number of wall types, L_{wi} is the penetration loss for a wall of type i, n_f is the number of floors crossed by the path, b and L_c are empirically derived constants and L_f is the loss per floor. Some recommended values are $L_w = 1.9$ dB (900 MHz), 3.4 dB (1800 MHz) for light walls, 6.9 dB (1800 MHz) for heavy walls, $L_f = 14.8$ dB (900 MHz), 18.3 dB (1800 MHz) and $b = 0.46$. The floor loss, i.e. the last term in Eq. (13.3), is shown in Figure 13.3. Notice that the additional loss per floor decreases with increasing number of floors. Section 13.3.3 examines potential reasons for this effect.

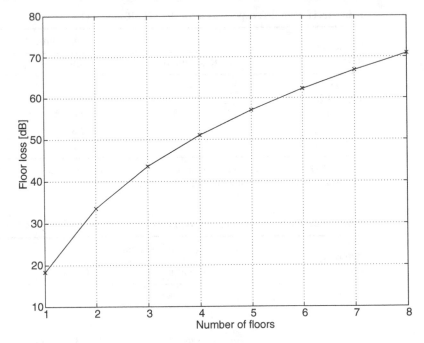

Figure 13.3: Floor loss for COST231 multi-wall model

13.2.3 Ericsson Model

In this model, intended for use around 900 MHz, the path loss including shadowing is considered to be a random variable, uniformly distributed between limits which vary with distance as indicated in Table 13.3 [Akerberg, 88]. The path loss exponent increases from 2 to 12 as the

Table 13.3: Ericsson indoor propagation model

Distance [m]	Lower limit of path loss [dB]	Upper limit of path loss [dB]
$1 < r < 10$	$30 + 20 \log r$	$30 + 40 \log r$
$10 \leq r < 20$	$20 + 30 \log r$	$40 + 30 \log r$
$20 \leq r < 40$	$-19 + 60 \log r$	$1 + 60 \log r$
$40 \leq r$	$-115 + 120 \log r$	$-95 + 120 \log r$

distance increases, indicating a very rapid decrease of signal strength with distance. A typical prediction from the model is shown in Figure 13.4. The model may be extended for use at 1800 MHz by the addition of 8.5 dB extra path loss at all distances.

13.2.4 Empirical Models for Wireless Lan

As wireless local area networks (WLAN) have increased in popularity, propagation modelling for these systems has received increasing attention. The main frequency bands of interest

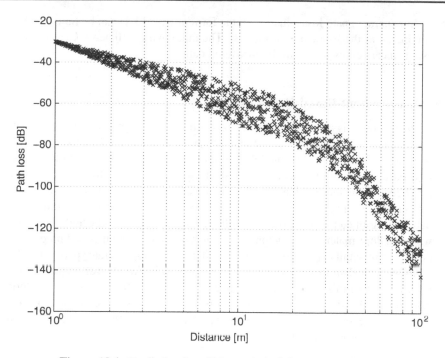

Figure 13.4: Prediction from Ericsson in-building path loss model

are the licence-exempt ('unlicensed') bands at around 2.4 and 5.2 GHz, typically supporting systems based on the Wi-Fi standards. The number of distinct channels available in these is limited and applications for these systems are increasingly critical, so coverage and interference modelling are both highly relevant. In one empirical model for use at WLAN frequencies [Tuan, 03], the path loss has the general form

$$L_T = k_1 + k_2 \log f + k_3 \log r + n_w(k_4 P_1 + k_5 P_2) + k_6 n_f \qquad (13.4)$$

In Eq. (13.4), P_1 and P_2 are associated with the angle of incidence θ to a wall. Various forms of P_1 and P_2 were proposed in [Tuan 03], and after validating the model with measurements, the path loss is given by

$$\begin{aligned} L_T = {} & 19.07 + 37.3 \log f + 18.3 \log r \\ & + n_w[21 \sin \theta + 12.2(1 - \sin \theta)] + 8.6 n_f \end{aligned} \qquad (13.5)$$

This model is given as valid for a frequency range between 900 MHz and 5.7 GHz. For the frequency range of interest, [Tuan, 03] reports a standard deviation of the error in the prediction of 6.7 dB. This model can be used in office environments, although the authors do not explicitly recommend other types of scenarios.

In general, a path loss equation of the form of Eq. (13.4) can be tuned with measurements conducted at the frequency of interest. The unknown coefficients k_1 to k_6 can be computed using linear regression from the measured data.

13.2.5 Measurement-Based Prediction

Empirical models are usually limited providing a rather general description of propagation, while higher accuracy for a specific site usually requires detailed physical models (see Section 13.3). Site-general models are not usually sufficient for an efficient system design in a particular building, whereas physical models are often too complex to implement in practice. As an intermediate approach, suitable for high-confidence designs, site-specific measurements may be used to determine the details of the propagation mechanisms and material parameters for a particular building without suffering the high cost of precise entry of wall and floor materials and geometries. Such a 'measurement-based prediction' approach is described in detail in [Aragón-Zavala, 06]. The use of appropriate spatial statistics enables the measurement data to be applied across the whole building, well beyond the measurement route. Reuse of this data, together with empirical models for the wall and floor-loss factors, also allows the system design to be optimised for antenna locations and types without re-measuring, even for different frequency bands than the original measurements. See Figure 13.5 for an example.

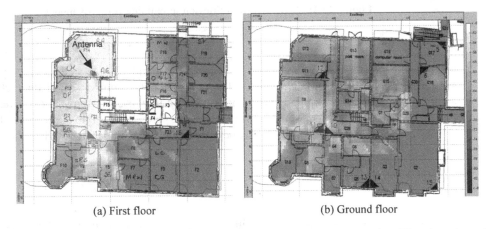

(a) First floor   (b) Ground floor

Figure 13.5: Measurement-based prediction model for picocell propagation. The plots show the predicted signal strength [dBm] received from an omni-directional antenna in the north-west area of the first floor. The predictions combine shadowing and correction information from the measurements with auto-tuned wall- and floor-factor modelling to achieve greater accuracy and flexibility than either source alone. (Reproduced by permission of Red-M Services Ltd.)

13.3 PHYSICAL MODELS OF INDOOR PROPAGATION WITHIN BUILDINGS

As with microcellular predictions, ray tracing and the geometrical theory of diffraction have been applied to deterministic prediction of indoor propagation, e.g. [Catedra, 98]. This can be used for site-specific predictions, provided that sufficient detail of the building geometry and materials is available. Building materials often need to be modelled in detail as multiple-layer structures with detailed internal construction to achieve high modelling accuracy. Fine detail of building geometry must also be obtained to account for the wave interactions with walls, floors and the edges of doors and windows. The number and complexity of these interactions is large and more advanced electromagnetic prediction techniques, such as the finite-difference time-domain (FDTD) approach, may also be useful in some cases since they avoid the need for an explicit ray tracing step. As with the physical models of microcellular propagation introduced in Chapter 12, however, these complexities mean that fully-detailed physical models are rarely used for practical system planning. These problems are particularly significant for picocells, where the influence of furniture and the movement of people can have a significant (and time-varying) effect on coverage. Yet some basic physical models can be used to yield insight into the fundamental processes and dependencies affecting building propagation as a precursor to simplified prediction and system design. They also yield detailed information concerning wave arrival angles, delays and the statistics of multipath propagation which would not otherwise be available.

13.3.1 Ray-Tracing Models for Picocells

The ray-tracing principles introduced in Chapter 3 may be applied to in-building situations, using geometrical optics and the geometrical theory of diffraction, if sufficient data and computing time is available. For example, a deterministic spatiotemporal propagation model has been proposed in [Lee, 01] based on ray launching techniques. Ray launching sends out test rays at a number of discrete angles from the transmitter. The rays interact with objects present in the environment as they propagate. The propagation of a ray is therefore terminated when its power falls below a predefined threshold. The model considers reflection, transmission and diffraction effects via UTD principles. Transmission and diffraction are considered in this model.

The complex electric field, E_i, associated with the ith ray path is determined by

$$E_i = E_0 f_{ti} f_{ri} L_{FSL}(r) \left[\prod_j \overline{R}_j \prod_k \overline{T}_k \prod_l \overline{D}_l A_l(S_l, S_l') \right] e^{-jkr} \qquad (13.6)$$

where E_0 represents the reference field, f_{ti} and f_{ri} are the transmitting and receiving antenna field radiation patterns, respectively, L_{FSL} is the free-space loss, R_j is the reflection coefficient for the jth reflection, T_k is the transmission coefficient for the kth transmission, D_l and A_l are the diffraction coefficient and the spreading attenuation for the lth diffraction, respectively and e^{-jkr} is the propagation phase factor, where r is the unfolded ray path length and k is the wave number. This is comparable to the formulation of Eq. (3.22) with the addition of diffraction effects and generalised to multiple interactions.

This model has been used with full three-dimensional data for both power delay and power azimuth profiles, departure and arrival angles, and coverage predictions as detailed in [Lee, 01].

13.3.2 Reduced-Complexity UTD Indoor Model

Given the complexity of models such as that given in the previous section, it is attractive to seek approaches which retain the physical principles while reducing the associated complexity. The model proposed in [Cheung, 98], validated for use around 900 MHz, incorporates much of the propagation phenomena suggested by electromagnetic theory, such as UTD, but still retains the straightforwardness of the empirical approach. Computation time is not as long as in pure ray tracing models, which will be discussed in more detail in Section 13.4. Likewise, the empirical factors required in the model can be closely related to theoretical derivations, so that model tuning or curve fitting to measured data may not be required.

Three propagation mechanisms have been incorporated in this model. The first factor is a dual-slope model for the main path, similar to those studied for microcells in Chapter 12. Around 900 MHz, a breakpoint distance of about 10 m is suggested, denoted as r_b. The second factor included is an angular dependence of attenuation factors. Less power is transmitted through walls when incidence is oblique as compared with normal incidence. Therefore, the wall attenuation factor L_{wi} (and likewise the floor attenuation factor L_{fi}) is made to depend on the angle of incidence. A simplified diffraction calculation is also included.

The resulting model is given by

$$L_\angle(r) = 10\log\left(\frac{r}{r_0}\right)^{n_1} u(r_b - r)$$
$$+ 10\left[\log\left(\frac{r_b}{r_0}\right)^{n_1} + \log\left(\frac{r}{r_b}\right)^{n_2}\right] u(r - r_b) \qquad (13.7)$$
$$+ \sum_{i=1}^{W}\frac{n_{wi}L_{wi}}{\cos\theta_i} + \sum_{j=1}^{F}\frac{n_{fi}L_{fi}}{\cos\theta_j}$$

In Eq. (13.12), θ_i and θ_j represent the angles between the ith wall and jth floor, respectively, and the straight line path joining the transmitter with the receiver. $u(\cdot)$ is the unit step function defined as

$$u(t) = \begin{cases} 0, & t < 0 \\ 1, & t \geq 0 \end{cases} \qquad (13.8)$$

To keep the diffraction model simple, the authors in [Cheung, 98] utilise only one level of diffraction from corners, including door and window frames in the building. To perform this, the field is calculated at each corner using Eq. (13.7) and the resulting diffracted field is determined using a diffraction coefficient. Thus, the total field at the receiver is computed as the summation of the field from the transmitter and all the corners.

Diffraction coefficients for perfect electrical conductors under UTD [Kouyoumjian, 74] are used in the model and are denoted as $D(r, \phi, r'r', \phi')$, where (r, ϕ) are the coordinates of the corner relative to the transmitter and (r', ϕ') are the coordinates of the receiver relative to the corner. Hence

$$L_T = -10\log\left[\sum_{m=1}^{M} l_\angle(r_m)l_\angle(r'_m) \times |D(r_m, \phi_m, r'_m, \phi'_m)|^2\right]$$
$$- 10\log[l_\angle(r)] \qquad (13.9)$$

where M is the number of corners in the building database, m refers to the mth corner and $l_\angle(\cdot)$ is a dimensionless quantity given by

$$l_\angle(\cdot) = 10^{-L_\angle(\cdot)/10} \qquad (13.10)$$

and L_ℓ is calculated from Eq. (13.12). In [Cheung, 98], at 900 MHz, the parameters $n_2 = 2.5$, $r_b = 10$ m, $L_w = 10$ dB (concrete block walls), $L_w = 5$ dB (hollowed plaster board walls) are obtained from measurements. n_1 is taken as 1. Good prediction accuracy is claimed.

13.3.3 Propagation Between Floors

Examining some particular cases from physical models enables us to gain some insight into the observed propagation phenomena for picocells. Figure 13.6 shows four distinct paths between a transmitter and receiver situated on different floors of the same building. Path 0 is the direct path, which encounters attenuation due to the building floors. Models such as those

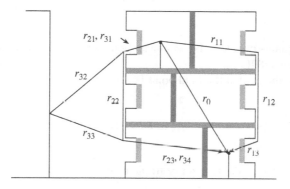

Figure 13.6: Alternative paths for propagation between floors

described in Section 13.2 implicitly assume this path is the dominant source of signal power, although the wall and floor loss factors applied may be modified to account for the average effect of other paths. Paths 1 and 2 encounter diffraction in propagating out of, and back in through, the windows of the building, but are unobstructed in propagating between the floors. Finally, path 3 is also diffracted through the windows of the building, although this is through a smaller angle than path 2. It is reflected from the wall of a nearby building before diffracting back into the original building [Honcharenko, 93].

In order to analyse the field strength due to paths 2 and 3, the geometry is approximated by the double-wedge geometry in Figure 13.7 , representing the building edges at the points where the rays enter and leave the building. The propagation is then analysed using the geometrical theory

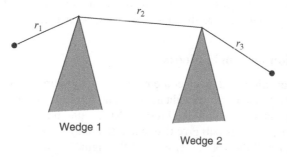

Figure 13.7: Double-wedge geometry

of diffraction, following the principles introduced in Chapter 3. The source is a point source and therefore radiates spherical waves. The field incident on wedge 1 is therefore

$$E_1 = \sqrt{Z_0 \frac{P_T}{4\pi r_1^2}} = \frac{1}{2r_1} \sqrt{\frac{Z_0 P_T}{\pi}} \qquad (13.11)$$

where P_T is the effective isotropic radiated power from the source. The diffraction process at wedge 1 then yields a field incident on wedge 2 which is approximated using GTD as

$$E_2 = E_1 \times D_1 \times \sqrt{\frac{r_1}{r_2(r_1 + r_2 + r_3)}} \qquad (13.12)$$

where the square-root factor is the spreading factor for spherical wave incidence on a straight wedge [Balanis, 89].

Similarly, the field at the field point is

$$E_3 = E_2 \times D_2 \times \sqrt{\frac{r_1 + r_2}{r_3(r_1 + r_2 + r_3)}} \qquad (13.13)$$

Hence the power available at an isotropic receive antenna is

$$P_r = P_T \frac{\lambda^2}{4\pi} \times \frac{E_3^2}{Z_0} = P_T \left(\frac{\lambda}{4\pi}\right)^2 \frac{D_1^2 D_2^2}{r_1 r_2 r_3 (r_1 + r_2 + r_3)} \qquad (13.14)$$

This result can be applied to both paths 1 and 2 by substitution of the appropriate distances. Path 3 also follows in the same way, but is multiplied by the reflection coefficient of the nearby building.

The sum of the power from the various contributions is shown in Figure 13.8. It is clear that two regimes are present; for small spacing between the transmitter and receiver, the signal drops rapidly as the multiple floor losses on path 0 accumulate. Eventually the diffracted paths (1 and 2) outside the building dominate, and these diminish far less quickly with distance. When a reflecting adjacent building is present, the diffraction losses associated with this path are less and this provides a significant increase in the field strength for large separations.

This analysis provides a physical justification for the reduction of path loss per floor exhibited by models such as that described in Section 13.2.2.

13.3.4 Propagation on Single Floors

When the transmitter and receiver are mounted on the same floor of a building, the dominant mode of propagation is line of sight as shown in Figure 13.9. However, the floor- and ceiling-mounted objects will result in the Fresnel zone around the direct ray becoming obstructed at large distances, and this will give rise to additional loss due to diffraction. The effective path loss exponent will then be increased and the signal strength will fall off very rapidly with distance, exhibiting path loss exponents which may be as steep as 8–10. The point at which

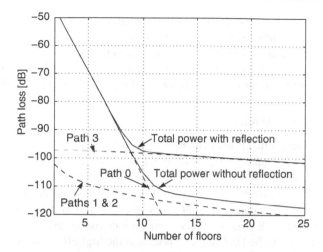

Figure 13.8: Variation of path loss with number of floors: here floor height is 4 m, building width is 30 m, distance to adjacent building is 30 m and the frequency is 900 MHz

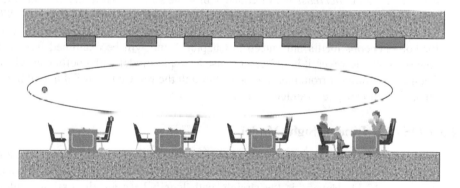

Figure 13.9: Propagation between antennas on a single floor

this occurs depends on the specific geometry, with the maximum unobstructed range being obtained when the antennas are mounted at the midpoint of the gap between the highest floor-mounted obstructions and the lowest point of the ceiling-mounted obstructions [Honcharenko, 92].

13.4 MODELS OF PROPAGATION INTO BUILDINGS

13.4.1 Introduction

There are two major motivations for examining signal penetration into buildings. First, since most cellular users spend most of their time inside buildings, the level of service which they perceive will depend heavily upon the signal strengths provided inside the buildings (the *depth* of coverage). When sufficient capacity exists within the macrocells and the microcells of the network, this indoor coverage is then provided by the degree of penetration into the buildings.

When, by contrast, it is necessary to serve very high densities of users within a building (e.g. in heavily populated office buildings, railway stations and airports), the indoor coverage must then be provided by dedicated picocells. It is inefficient to allocate distinct frequencies to them, so it is necessary to reuse frequencies already allocated to macrocells and microcells, based on clear knowledge of the extent to which the two cell types will interfere within the building.

13.4.2 Measured Behaviour

Surveys of measurements in the literature produce a rather confusing outcome, highlighted by examining the frequency dependence reported for penetration loss. If specific building materials are measured in isolation, the general trend is for the attenuation to increase with frequency. See [Stone, 97] for a particularly thorough set of measurements, which examines attenuation through many different types of construction materials in the frequency range 0.5–2.1 GHz and 3–8 GHz and finds a general increase in loss. This trend is to be expected, as the skin depth described in Chapter 2 is least at the highest frequencies, so the current density is greatest at highest frequencies and the losses arising from conversion from electromagnetic to thermal energy are greater.

When the loss is examined in practical buildings, however, the building penetration loss has been found to *decrease* with frequency in some studies, but to increase in others. This is clear from the compilation of several studies shown in Figure 13.10 [Davidson, 97]. Clearly the mechanisms involved are more complex than the simple Fresnel transmission coefficients for homogeneous media introduced in Chapter 2 and will be examined further in Section 13.4.6. Often, however, it is sufficient to use a simple model with coefficients chosen for the frequency of interest from measurements through the relevant material in real buildings, and three such models are examined below.

13.4.3 COST231 Line-of-Sight Model

In cases where a line-of-sight path exists between a building face and the external antenna, the following semi-empirical model has been suggested [COST231, 99], with geometry defined by Figure 13.11. Here r_e is the straight path length between the external antenna and a reference point on the building wall; since the model will often be applied at short ranges, it is important to account for the true path length in three dimensions, rather than the path length along the ground. The loss predicted by the model varies significantly as the angle of incidence, $\theta = \cos^{-1}(r_p/r_e)$, is varied.

The total path loss is then predicted using

$$L_T = L_F + L_e + L_g(1 - \cos\,\theta)^2 + \max(L_1, L_2) \tag{13.15}$$

where L_F is the free space loss for the total path length $(r_i + r_e)$, L_e is the path loss through the external wall at normal incidence $(\theta = 0°)$, L_g is the additional external wall loss incurred at grazing incidence $(\theta = 90°)$ and

$$L_1 = n_w L_i \quad L_2 = \alpha(r_i - 2)(1 - \cos\,\theta)^2 \tag{13.16}$$

where n_w is the number of walls crossed by the internal path r_i, L_i is the loss per internal wall and α is a specific attenuation $[dBm^{-1}]$ which applies for unobstructed internal paths. All distances are in metres.

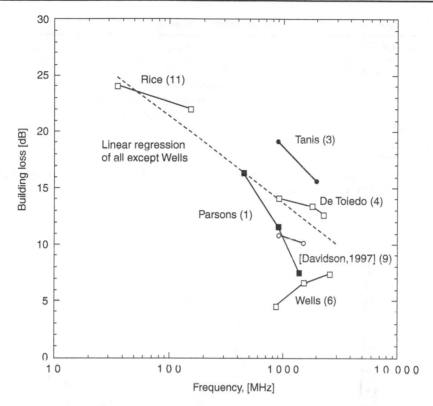

Figure 13.10: Measured building penetration loss versus frequency, with number of buildings measured in brackets based on a compilation of several studies (from [Davidson, 97]) (Reproduced by permission of IEEE, © 1997 IEEE)

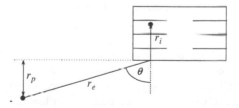

Figure 13.11: Geometry for COST231 line-of-sight building penetration model

The model is valid at distances up to 500 m and the parameter values in Table 13.4 are recommended for use in the 900–1800 MHz frequency range. They are in good agreement with measurements from real buildings and implicitly include the effects of typical furniture arrangements.

13.4.4 Floor Gain Models

In most macrocell cases, no line-of-sight path exists between the base station and the face of the building. Empirical models of this situation are then most usually based on comparing the path loss encountered in the street outside the building (L_{out} in Figure 13.12) to the path loss

Table 13.4: Parameters for COST231 line-of-sight model

Parameter	Material	Approximate value
L_e or L_i [dB m^{-1}]	Wooden walls	4
	Concrete with non-metallised windows	7
	Concrete without windows	10–20
L_g [dB]	Unspecified	20
α [dB m^{-1}]	Unspecified	0.6

$L_f(n)$ within the building at various floor levels (where n is the floor number defined in Figure 13.12). It is then possible to define a penetration loss as

$$L_p = L_f(n) - L_{out} \tag{13.17}$$

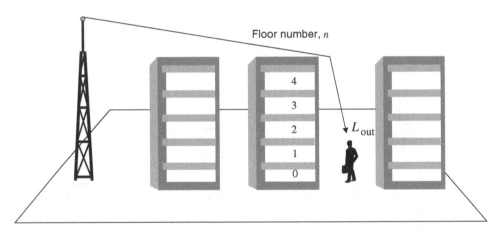

Figure 13.12: Geometry for building penetration in non-line-of-sight conditions (reproduced by permission of Jaybeam Limited)

The penetration loss has been found to *decrease* with frequency in [Turkmani, 92]; typical values for the ground floor penetration loss $L_f(0)$ are 14.2, 13.4 and 12.8 dB measured at 900, 1800 and 2300 MHz, respectively. It does not necessarily indicate that the actual wall attenuations follow this trend, since the penetration loss defined this way makes no attempt to isolate effects due to individual waves. The loss decreases with height from the ground floor upwards at a rate of around 2 dB per floor and then starts to increase again with height beyond about the ninth [Turkmani, 87] or fifteenth [Walker, 92] floor. The precise variation is likely to be very dependent on the specific geometry of the surrounding buildings.

13.4.5 COST231 Non-line-of-Sight Model

This model [COST231, 99] relates the loss inside a building from an external transmitter to the loss measured outside the building, on the side nearest to the wall of interest, at 2 m above ground level. The loss is given by

$$L_T = L_{out} + L_e + L_{ge} + \max(L_1, L_3) - G_{fh} \tag{13.18}$$

where $L_3 = \alpha r_i$ and r_i, L_e, α and L_1 are as defined in the COST231 line-of-sight model (Section 13.4.3), and the floor height gain G_{fh} is given by

$$G_{fh} = \begin{cases} nG_n \\ hG_h \end{cases} \qquad (13.19)$$

where h is the floor height above the outdoor reference height [m] and n is the floor number as defined in Figure 13.12. Shadowing is predicted to be log-normal with location variability of 4–6 dB. Other values are as shown in Table 13.5.

Table 13.5: Parameters for COST231 non-line-of-sight model

Parameter	Approximate value
L_{ge} [dB] at 900 MHz	4
L_{ge} [dB] at 1800 MHz	6
G_n [dB per floor] at 900/1800 MHz	1.5–2.0 for normal buildings
	4–7 for floor heights above 4 m

Both the line-of-sight and non-line-of-sight models of COST-231 rely on the dominant contribution penetrating through a single external wall. A more accurate estimation may be obtained by summing the power from components through all of the walls.

13.4.6 Propagation Mechanisms

As indicated in Section 13.4.2, the mechanisms involved in building penetration are rather complicated. At first, it might be imagined that the exterior wall of a building could be modelled as a simple slab of lossy dielectric material, with the penetration loss predicted simply via the Fresnel transmission coefficient from Chapter 3, given appropriate constitutive parameters to represent the material. This would suggest that the penetration loss would be a rather slowly-changing, smooth, increasing function of frequency. Given that practical measurements suggest a far more complex behaviour, particularly at VHF and above, several mechanisms must be present.

First, the constitutive parameters of materials are themselves frequency dependent, even for relatively uniform walls, due to the specific molecular structure of the materials used. This can be compared with the atmospheric absorption effects described in Section 7.2.3, where molecular resonance effects cause loss maxima at particular frequencies. Second, the wall structure frequently has several layers, setting up multipath interference and associated resonances within the structure. These can be analysed by treating each layer as a section of transmission line, with a characteristic impedance determined by the wave impedance, the frequency and the angle of incidence. At each boundary reflections are created leading to a set of multiple reflections with interference between each contribution.

For example, Figure 13.13 shows how a double-glazed window can be approximated by a five-section transmission line, with the characteristic impedance of each section determined from the wave impedance in the corresponding medium and the angle of incidence [Stavrou, 03a]. The result of such analysis is a loss which varies widely with

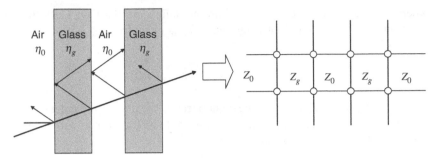

Figure 13.13: Transmission line analogy for double-glazed window

frequency and angle as shown in Figure 13.14. Comparisons of loss with frequency over a limited range can therefore produce a loss which either rises or falls depending on the specific range examined.

Additionally, the presence of windows within the walls creates an aperture, whose loss depends on the precise relation between the Fresnel zone size of the penetrating ray and the window size as well as the glass material. At higher frequencies, the Fresnel zone reduces in size and is more likely to fit through the window unobstructed, so overall the excess loss reduces with frequency. However, the diffraction contributions from each window edge sum together to produce variation in the loss at particular locations which may reverse the trend as illustrated in Figure 13.15.

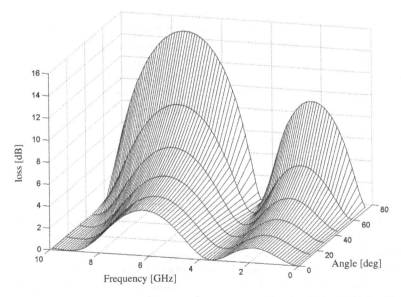

Figure 13.14: Transmission loss for double-glazed window. The glass is 8 mm thick with a 4 mm air gap. The glass has a relative permittivity of 4 and a loss tangent of 0.0012. From [Stavrou, 03a] (reproduced by permission of IET)

Table 13.6: Complex permittivity of typical construction materials

	1 GHz	57.5 GHz	78.5 GHz	95.9 GHz
Concrete	$7.0 - j0.85$	$6.50 - j0.43$	—	$6.20 - j0.34$
Lightweight concrete	$2.0 - j0.50$	—	—	—
Floorboard (synthetic resin)	—	$3.91 - j0.33$	$3.64 - j0.37$	$3.16 - j0.39$
Plasterboard	—	$2.25 - j0.03$	$2.37 - j0.10$	$2.25 - j0.06$
Ceiling board (rock wool)	$1.2 - j0.01$	$1.59 - j0.01$	$1.56 - j0.02$	$1.56 - j0.04$
Glass	$7.0 - j0.10$	$6.81 - j0.17$	—	—
Fibreglass	$1.2 - j0.10$	—	—	—

13.5 CONSTITUTIVE PARAMETERS OF BUILDING MATERIALS FOR PHYSICAL MODELS

All physical models require both the geometry and constitutive parameters of the buildings as input. Representative values of the complex permittivity at various frequencies are given in Table 13.6 [ITU, 1238].

The constitutive parameters of building materials have been summarised over a wide frequency range in [Stavrou, 03b]. Table 13.7 shows a summary of the findings. See also [Stone, 97].

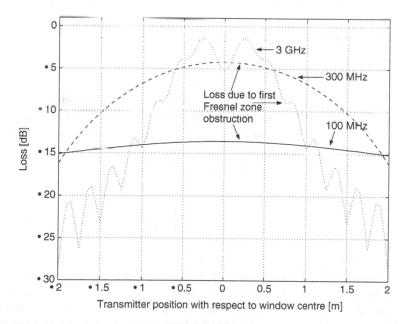

Figure 13.15: Transmission loss for a 0.8 m square single-glazed window set in a brick wall. The glass is 8 mm thick and the brick is 30 cm thick. The transmitter and receiver are 30 m and 2 m from the centre of the window respectively. The glass has a relative permittivity of 4 and a loss tangent of 0.0012. From [Stavrou, 03a] (reproduced by permission of IET)

Table 13.7: Review of constitutive parameters of building materials [Stavrou, 03b]

Material	Frequency range	Permittivity	Conductivity
Brick	1.7–18 GHz	4.62–4.11	0.0174–0.0364
Concrete	3–24 GHz	5–7	0.0138–0.025
Silica glass	VHF – microwave	4	0.00005–0.035
Commercial glass	VHF – microwave	4–9	0.00005–0.035
Wood	20 Hz–100 GHz	1.2–6.8	0.005–0.063

Table 13.8: Measured values for location variability in indoor environments

Frequency [GHz]	Location variability [dB]			
	Residential	Office	Commercial	Laboratory
0.8–1.0	—	3 [Keenan, 90]	—	6 [Keenan, 90]
1.7–2.0	8 [ITU, 1238]	10 [ITU, 1238] 4 [Keenan, 90]	10 [ITU, 1238]	—

13.6 SHADOWING

It is usual to model shadowing in indoor environments as log-normal, just as in other cell types (Chapter 9). However, there is some evidence that the location variability is itself more environment-dependent. Reported values are shown in Table 13.8.

13.7 MULTIPATH EFFECTS

In Chapter 10, it was assumed that waves arrived with uniform probability from all horizontal angles, leading to the classical Doppler spectrum. By contrast, a more reasonable assumption for the indoor environment, particularly when propagation occurs between floors, is that waves arrive with uniform probability from *all* angles. The resulting Doppler spectrum is then relatively uniform, so for simulation purposes, it is reasonable to assume a flat Doppler spectrum given by

$$S(f) = \begin{cases} 1/2f_m & |f| \leq f_m \\ 0 & f > f_m \end{cases} \qquad (13.20)$$

where f_m is the maximum Doppler frequency.

As regards the RMS delay spread of the channel, values encountered in most cases are very much lower than those found in either micro- or macrocells. But the variability around the median value is large, although there is a strong correlation with the path loss [Hashemi, 93a] and there are occasionally cases where the delay spread is very much larger than the median. In order to provide reasonably realistic simulations both situations must be considered. Tables 13.9 and 13.10 give suitable channels for an indoor office scenario and for an outdoor-to-indoor scenario, respectively, intended for evaluation purposes at around 2 GHz [ETSI, 97]. Values

Table 13.9: Indoor office test environment wideband channel parameters[a]

Median channel $\tau_{RMS} = 35$ ns		Bad channel $\tau_{RMS} = 100$ ns	
Relative delay τ [ns]	Relative mean power [dB]	Relative delay τ [ns]	Relative mean power [dB]
0	0	0	0
50	−3.0	100	−3.6
110	10.0	200	−7.2
170	−18.0	300	−10.8
290	−26.0	500	−18.0
310	−32.0	700	−25.2

[a]Doppler spectrum for all taps is flat.

Table 13.10: Outdoor-to-indoor test environment: wideband channel parameters[a]

Median channel $\tau_{RMS} = 45$ ns		Bad channel $\tau_{RMS} = 750$ ns	
Relative delay τ [ns]	Relative mean power [dB]	Relative delay τ [ns]	Relative mean power [dB]
0	0	0	0
110	−9.7	200	−0.9
190	−19.2	800	−4.9
410	−22.8	1200	−8.0
—	—	2300	−7.8
—	—	3700	−23.9

[a]Doppler spectrum for all taps is classical.

for the RMS delay spread for indoor-to-indoor environments are also shown in Table 13.11 [ITU, 1238]; case A represents low but frequently occurring values, case B represents median values and case C gives extreme values which occur only rarely. The very high cases can occur particularly if there are strong reflections from buildings situated a long way from the building under test.

More details of the statistics and structure of the indoor wideband channel are available in references such as [Hashemi, 93b], [Hashemi, 94], [Rappaport, 91] and [Saleh, 87a]. In particular, [Saleh, 87a] proposes that the power delay profile tends to follow a doubly exponential

Table 13.11: RMS delay spread in nanoseconds in indoor-to-indoor environments

Environment	Case A	Case B	Case C
Indoor residential	20	70	150
Indoor office	35	100	460
Indoor commercial	55	150	500

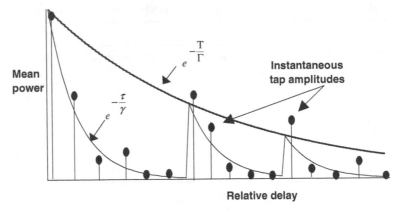

Figure 13.16: Doubly exponential power delay profile in indoor channels after [Saleh, 87a]

distribution (Figure 13.16), where the peaks of the individual exponentials can be reasonably accurately predicted from ray-tracing models, but where the associated weaker signals result from rough scattering and fine detail which cannot easily be predicted from a deterministic physical model. It has also been observed that the number of multipath components follows a Gaussian distribution, with a mean value which increases with antenna separation [Hashemi, 93a].

13.8 ULTRA-WIDEBAND INDOOR PROPAGATION

In ultra-wideband (UWB) systems, the transmission bandwidth is very much greater than the required information bandwidth, with practical systems occupying signal bandwidths as wide as 7.5 GHz. This reduces the power spectral density to very low levels, so that the interference to other systems can be made negligible. Such systems are therefore one means of providing very high data rates over short distances for applications such as video distribution. Although, in principle, UWB systems experience all the same propagation mechanisms as other in-building systems, they differ in detail because their bandwidth may be a large fraction of the central frequency, resulting in significant differences in propagation effects across the operating bandwidth. Additionally, they always experience a dispersive channel, so wideband and narrowband propagation effects cannot usually be separated in the conventional fashion.

For example, some studies have shown a significant difference in the path loss exponent experienced for the total power in a UWB system to that exhibited by the strongest pulse (which usually has the shortest delay). In [Siwiak, 03], for example, measurements were made in a large single-story building. While the total power was found to have a path loss exponent of 2, as would be expected from a broadly free-space situation for short distances, the strongest path was found to diminish in power with range cubed, while the RMS delay spread increased linearly with range at the rate of 3 ns per metre.

A simple theory to account for this observation follows [Siwiak, 03]. Model the total power, P_r, received with a transmit-receive separation of r, as having a free-space loss component and a dissipative attenuation coefficient γ [dB/m] as follows:

$$\frac{P_r(r)}{P_t} = \frac{1}{4\pi r^2} e^{\frac{-r\gamma}{10}} \qquad (13.21)$$

If the channel dispersion is exponential with delay, as suggested in section 13.7, with an RMS delay spread of τ_{RMS}, then the power delay profile is:

$$P(\tau) = \frac{1}{\tau_{RMS}} e^{-\frac{\tau}{\tau_{RMS}}} \tag{13.22}$$

The channel impulse response is in practice not continuous, but is made up of a number of discrete rays. Assuming these arrive at the receiver at an average rate $1/t_0$ rays per second, then the fractional power in the first arriving ray, P_1, is the integral of the power delay profile over the first t_0 seconds:

$$P_1 = \int_0^{t_0} P(\tau) d\tau = \int_0^{t_0} \frac{1}{\tau_{RMS}} e^{\frac{\tau}{\tau_{RMS}}} d\tau = 1 - e^{-\frac{t_0}{\tau_{RMS}}} \tag{13.23}$$

The power in this delay range is then a fraction of the total power in the path loss model of Eq. (13.22) as follows:

$$\frac{P_1(r)}{P_t} = \frac{1}{4\pi r^2} e^{-\frac{r}{10}} \left(1 - e^{-\frac{t_0}{\tau_{RMS}}} \right) \tag{13.24}$$

If the UWB channel bandwidth is large compared with the delay spread, which would be the usual case, then the exponential term can be replaced with the approximation $e^x \approx 1 + x$, so:

If we further assume, following the measured observations, that the delay spread has a linear dependence on distance with a slope of τ_0 so that $\tau_{RMS} = \tau_0 r$, then

$$\frac{P_1(r)}{P_t} = \frac{1}{4\pi r^2} e^{-\frac{r}{10}} \frac{t_0}{\tau_{RMS}}, \text{ for } \frac{t_0}{\tau_{RMS}} \ll 1 \tag{13.25}$$

$$\frac{P_1(r)}{P_t} = \frac{1}{4\pi r^2} e^{-\frac{r}{10}} \frac{t_0}{\tau_0 r} = \frac{1}{4\pi r^3} e^{-\frac{r}{10}} \frac{t_0}{\tau_0}; \text{ for } \frac{t_0}{\tau_{RMS}} \ll 1 \tag{13.26}$$

This is consistent with the path loss exponent observed in the measurements.

It is clearly essential to model channel dispersion in an ultra-wideband channel. The group standardising the IEEE 802.15.3a UWB system have proposed such a model for comparing the performance of various schemes [Molisch, 03]. The model is based on a modified version of the [Saleh, 87a] model described in Section 13.7. Differences arise because of the very high bandwidths, leading to a larger number of discrete multipaths than would be apparent at narrower bandwidths. For example, given a 7.5 GHz signal bandwidth, paths arriving just 133 ps separated in time are resolvable, corresponding to a path length difference as small as 4 cm. The arrival rates of rays need to be set appropriately. Also, the individual multipaths do not tend to follow Rayleigh fading distributions since the usual assumptions of a large number of component waves no longer hold, so the model instead uses lognormal statistics for both the short term variations of the individual waves as well as for the overall bulk shadowing component. The model sets parameters for a variety of short-range (< 10 m) situations, with RMS delay spreads off from 5–14 ns, ray arrival rates from 0.5 to 2.5 ns^{-1}, cluster arrival rates from 0.02 to 0.4 ns^{-1}, and a requirement to generate more than 60 individual paths to fully represent the channel energy. See [Molisch, 03] for full details.

13.9 PROPAGATION IN TUNNELS AND OTHER ENCLOSED SPACES

A special indoor environment, which has important practical significance for some systems, is in tunnels. Tunnels are relevant for systems serving both roads and railways, and the mechanisms are also very similar for other enclosed spaces such as underground mines.

13.9.1 Measured Behaviour

Some example measurements are shown in Figure 13.17 for a system operating at 900 MHz with an external mast radiating into a train tunnel. The measurements are taken from within the train, so they include the loss involved in penetrating into the train itself.

The main features of such measurements are summarised diagrammatically in Figure 13.18 (see [Zhang, 00] for further examples). Propagation along the track outside the tunnel is strongly influenced by the shape of the surrounding terrain which may often be a steep cutting or an elevated embankment, by the presence of clutter around the train track such as trees and overhead gantries and by the existence of strong reflections which may produce a situation similar to a plane earth loss model.

Notice how the field drops rapidly in the first few metres following entrance of the train into the tunnel. This drop is dependent on the distance of the transmitting antenna from the tunnel mouth, the angle of arrival of waves into the tunnel relative to its axis, its cross-sectional area and the frequency of operation. Propagation within the tunnel is further influenced by the tunnel shape, the extent to which it is filled by the train and by the tunnel construction materials.

13.9.2 Models of Tunnel Propagation

In order to account for the measured effects, it is common to account for in-tunnel propagation via waveguide theory, in which the tunnel acts as a waveguide supporting multiple lossy

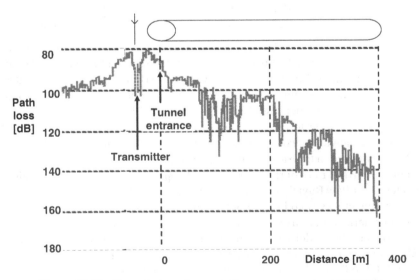

Figure 13.17: Example tunnel propagation measurements at 900 MHz

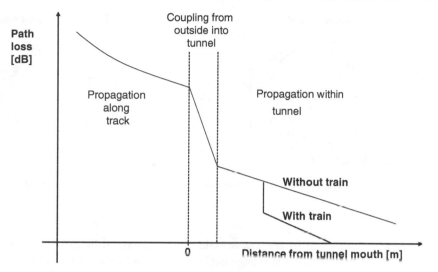

Figure 13.18: Main features of tunnel propagation

modes as illustrated in Figure 13.19. *Modes* are the waves representing solutions of Maxwell's equations subject to the boundary conditions created by the interfaces between the interior of the tunnels and the wall materials. *Higher order* modes are those which propagate at large angles to the tunnel axis and therefore experience more reflections per unit distance. See [Balanis, 89] for more details. At VHF/UHF frequencies and for typical materials, the dominant loss is refraction through the tunnel walls, rather than ohmic losses. All of the modes produce exponential losses, where the loss increases linearly with distance. The tunnel thus acts as a high-pass filter, with the loss decreasing with frequency above a cut off

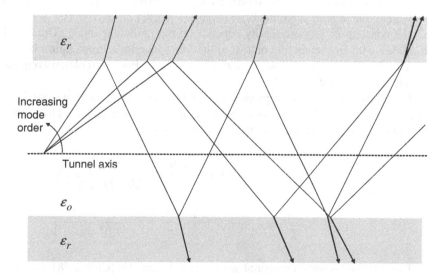

Figure 13.19: Tunnel propagation via multiple modes

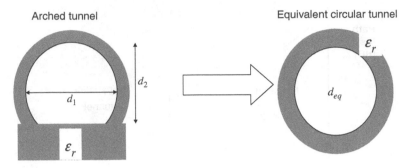

Figure 13.20: Equivalent circular geometry for in-tunnel propagation

frequency determined by the lowest-order mode supported by the tunnel. The high order modes (those propagating at a large angle to the tunnel axis) suffer highest attenuation per distance from the multiple reflections, so the low order modes become the dominant contributions to the field observed at large distances.

Real tunnels are frequently arched, with a cross-section similar to a truncated circle (see Figure 13.20). In order to analyse them as waveguides, an approximation needs to be made.

One simple approach is to assume that the tunnel can be approximated by an equivalent circular tunnel made from the same material, with a diameter d_{eq} chosen to produce the same overall area, as shown in Figure 13.20. In this case, it can be shown [Chiba, 78] that each mode of propagation produces an exponential loss of the following general form:

$$L_{\mathrm{mode}} \propto r \frac{1}{f^2 d_{eq}^3} \tag{13.27}$$

where r is distance along the tunnel axis, f is carrier frequency and d_{eq} is the circular-equivalent diameter of the tunnel.

The constant of proportionality depends on the particular mode. Thus the loss actually decreases with frequency, in contrast to the free space loss. Propagation is essentially non-existent for frequencies below the cut-off frequency. For a circular waveguide the cut-off wavelength is $1.71 \, d_{eq}$.

The dominant mode of propagation depends on the dielectric constant: for $\varepsilon_r < 4.08$ the mode producing the lowest loss is the EH_{11} mode; for denser dielectrics, the $\mathrm{TE}01$ mode dominates. For these two modes the associated loss is

$$
\begin{aligned}
L_{EH_{11}} &= 8.686r \left(\frac{2.405}{2\pi} \right)^2 \frac{\lambda^2}{(d_{eq}/2)^3} \frac{(1 + \varepsilon_r)}{\sqrt{\varepsilon_r - 1}} \\[2ex]
L_{TE_{01}} &= 8.686r \left(\frac{3.832}{2\pi} \right)^2 \frac{\lambda^2}{(d_{eq}/2)^3} \frac{1}{\sqrt{\varepsilon_r - 1}}
\end{aligned}
\tag{13.28}
$$

These formulas were compared with measurements by [Chiba, 78] and good agreement was obtained in an empty, straight tunnel. In more complex situations, however, these formulas alone tend to substantially underestimate the measured loss and other effects need to be considered.

A good example is the work by [Emslie, 75] which was later extended and corrected by [Zhang 97]. This work was originally intended for tunnels of approximately rectangular cross-section, such as coal mines, where there is a considerable degree of roughness and the tunnel axis may not be straight. Taking the tunnel width and height as a and b, respectively, the loss for vertical and horizontal polarisations is given as

$$L_{mn}^h = 4.343 r \lambda^2 \left(\frac{m^2 \varepsilon_{r1}}{a^3 \sqrt{\varepsilon_{r1} - 1}} + \frac{n^2}{b^3 \sqrt{\varepsilon_{r2} - 1}} \right) \tag{13.29}$$

$$L_{mn}^h = 4.343 r \lambda^2 \left(\frac{m^2}{a^3 \sqrt{\varepsilon_{r1} - 1}} + \frac{n^2 \varepsilon_{r2}}{b^3 \sqrt{\varepsilon_{r2} - 1}} \right)$$

where ε_{r1} and ε_{r2} are the relative permittivities of the side walls and ceiling/floor, respectively. The mode numbers – the number of half-wavelengths in the vertical and horizontal directions – are m and n, respectively. For long distances, the $m = 1$, $n = 1$ mode dominates. For distances less than the breakpoint distance r_{bp}, the effects of multiple higher-order modes must be accounted for, where

$$r_{bp} = \max \left(\frac{a^2}{\lambda}, \frac{b^2}{\lambda} \right) \tag{13.30}$$

The higher order modes fall off very rapidly, accounting for the rapid reduction in signal near the tunnel mouth observed in measurements such as Figures 13.17 and 13.18. Nevertheless, such a model still produces lower losses than observed in practice beyond the breakpoint. One effect which increases the observed loss is the impact of roughness. This adds to the previous losses depending on the RMS surface roughness, h_{RMS} as

$$L_r = 8.686 \pi^2 h_{RMS}^2 \lambda \left(\frac{1}{a^4} + \frac{1}{b^4} \right) r \tag{13.31}$$

If the tunnel also has minor variations from being straight, there is a further loss component dependent on the RMS tilt angle, θ_{RMS}, defined by the variation of the tunnel axis from a straight line,

$$L_t = \frac{17.372 \pi^2 \theta_{RMS}^2}{\lambda} r \tag{13.32}$$

Overall, then, the model predicts the path loss as

$$L = \left. \begin{array}{l} L_{mn} + L_r + L_t \\ L_{11} + L_r + L_t \end{array} \right\} \begin{array}{l} r < r_{bp} \\ r \geq r_{bp} \end{array} \tag{13.33}$$

where the L_{mn} term is the sum of powers arising from multiple modes until the associated power is too small to be significant.

Two other sources of loss also need to be considered to provide a reasonable prediction of measured effects. First, a practical antenna excites the tunnel waveguide modes only imperfectly due to its finite size and the distribution of current across it. For the E_h (1,1) mode the result for a dipole placed at the point (x_0, y_0) on the tunnel cross section is [Emslie, 75]

$$L_{ant} = -10\log\left(\frac{Z_{0,1,1}\lambda^2}{\pi^2 R_r d_1 d_2}\cos^2\frac{\pi x_0}{d_1}\cos^2\frac{\pi y_0}{d_2}\right) \qquad (13.34)$$

where $Z_{0,1,1}$ is the characteristic impedance of the $E_h(1,1)$ mode and R_r is the radiation resistance of the antenna, which are approximately the free space values 377 and 73 Ω respectively, provided that λ is small compared with d_1 and d_2. For example, a tunnel-centred dipole produces a loss of 26.3 dB for 900 MHz and $d_1 = d_2 = 5$ m. This effect should be carefully considered when siting antennas specifically to excite a tunnel.

Finally, when the tunnel is illuminated from an antenna external to the tunnel, the power available to excite modes within the tunnel depends on the ratio between tunnel area and antenna beamwidth projected to area of the tunnel mouth as illustrated in Figure 13.21. Under these circumstances an additional loss is experienced, which can be estimated as

$$L_{mouth} = 10\log\frac{A_{ill}}{A\cos\phi} \qquad (13.35)$$

where A is the area of the tunnel mouth, A_{ill} is the area of the 3 dB beam incident on the tunnel mouth and ϕ is the angle between the tunnel axis and the direct path from the transmit antenna to the centre of the tunnel mouth.

Combining the effects described so far, an overall prediction such as that shown in Figure 13.22 can be obtained. This exhibits most of the behaviour observed in measurements and provides an overall model useful for initial estimation of effects. Other effects should also be considered in detailed work, however. These include the impact of trackside clutter outside the tunnel, particularly trees, buildings and overhead bridges and power lines. It is often important to also consider the effect of the presence of the train within a railway tunnel, which leads to a much higher cut-off frequency in the gap between tunnel and train and hence greater loss rate with distance. The penetration of signals into the train is affected by the mode

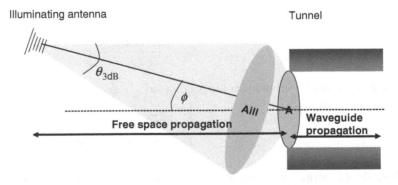

Figure 13.21: Transition from free-space to waveguide propagation at a tunnel mouth

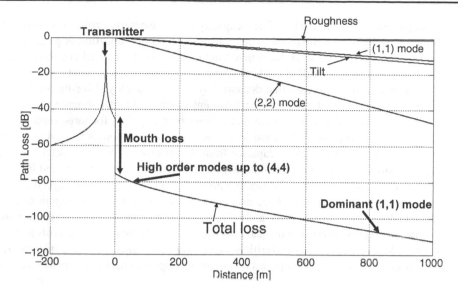

Figure 13.22: Overall prediction of sources of loss in a tunnel; here $f_c = 900\,\text{MHz}$, $d_{eq} = 5\,\text{m}$, $\varepsilon_r = 5.5$, $h_{rough} = 0.1\,\text{m}$, $\theta_{tilt} = 0.3°$ and the illuminating antenna is 30 m from the tunnel mouth with a gain of 12 dB

number; the angle of incidence is very shallow for the dominant so the loss may be much larger than would be observed in an open environment. Finally, although fast multipath effects are usually considered statistically as fast fading, as described in chapter 10, the interference between waveguide modes with similar angles of arrival may produce destructive fading at certain locations within the tunnel which persist for much longer distances than the typical fast fading observed in an open situation and should be calculated explicitly.

13.10 DISCUSSION

Propagation effects in enclosed spaces such as buildings and tunnels are even more geometry-dependent than in microcells, placing even greater burdens on the quality of data and computational requirements if deterministic physical models are to produce useful predictions. In the near future, practical picocell system design is more likely to rely on empirical models and engineering experience. In the longer term, however, combinations of physical models with statistics are expected to yield significant benefits.

13.11 DISTRIBUTION SYSTEMS FOR INDOOR AND ENCLOSED SPACE APPLICATIONS

The first choice approach to providing coverage inside a building, at least for cellular systems, has usually been to illuminate the building from external macrocells and microcells. Each cell then covers a large number of buildings and the cell capacity is shared amongst users across the whole cell area. However, the large variability of penetration losses reported in Section 13.4 implies that large numbers of buildings, particularly near the cell edge, may be poorly covered using such an approach. It also imposes a further constraint in the amount of channels that can be allocated to the building – capacity from the external cell is taken to handle the in-building traffic at the expense of an increase in blocking probabilities. A general trend towards higher frequencies and to systems involving CDMA where the cell coverage decreases with the traffic handled (so-called 'cell breathing') exacerbates these issues.

It is increasingly the case that dedicated systems are deployed for indoor coverage. These have long been used in large public environments, such as airports, shopping centres (malls) and railway stations and hotels, but are increasingly being applied for large multi-tenanted office environments. Such picocells (and even *femtocells* in the domestic environment) not only improve coverage and offload capacity from the macrocells but also offer the capability of services tailored to individual buildings with special tariffs and services and higher data rates than would be available from the outdoor network. Picocells must be designed to provide these services throughout the building, accounting carefully for the propagation characteristics described earlier in this chapter and selecting an appropriate means of distributing transmit power and capacity economically and effectively across the building. The costs and performance of the indoor system will depend critically on making the choice; therefore it is important to have a clear understanding of the options before attempting to optimise any indoor system.

13.11.1 Distributed Antenna Systems – General Considerations

A particularly useful application of antennas in indoor systems is the idea of *distributed antennas*. As illustrated in Figure 13.23, the idea is to split the transmitted power among several antenna elements, separated in space so as to provide coverage over the same area as a single antenna but with reduced total power and improved reliability [Saleh, 87b]; [Chow, 94]. This is possible because less power is wasted in overcoming penetration and shadowing losses, and because a line-of-sight channel is present more frequently, leading to reduced fade depths and reduced delay spread.

A distributed antenna system (DAS) can be implemented using passive splitters and feeders, or active repeater amplifiers can be included to overcome the feeder losses. These distribution techniques will be explained in the following sections. Since the antennas are all radiating the same signals, they effectively act as artificial sources of multipath. Although this would increase

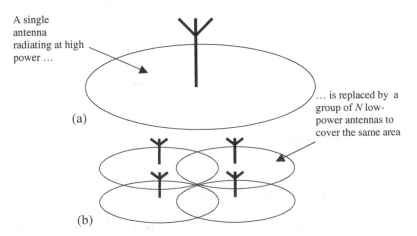

Figure 13.23: Indoor coverage: (a) single antenna, (b) distributed antenna

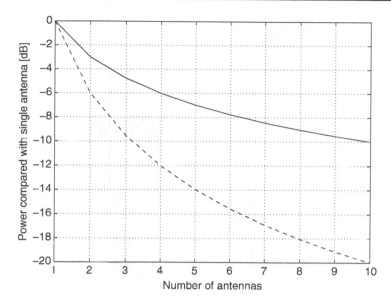

Figure 13.24: Distributed antenna power compared with single antenna: (—) total radiated power; (- - -) power for each antenna. Path loss exponent $n = 4$

fading in principle, in practice, most indoor environments are already close to Rayleigh environments so this does not create degradation. Indeed, in systems where equalisation or other active means of benefiting from a wideband channel (Chapter 17) is applied, it may be desirable to introduce delays between the antenna elements. This artificially increases delay spread in areas of overlapped coverage, permitting quality improvements via time diversity (Chapter 16).

It can be shown that if a given area is to be covered by N distributed antenna elements rather than a single antenna, then the total radiated power is reduced by approximately a factor $N^{1-n/2}$ and the power per antenna is reduced by a factor $N^{n/2}$ where a simple power law path loss model with path loss exponent n is assumed. This is shown in Figure 13.24. As an alternative, the total area covered could be extended for a given limit of effective radiated power, which may be important to ensure compliance with safety limits on radiation into the human body (see Section 16.2.8).

13.11.2 Passive Distributed Antenna Systems

The distributed antenna concept may be implemented using a purely passive infrastructure via conventional coaxial-based components. A list of components for a passive system varies by design, but usually includes coaxial cable, signal combiners, power splitters, diplexers, antennas and attenuators. These components require no external power to operate, only an RF signal input.

An example of such a system is shown in Figure 13.25. The base station is located in a central equipment room, connected via wired or wireless backhaul to the wider network. The base station antenna connector acts as the point of interface to the DAS. The coaxial cable used to distribute power from the base station to different floors (*vertical distribution*) is usually of large diameter, minimising the loss in distributing the power to higher floors, while narrower, higher

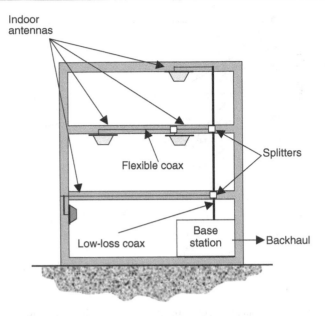

Figure 13.25: Passive distributed antenna system

loss cable which is easier to install is used for connecting the antennas on each floor to this 'backbone' (*horizontal distribution*). Differing coaxial splitters and attenuators can be used to provide approximate equalisation of the losses from the central location to each antenna. Eventually these losses become too large to provide useful coverage extension, and the building has to be served either using a further base station in another location or via the active approach in the following section. Thus passive systems tend to be applied mostly for small-to-medium sized buildings. Where coverage is provided by multiple operators in the same building, multiple base stations are located in the equipment room and combined together, so that the costs of the installation of the DAS can be shared amongst the operators. In this case, it is important to ensure that the combiner provides sufficient isolation amongst operators to avoid interference between transmitters and receivers due to blocking or intermodulation effects.

The advantages and disadvantages of passive distributed antenna systems are summarised in Table 13.12.

Table 13.12: Advantages and disadvantages of passive distributed antenna systems

Advantages	Disadvantages
Low hardware cost	Slow deployment
Simple architecture	Difficult installation of large low-loss cables due to limited bending radius
Very wide bandwidth	Rigid solution - little scope for upgrading
	Difficult to 'zone' or sectorise if not planned from the beginning
	Losses restrict uplink sensitivity

13.11.3 Active Distributed Antenna Systems

When the losses inherent in passive systems become excessive, active amplification can be used to overcome the losses in both the uplink and the downlink. Uplink amplification tends to be placed closest to the antennas to minimise the overall system noise figure, while downlink amplification can be distributed across the system. On-frequency amplifiers can be used to directly extend the range of simple coaxial cables, but the gain of such systems is limited by the need to avoid feedback and hence oscillation. It is more common to use frequency-translating repeaters, which also permit the use of other cabling technologies such as optical fibre and twisted pair copper cable. These can be cheaper and easier to handle than large-diameter coaxial cable. On the contrary, insertion of the amplifiers requires the use of band filtering to define the signals amplified and hence restricts the range of technologies available. Also, the amplifier power can limit the number of operators or capacity handled since the power per carrier reduces with the number of carriers.

One common type of active distributed antenna system uses a combination of RF and optical fibre technology as shown in Figure 13.26. The RF power from the base station is converted to optical by a main fibre unit, which is transmitted to remote fibre units using single or multi-mode fibre. Multi-mode fibre is somewhat cheaper and is more widely installed in existing buildings, but suffers greater loss and usually more restricted bandwidth than single-mode. After the optical signal has been transmitted from the main fibre unit, the signal is reconverted to RF and connected to one or multiple antennas using passive distribution. Thus the optical signal avoids the unwieldy coaxial backbone.

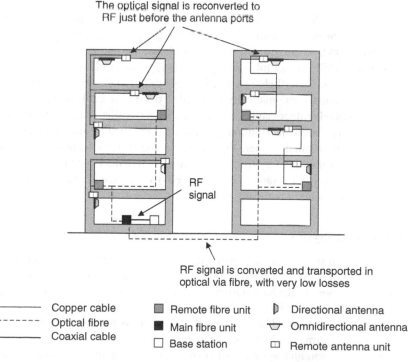

Figure 13.26: Active distributed antenna system based on RF-over-fibre technology

Table 13.13: Advantages and disadvantages of active distributed antenna systems

Advantages	Disadvantages
Easy zoning	Intermodulation interference, leading to a rapid decrease of available power with increasing number of carriers
Large coverage areas	Local power is required at remote units
Flexibility	Hardware and maintenance costs
Easier power adjustments per antenna	Uplink noise contributed by active elements, increasing with the number of elements
Easy and rapid installation	

This allows very low losses between the two fibre units, and hence the distance between an antenna element and the base station is maximised, allowing fibre lengths of typically from 2 to 20 km. For example, the losses for certain types of single-mode optical fibre are approximately 0.5 dB km^{-1} @ 1300 nm wavelength, which for 2 km equates to 2 dB total loss. A typical 900 MHz coaxial cable of the type used for picocell installations, has a loss of 11.5 dB per 100 m, which for 2 km would equate to 230 dB. Fibre systems are therefore commonly used to interconnect multiple buildings separated across a campus.

In extreme cases a large number of buildings across a city can be connected to a centralised bank of base stations to produce a *base station hotel*. This makes channel resources available to large numbers of users, increasing trunking gains, while allowing dense frequency reuse by making good use of inter-building shielding.

However, the uplink of an active DAS can represent a performance imitation. Since the signals from multiple antenna units are incoherently summed, the noise power generated by each unit adds, increasing the noise at the base station receiver continuously with the number of units, while the useful signal power is typically only contributed to by the nearest 2–3 antenna units. Nevertheless, active DAS are easier to design and deploy because the amplifier gains are usually automatically adjusted to overcome cable losses.

A list of advantages and disadvantages for active distributed antenna systems is included in Table 13.13. These distribution systems are employed frequently in large coverage areas and campus environments, where the base station needs to be located close to some antennas but a hundreds of metres away from others. Also, active distributed antenna systems are preferred when zoning (sectorisation of indoor cells) is required, and offer a better range of power adjustments when optimising the design.

13.11.4 Hybrid Systems

Elements of passive and active distribution systems may be combined to achieve an optimum design, For example, passive distribution can be used for antenna elements which are relatively close to the base station, with active used over longer distances. Passive distribution can also be used to increase the number of antennas served per remote antenna unit. These systems are most used in situations that encompass requirements typical for both distributed antenna systems, in which specific areas of the building may be more suitable for one type of distributed antenna systems whereas another area may be better implemented using a different distributed antenna system.

13.11.5 Radiating Cables

The ultimate form of a passive distributed antenna is a radiating cable (commonly known as a leaky feeder) which is a special type of coaxial cable where the screen is slotted to allow radiation along the cable length. With careful design, such cables can produce virtually uniform coverage, i.e. an effective path loss exponent of unity.

Radiating cables were originally conceived to provide subterranean radio propagation, for example, in railway tunnels and coal mines [Monk, 56]. They are ideal for providing uniform coverage in such locations, overcoming the complexity and potential uncertainty arising from the propagation considerations described in Section 13.9. Radiating cables have also been used as an alternative distribution system for in-building scenarios [Saleh, 87b]; [Zhang, 01]; [Bye, 88], especially for areas which are difficult to cover with conventional antennas such as long corridors and airport piers. An example of a radiating cable is shown in Figure 13.27.

Figure 13.27: Radiating cable

Radiating Cable Parameters

A radiating cable functions as both a transmission line and as an antenna. The radiating slots are effectively individual antennas, and propagation can be analysed as the summation of all of these antennas individually (e.g. [Park, 00]). It is usually more convenient, however, to analyse radiation from radiating cables as if the cable radiated continuously along its length. The amount of radiation is quantified by two main parameters. The *coupling loss* is defined as the ratio between the power transmitted into the cable and the power received by a half-wavelength dipole antenna located at a fixed reference distance from the cable in free space. The *insertion loss* is a measure of the longitudinal attenuation of the cable, usually expressed in decibels per metre at a specific frequency. For a given cable size, the insertion loss increases as the frequency of operation increases.

Propagation Modelling for Radiating Cables

A radiating cable model for indoor environments is proposed in [Zhang, 01]. This model is an empirical one, configured for frequencies of around 2 GHz. It takes into consideration both coupling and longitudinal losses of the cable. The equation that shows the overall link loss, combining the cable losses and propagation losses is

$$L = \alpha z + L_c + L_v + L_b + 10\, n \log r \qquad (13.36)$$

where z is the longitudinal distance along the cable to the point nearest the receiver [m], α is the attenuation per unit length of the cable [dB m^{-1}], L_c is the coupling loss referenced to 1 m

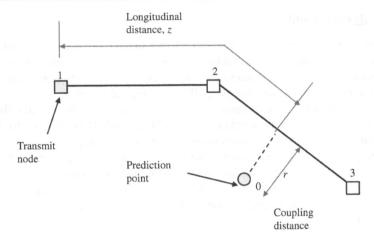

Figure 13.28: Geometry for radiating cable propagation model

radial distance from the cable [dB], L_v is the variability in coupling loss [dB], L_b is the loss factor due to blockage [dB], r is the shortest distance between the cable and the receiver [m] and n is the path loss exponent (see Figure 13.28). The particular situation analysed by the author was an academic building with reinforced concrete floors, 3.83 m ceiling height with a suspended false ceiling at 2.83 m, internal brick walls and external glass walls, with rooms interconnected via a corridor, along which ran the leaky feeder. In this case the model parameters were $L_c = 70.9$ dB, $L_v = 3.4$ dB, $L_c = 9.9$ dB, $\alpha = 0.12$ dB m^{-1} and $n = 0.6$. Note the very small value of the path loss exponent, indicating how the cable produces a very consistent level of coverage, although the particular parameters would depend strongly on the cable type and the environment. Coupling loss, longitudinal loss and variability in coupling loss are obtained from the cable's manufacturer.

Practical Considerations

Leaky feeders are an attractive solution for cellular coverage in areas difficult to illuminate with other distribution systems, such as tunnels, airport piers and long corridors. For multi-operator solutions, a leaky feeder often offers a viable solution, since its reasonably large bandwidth can accommodate many frequency bands to support multi-operator designs.

When installing radiating cables, though, special care must be taken to ensure that the cable is suspended carefully with reasonable separation from metal cabling trays and other cables to avoid major impacts to its radiation characteristics. Compliance with fire and cable installation standards must also be taken into account.

For large tunnels, leaky feeders are also accompanied by bi-directional amplifiers or repeaters, which amplify the signal when longitudinal losses are beyond acceptable limits. Figure 13.29 shows an example of this configuration.

Research has shown that the coverage provided by radiating cables, when compared to distributed antenna system elements, is smoother and better controlled, although more input power is often required for the former [Stamopoulos, 03]. Despite this constraint, leaky feeders are still preferred in environments where leakage out of the building needs to be controlled more accurately.

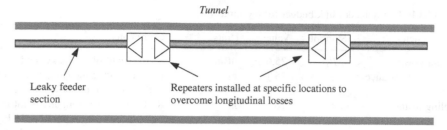

Figure 13.29: Leaky feeder installation in a tunnel using repeaters

When performing radiating cable signal strength predictions, it is important to understand each of the technical parameters given by manufacturers, to be able to apply Eq. (13.36) properly. The approach used is explained in Example 13.1.

Example 13.1

Cellular coverage at 1800 MHz is desired for mobile phone users in an airport pier. Due to the characteristics of the airport pier, a distributed antenna system would not be suitable, since leakage out of the pier would be excessive at locations where antenna elements can be deployed. Therefore, wireless system designers have chosen a leaky feeder to provide this coverage. All the leaky feeder specifications as given by the manufacturer are shown in Table 13.14.

Propagation measurements have shown that the airport pier environment has a location variability of 12 dB. Coverage is intended to be provided for 90% of the locations in the airport pier, and connector losses are assumed to add up to 1 dB in excess. A transmit power of 25.6 dBm has been considered for this application, which is in line with typical values for transmit antenna power levels in passive distributed antenna systems. Wall losses of 17 dB are assumed for concrete walls, as reported in Table 13.4 for 1800 MHz. A path loss exponent of 0.6 is assumed.

Find the received power at a point located in a longitudinal distance of 25 m and a coupling distance of 10 m, according to Figure 13.28.

Solution

Referring to Eq. (13.36), it is necessary to determine the coupling loss variability L_v. From the manufacturer's data provided, the coupling loss has been specified at 50 and 95% of the area of the feeder. Hence, the coupling loss location variability is

$$L_v = \frac{L_{C,95} - L_{C,50}}{Q^{-1}(0.95)} \tag{13.37}$$

$Q^{-1}(\cdot)$ is the inverse of the complementary cumulative normal distribution. Therefore the leaky feeder total coupling loss is given by the sum of the median coupling loss and the variability,

$$L_C = L_{C,50} + L_v \tag{13.38}$$

Table 13.14: Leaky feeder link budget for Example 13.1

User parameter	Symbol	Value	Units	Comments
Transmit power	P_T	25.6	dBm	At the input of the leaky feeder
Longitudinal distance	z	25	m	Distance along the cable to the point nearest the receiver, as shown in Fig. 13.19
Coupling distance	r	10	m	Radial distance from any point in the feeder to the prediction point, as in Fig. 13.19
Additional losses	L_e	1	dB	Connector, adaptor, etc after leaky feeder transmit node

Manufacture data	Symbol	Value	Units	Comments
Longitudinal loss	α	9.4	dBm/100m	As given by the manufacturer
Coupling loss @ 50%	$L_{C,50}$	60	dB	As given by the manufacturer, in 50% of locations in the feeder
Coupling loss @ 95%	$L_{C,95}$	66	dB	As given by the manufacturer, in 95% of locations in the feeder
Coupling loss variability margin	L_v	3.6	dB	Margin given to account for variations in coupling loss along the feeder
Coupling loss	L_C	63.6	dB	Loss at 1m radial distance from the leaky feeder

Propagation	Symbol	Value	Units	Comments
Location variability	σ_s	12	dB	Variation in path loss around the median value. This accounts for no measurements
Area availability		90	%	Coverage requirement area, which indicates the percentile of locations to be covered
Shadowing margin	L_S	15	dB	Margin to account for slow fading variations along the mean estimated value
Wall loss	L_w	17	dB	Assumed wall loss for a certain material at a specific frequency
Number of walls	n_w	1	walls	Number of walls (radially) between the leaky feeder and the prediction point
Path loss exponent	n	0.6		Assumed path loss exponent for leaky feeder propagation
Received power	P_R	−80	dBm	Expected received power at prediction point

The shadowing margin L_s is computed from the values of location variability and percentile of coverage area, assuming a lognormal distribution, as explained in Chapter 9. The received power P_R is then calculated from Eq. (13.36), as follows:

$$
\begin{aligned}
P_R &= P_T - L \\
&= P_T - \alpha z - n_w L_w - L_c - L_e - L_s - 10\,n \log(r)
\end{aligned}
\tag{13.39}
$$

Table 13.14 summarises the calculations performed to compute the received power at the given point.

Such calculations assume that the receiver point is central to the radiating cable. For points close to the ends of the radiating cable coverage is more complex to compute and may exhibit nulls at specific points [Stamopoulos, 03]. It is common to terminate the far end of cable in an antenna to minimise the impact of these nulls and make efficient use of the power remaining – otherwise it is necessary to terminate the feeder in a resistive matched load.

13.11.6 Repeaters

The distribution techniques described early have to be fed by an appropriate signal. This may be done using a base station, either on-site or remotely from some centralised location ('base-station hotel'). This may often produce unwarranted expense, however, so an alternative is to feed the system from a repeater as shown in Figure 13.30. A repeater is essentially a bi-directional amplifier with a high-gain, narrow-beam antenna directed at a 'donor' base station. The RF signal received by the repeater can be distributed across the building using any of the passive or active distribution techniques described previously. Thus, the repeater gain overcomes the building penetration loss, improving the indoor coverage.

Repeater systems must be carefully designed to ensure sufficient isolation between the donor and distribution antennas to avoid feedback, which limits the practical gain available. Although repeaters extend coverage, they do not increase network capacity, as they 'steal' the resources from the donor cell to be used indoors. The uplink from the repeater contains noise due to the active amplification of the uplink, which desensitises the uplink of the donor macrocell and reduces its range and its capacity in the case of a CDMA system. This is effect is cumulative for all repeaters under the coverage of a single cell. This is a major limitation, as for dense-populated areas, a sacrifice in the capacity of the donor cell may lead to cell congestion, if the resources are not properly allocated. It can also make it difficult to manage and optimise

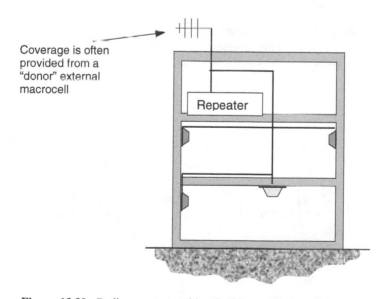

Coverage is often provided from a "donor" external macrocell

Repeater

Figure 13.30: Radio repeater used in a building to feed a passive DAS

coverage in the macro network because the need to retain appropriate donors acts as an additional constraint for the optimiser. Nevertheless, this option is still in use for low-cost easy-deployment situations with low-capacity demand and small coverage areas.

13.11.7 Digital Distribution

The signal distribution options examined so far all distribute analogue signals in essentially the same form as emerges from the antenna connector of the base station, although some of them do so in a frequency-translated form. Although this approach is flexible and relatively low-cost, these approaches all introduce signal distortions in the form of loss, noise and distortion at various levels and require specialist cabling and design works. It is now increasingly feasible to transport radio signals in digital form across a local area network (LAN), potentially making use of existing cabling and switching elements. In principle this could be done by digitising the RF signal directly, but this would require enormous LAN bandwidth for little benefit, so in practice it is more reasonable to digitise the signal at or near to baseband, so the sample rate can be reduced to a low multiple of the signal bandwidth of the signal. This is illustrated in Figure 13.31, where several antenna units provide RF processing and baseband digitisation, are connected to a LAN and are controlled by a centralised controller which provides the remainder of the conventional base station functions. Thus the controller/LAN/antenna unit combination constitute a distributed base station. In principle, the distances between units are not limited, provided the LAN bandwidth and switching capability are sufficient, which typically implies the use of gigabit-ethernet networks. The remote units need not simply replicate signals, but can be controlled intelligently by the controller to direct power and capacity appropriately, to coherently combine uplink signals and to reject signals containing only noise, thereby avoiding the uplink noise problems referred to in Section 13.11.3. This also allows the system to produce diversity and adaptive antenna benefits (as described in Chapters 16 and 18), yielding *intelligent picocells* [Fiacco, 01]. Standard interfaces are being produced to allow the antenna units and controller to be sourced from different manufacturers [CPRI], [OBSAI].

Such approaches are comparable to the architectures of many wireless LAN systems, and are expected to increasingly represent the standard approach to indoor coverage in the future as the economics of such systems become increasingly attractive.

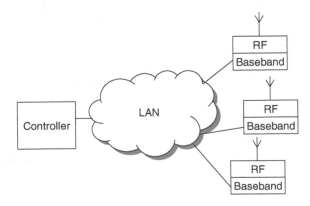

Figure 13.31: Digital indoor signal distribution

13.11.8 Selecting the Most Appropriate Distribution System

Several options of distribution system were presented in this section. The most appropriate choice depends on the building-specific capacity and coverage requirements and the financial budget. The best flexibility is provided by active distribution systems, but these are typically more expensive than passive systems for small- and medium-scale applications. Base station feeding provides the best capacity level and ease of management, but for rapid and simple indoor coverage, a combination of repeaters and macrocell penetration are common. The latter comes at the expense, however, of a reduction in capacity resulting from the amplification of the noise in the repeater, which impact on capacity in a CDMA system.

The selection of the distribution system also affects the optimisation method. With passive systems, individual antenna power reduction and zoning is difficult, although possible in special circumstances. Active systems can be adjusted more easily, and radiating cable installations are the best to contain leakage and for long corridors or tunnels. A hybrid system, combining the best of all these distribution systems is often the best choice for special installations, with demanding requirements of coverage and capacity.

13.12 INDOOR LINK BUDGETS

In-building design is one of the most demanding cellular engineering tasks. As we have seen in this chapter, indoor propagation is complex in nature, due to the high variability of clutter and materials. Accurate predictions to estimate path loss are required, combined with accounting for losses and other performance degradations in the distribution system.

In Chapter 5, link budgets were introduced as a useful approach to the basic design of a complete communication system. For indoor scenarios, link budgets are often constructed based on the specific distribution system in use. Losses are accounted for in a different way, as active distribution systems usually compensate for such losses with the use of amplification. The approach is best illustrated via examples:

Example 13.2: Passive System Link Budget

A cellular operator in the 900 MHz band has been asked to provide coverage in a two-storey Victorian-style office building, with thick brick and concrete walls. After an initial survey, the design engineer in charge has determined that a passive distribution system is the most viable alternative, as the size of the building and budget constraints do not justify the use of an active system. A summary of the specifications of the passive elements to be used in the design is as follows:

2-way splitter: 3.5 dB insertion loss.
Coaxial cable: 24.5 dB loss per 100 m @ 900 MHz.
Directional indoor antenna gain: 5.7 dBi @ 900 MHz.
Omnidirectional indoor antenna gain: 2 dBi @ 900 MHz.
Connector losses: 0.5 dB each.
Pico base station maximum transmit power: 33 dBm.

The schematic for this system is shown in Figure 13.32. Directional antennas have been selected for use in corridors whereas omnidirectional antennas were preferred for high-ceiling rooms. Propagation modelling validated with measurements has shown that

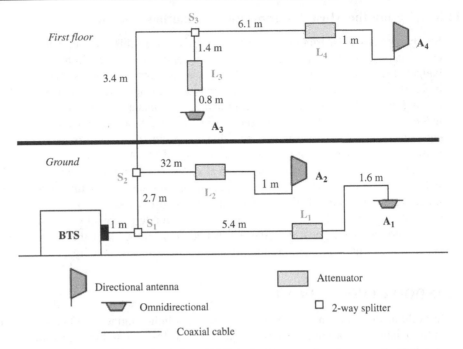

Figure 13.32: Passive indoor distributed antenna system for example 13.2

high-ceiling rooms. Propagation modelling validated with measurements has shown that in order to achieve sufficient received power, the required transmit power for directional antennas is 11.8 dBm, whereas for the omnidirectional antennas it is 17 dBm. Determine the required attenuation values L_1, L_2, L_3 and L_4 for this system.

Solution

Referring to Figure 13.31, the required transmit power per antenna P_{ant} is given by

$$P_{ant} = P_{BTS} - L_{cable} - L_{att} - L_{conn} - L_{splitter} + G_{ant} \qquad (13.40)$$

where P_{BTS} is the indoor BTS transmit power, L_{cable} represents the total cable losses, L_{att} are the attenuator losses, L_{conn} are connector losses, $L_{splitter}$ are splitter losses and G_{ant} is the maximum antenna gain.

For indoor link budgets, losses need to be computed for each individual antenna. For instance, Antenna 1, as shown in Figure 13.20, is an omnidirectional antenna intended to provide coverage in all directions, perhaps in a suspended ceiling mounting. The cable losses $L_{cable,A1}$ are then calculated by adding the cable lengths along the cable route as follows:

$$L_{cable,A1} = \frac{(1 + 5.4 + 1.6) \times 24.5}{100} = 1.96\,\text{dB} \qquad (13.41)$$

Therefore, the total losses that need to be accounted for in the link budget for antenna 1 include cable, splitter, attenuator and connector losses, as shown in Table 13.15.

Table 13.15: Passive indoor link budget

	Antenna Number			
	A1	A2	A3	A4
BTS transmit power [dBm]	33	33	33	33
Cable losses [dB]	2.0	1.9	2.3	3.5
Connector losses [dB]	3	4	5	5
Splitter losses [dB]	3.5	7	10.5	10.5
Antenna gain [dBi]	2	5.7	2	5.7
Desired transmit power [dBm]	11.8	11.8	11.8	11.8
Required attenuation [dB]	14.7	14.0	5.4	7.9

Hence, the required attenuation L_1 for antenna 1 is given by

$$L_1 = P_{BTS} - L_{cable,A1} - L_{splitter,A1} - L_{conn,A1} - P_{A1} + G_{A1}$$
$$= 33 - 1.96 - 3.5 - (6)(0.5) - 11.8 + 2 = 14.74 \, \text{dB} \qquad (13.42)$$

The required attenuation for the other antennas A2 to A4 can be computed in a similar way. The results are summarised in Table 13.15. The attenuation values would usually be rounded to the nearest decibel for practical implementation.

Example 13.3: Active System Link Budget

An active distribution system has been deployed in an airport to provide cellular coverage in the 1800 MHz band, as shown in Figure 13.21. The airport cell is comprised of the main terminal building and two piers, one for arriving passengers and the other for departing passengers. Sufficient field strength level in the piers is particularly important for cellular operators to guarantee that roamers have enough coverage. The following equipment specifications apply for this active system:

Coaxial cable: 29.3 dB loss per 100 m @ 1800 MHz.
Directional indoor antenna gain: 5.2 dBi @ 1800 MHz.
Omnidirectional indoor antenna gain: 1.7 dBi @ 1800 MHz.
Connector losses: 0.5 dB each.
Pico-base station maximum transmit power: 37 dBm.
Optical fibre losses: 0 dB for multi-mode fibre length up to 2 km.
Maximum power per carrier in main fibre unit: see Table 13.16.

Traffic forecasts have shown that two sectors within the indoor cell are required to provide the required capacity: sector A, formed by antennas 1,2 and 5, will use three carriers; and sector B, with antennas 3,4 and 6, needs four carriers (refer to Chapter 1 to check how this capacity dimensioning is obtained). Perform the required downlink link budget calculations to fill in all the parameters in Table 13.15 for the six antennas indicated in Figure 13.33.

Table 13.16: Main fibre unit maximum power per carrier

Number of 1800 MHz carriers	Maximum power per carrier [dBm]
1	13
2	10
3	8.2
4	7
5	6.8
6	5.1
7	4.5
8	4
9	4
10	3.8

Solution

The losses for an active distribution system are accounted for in a very similar way to a passive system, except for those losses between the main fibre unit and the remote fibre unit, which can be considered negligible. Therefore Eq. (13.23) still applies for this system, but L_{cable} needs to be recalculated as follows:

$$L_{cable} = L_{fibre} + L_{coaxial} = 0 + L_{coaxial} = L_{coaxial} \qquad (13.43)$$

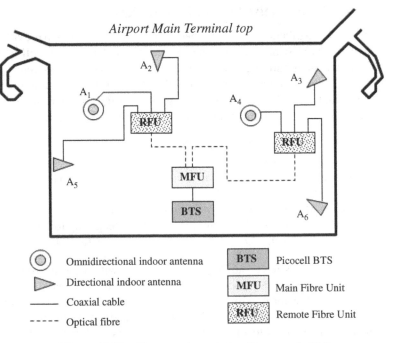

Figure 13.33: Airport active solution for example 13.3

Table 13.17: Summary of active link budget results for Example 13.3

	Antenna Number					
	A1	A2	A3	A4	A5	A6
BTS transmit power [dBm]	37	37	37	37	37	37
Maximum power per carrier [dBm]	8.2	8.2	7	7	8.2	7
Distance BTS to MFU [m]	1.5	1.5	1.5	1.5	1.5	1.5
Distance RFU to antenna [m]	1.0	1.2	1.8	1.7	2.1	2.5
Cable losses [dB]	0.7	0.8	1.0	0.9	1.1	1.2
Connector losses [dB]	2	2	2	2	2	2
Splitter losses [dB]	0	0	0	0	0	0
Optical fibre losses [dB]	0	0	0	0	0	0
Antenna gain [dBi]	1.7	5.2	5.2	1.7	5.2	5.2
Available eirp [dBm]	7.2	10.6	9.2	5.8	10.3	9.0
Attenuation BTS to MFU [dB]	28	28	30	30	28	30

$L_{coaxial}$ are the losses of any coaxial cables *after* the remote fibre unit. Therefore, only these losses need to be accounted for in the link budget.

For the case of active systems, there is an additional restriction to take into account when working out power levels in the link budget: the maximum output power per carrier in the main fibre unit. As active systems often have a non-linear behaviour when approaching saturation or when are operated close to maximum output power, intermodulation distortion is added to the system and hence special care should be taken. For this reason, as the number of employed carriers is increased in the main fibre unit, less power per carrier can be obtained at the output, as shown in Table 13.14. Manufacturers of active distribution system equipment often provide this information for the specified operating frequencies, at various number of carriers requirements. For this example, as 4 carriers are employed, then a maximum power per carrier of 7 dBm can be obtained and this figure should be used in the link budget as parameter. Practical calculation may also need to account for the uplink noise introduced by all of the active elements in the uplink.

Table 13.17 summarises the link budget results for the six antennas specified in Figure 13.21. The distances from RFU to each antenna have been included in the link budget results.

13.13 CONCLUSION

Picocell propagation is affected by a wide range of mechanisms operating on a complex, three-dimensional environment, the details of which are rarely available for propagation predictions. Simple models can give some useful estimates of in-building propagation, however, and further progress in these areas is strongly motivated by the growing importance of in-building communication, particularly for very high data rates. The use of appropriate distributed antenna structures or radiating cables helps considerably in providing controlled coverage around buildings, and it is expected that the provision of intelligence within such units will further allow systems to be installed without the need for detailed propagation predictions in every case (Chapter 20).

REFERENCES

[Akerberg, 88] D. Akerberg, Properties of a TDMA picocellular office communication system, In *Proceedings of. IEEE Global Telecommunications Conference Globecom* 1988, Hollywood, 1988, pp. 1343–1349.

[Aragón, 06] A. Aragón-Zavala, B. Belloul, V. Nikolopoulos and S. R. Saunders, Accuracy evaluation analysis for indoor measurement-based radio-wave-propagation predictions, *IEE Proceedings-Microwaves Antennas and Propagation.*, 153 (1), 67–74, 2006.

[Balanis, 89] C. A. Balanis, *Advanced Engineering Electromagnetics*, John Wiley & Sons, Inc., New York, ISBN 0-471-62194-3, 1989.

[Bye, 88] K. J. Bye, Leaky feeders for cordless communication in the office, In *Proceedings of European Conference on Electrotechnics* (*Eurocon '88*), Stockholm, June pp. 387–390, 1988.

[Catedra, 98] M. F. Catedra, J. Perez, F. S. de Adana and O. Gutierrez, Efficient ray-tracing techniques for three-dimensional analyses of propagation in mobile communications: application to picocell and microcell scenarios, *IEEE Antennas and Propagation Magazine*, 40 (2), 15–28, 1998.

[Cheung, 98] K. W. Cheung, J. H. M. Sau, and R. D. Murch, A new empirical model for indoor propagation prediction, *IEEE Transactions on Vehicular Technology*, 47 (3), 996–1001, 1998.

[Chiba, 78] J. Chiba, T. Inaba, Y. Kuwamoto, O. Banno and R. Sato, Radio Communication in Tunnels, *IEEE Transactions on Microwave Theory and Technology*, 26 (6), 439–443, 1978.

[Chow, 94] P. Chow, A. Karim, V. Fung and C. Dietrich, Performance advantages of distributed antennas in indoor wireless communication systems, *Proceedings of IEEE Vehicular Technology Society Conference*, Vol. 3, 1994, pp. 1522–1526.

[COST231, 99] COST231 final report, COST Action 231, *Digital Mobile Radio Towards Future Generation Systems*, European Commission/COST Telecommunications, Brussels, Belgium, 1999.

[CPRI] Common Public Radio Initiative: http://www.cpri.info/

[Davidson, 97] A. Davidson and C. Hill, Measurement of building penetration into medium buildings at 900 and 1500 MHz, *IEEE Transactions on Vehicular Technology*, 46 (1), 1997.

[Emslie, 75] A. G. Emslie, R. L. Lagace and P. F. Strong, Theory of the propagation of UHF radio waves in coal mine tunnels, *IEEE Antennas and Propagation Magazine*, AP-23, 192–205, 1975.

[ETSI, 97] European Telecommunication Standards Institute, *Selection procedures for the choice of radio transmission technologies of the Universal Mobile Telecommunications System* (*UMTS*), DTR/SMG-50402, 1997.

[Fiacco, 01] M. Fiacco, Intelligent picocells for adaptive indoor coverage and capacity, *Ph.D. thesis*, Centre for Communication Systems Research, University of Surrey, UK, 2001.

[Hashemi, 93a] H. Hashemi, Impulse response modelling of indoor radio propagation channels, *IEEE Journal on Selected Areas in Communications*, 11 (7), 1993.

[Hashemi, 93b] H. Hashemi, The indoor radio propagation channel, *Proceedings of IEEE*, 81 (7), 943–967, 1993.

[Hashemi, 94] H. Hashemi and D. Tholl, Statistical modelling and simulation of the RMS delay spread of the indoor radio propagation channel, *IEEE Transactions on Vehicular Technology*, 43 (1), 110–120, 1994.

[Honcharenko, 92] W. Honcharenko, H. L. Bertoni, J. L. Dailing, J. Qian and H. D. Yee, Mechanisms governing UHF propagation on single floors in modern office buildings, *IEEE Transactions on Vehicular Technology*, 41 (4) 496–504, 1992.

[Honcharenko, 93] W. Honcharenko, H. L. Bertoni and J. Dailing, Mechanisms governing propagation between different floors in buildings, *IEEE Antennas and Propagation Magazine*, 41 (6), 787–790, 1993.

[ITU, 1238] International Telecommunication Union, ITU-R Recommendation P.1238, *Propagation data and prediction models for theplanning of indoor radio communication systems and radio local area networks in the frequency range 900 MHz to 100 GHz*, Geneva, 1997.

[Keenan, 90] J. M. Keenan and A. J. Motley, Radio coverage in buildings, *BT Technology Journal*, 8 (1), 19–24, 1990.

[Kouyoumjian, 74] R. G. Kouyoumjian and P. H. Pathak, A uniform geometrical theory of diffraction for an edge in a perfectly conducting surface, *Proceedings of IEEE*, 62 (11), 1148–1161, 1974.

[Lee, 01] B. S. Lee, A. R. Nix and J. P. McGeehan, Indoor space-time propagation modelling using a ray launching technique, IEE *Proceedings of International Conference of Antennas and Propagation*, pp. 279–324, 2001.

[Molisch, 03] A. F. Molisch, J. R. Foerster, M. Pendergrass, Channel models for ultrawideband personal area networks, *IEEE Wireless Communications*, 14–21 December 2003.

[Monk, 56] N. Monk and H. S. Wingbier, Communications with moving trains in tunnels, *IRE Transactions on Vehicular Communications*, 7, 21–28, 1956.

[Stone, 97] W. C. Stone, National Institute of Standards and Technology, Electromagnetic Signal Attenuation in Construction Materials, NIST study NISTIR 6055, NIST Construction Automation Program, Report No. 3, October 1997.

[OBSAI] The Open Base Station Architecture Initiative: http://www.obsai.org

[Park, 00] J. K. Park and H. J. Eom, Radiation from multiple circumferential slots on a coaxial cable, *Microwave and Optical Technology Letters*, 26 (3), 160–162, 2000.

[Rappaport, 91] T. S. Rappaport, S. Y. Seidel and K. Takamizawa, Statistical channel impulse response models for factory and open plan building radio communication system design, *IEEE Transactions on Communications*, 39 (5), 794–806, 1991.

[Saleh, 87a] A. A. M. Saleh and R. A. Valenzuela, A statistical model for indoor multipath propagation, *IEEE Journal on Selected Areas in Communications*, 5 (2), 28–137, 1987.

[Saleh, 87b] A. A. M. Saleh, A. J. Rustako, Jr and R. S. Roman, Distributed antennas for indoor communication, *IEEE Transactions on Communications*, 35 (11), 1245–1251, 1987.

[Siwiak, 03] K. Siwiak, H. L. Bertoni, and S. M. Yano, Relation between multipath and wave propagation attenuation, *Electronic Letters*, 39 (1), 142–143, 2003.

[Stamopoulos, 03] I. Stamopoulos, A. Aragón-Zavala and S. R. Saunders, Performance comparison of distributed antenna and radiating cable systems for cellular indoor environments in the DCS band, *IEE International Conference on Antennas and Propagation*, ICAP 2003, April 2003.

[Stavrou, 03a] S. Stavrou and S. R. Saunders, Factors influencing outdoor to indoor Radio-wave Propagation, *IEE Twelfth International Conference on Antennas and Propagation (ICAP)*, University of Exeter, UK, 31st March–3rd April 2003a.

[Stavrou, 03b] S. Stavrou and S. R. Saunders, Review of Constitutive Parameters of Building Materials, *IEE Twelfth International Conference on Antennas and Propagation (ICAP)*, University of Exeter, UK, 31st March–3rd April 2003b.

[Tuan, 03] S. C.Tuan, J. C. Chen, H. T. Chou and H. H. Chou, Optimization of propagation models for the radio performance evaluation of wireless local area networks, *IEEE Antennas and Propagation Society International Symposium*, Vol. 2, pp. 146–149, 2003.

[Turkmani, 87] A. M. D. Turkmani, J. D. Parsons and D. G. Lewis, Radio propagation into buildings at 441, 900 and 1400 MHz, *IEE Proceedings of 4th International Conference on Land Mobile Radio*, 1987.

[Turkmani, 92] A. M. D. Turkmani and A. F. Toledo, Propagation into and within buildings at 900, 1800 and 2300 MHz, *IEEE Vehicular Technology Conference*, 1992.

[Walker, 92] E. H. Walker, Penetration of radio signals into buildings in cellular radio environments, *Proceedings of IEEE Vehicular Technology Society Conference*, 1992.

[Zhang, 97] Y. P. Zhang and Y. Hwang, Enhancement of rectangular tunnel waveguide model, *IEICE Asia Pacific Microwave Conference*, 197–200, 1997.

[Zhang, 00] Y. P. Zhang, Z. R. Jiang, T. S. Ng and J. H. Sheng, Measurements of the propagation of UHF radio waves on an underground railway train, *IEEE Transactions on Vehiculatr Technology*, 49 (4), 1342–1347, 2000.

[Zhang, 01] Y. P. Zhang, Indoor radiated-mode leaky feeder propagation at 2.0 GHz, *IEEE Transactions on Vehiculatr Technology*, 50 (2), 536–545, 2001.

PROBLEMS

13.1 Calculate the maximum number of floors which can separate an indoor base station and a mobile at 900 MHz, with a maximum acceptable path loss of 110 dB, using the ITU-R model in Eq. (13.2). Assume that shadowing effects are negligible initially, then recalculate with a location variability of 10 dB.

13.2 Use Table 13.6 to calculate the reflection coefficient for a smooth concrete wall at 1 GHz for normal incidence.

13.3 Estimate, using 13.2.2, the number of floors required to separate two co-channel transmitters in an isolated building if a median signal-to-interference ratio of 12 dB is required.

13.4 Repeat Problem 13.3 in the case where the signal-to-interference ratio is required to be 12 dB at 90% of locations with a location variability of 5 dB.

13.5 Prove the expression for the power reduction for a given coverage area using distributed antennas given in Section 13.11.1. Derive an expression for the extension of coverage area which is obtained using a distributed antenna solution with the same radiated power per antenna as the single antenna solution. Calculate the result from your expression for eight antennas and a path loss exponent of 4. What practical limitations would there be on achieving this potential?

13.6 A radio consultant has been asked to make recommendations on the installation of a new cellular infrastructure to provide coverage in an underground station. Explain what type of distribution system you would install, giving reasons for your choice.

13.7 Explain the characteristics of a hybrid distribution system. Is it possible to construct a hybrid DAS with antennas and radiating cables? Sketch a diagram for this architecture.

13.8 An RF-over-fibre active distribution system has been designed to provide 11 dBm of power per carrier if three carriers at 900 MHz are used. What is the composite power that you would expect if all carriers are added together?

13.9 Zoning is a technique similar to sectorisation in macrocells which allows that indoor antennas be grouped together and channels assigned to each zone independently. Explain how you would achieve this in active and passive distribution systems, stating any advantages and disadvantages that you might envisage.

13.10 A radiating cable has been deployed in an office corridor to provide smooth and controlled VHF-band coverage to employees in the company. The characteristics of this cable are as follows:

Longitudinal loss @ 102 MHz: 4.2 dB per 100 m.
Coupling loss 50%: 45 dB.
Coupling loss 90%: 49 dB.

Propagation measurements show that a location variability of 9.8 dB for office environments is typical. Coverage is intended for a 90% of the area in the corridor. The 'common room' is a place where employees gather during their breaks, and coverage is especially desired in this place. Two wooden walls with 7 dB loss each are separating the leaky feeder with this room. Estimate the received signal strength at a point located at a longitudinal distance of 4 m and a coupling distance of 7 m, if a transmit power of 6 dBm is fed into the radiating cable. State and justify any assumptions made in the process.

13.11 Estimate the coverage area of each of the omnidirectional antennas in Example 13.2, assuming an appropriate path loss model from this chapter and a maximum acceptable path loss of 70 dB.

14 Megacells

*'It's like a big parking lot up there. There are lots of empty places,
and you can park in one as long as it doesn't belong to someone else'.*
Robert Grove Jr, Satellite engineer

14.1 INTRODUCTION

Mobile systems designed to provide truly global coverage using constellations of low and medium earth orbit satellites are now in operation. These form *megacells*, consisting of the footprint from clusters of spotbeams from each satellite, which move rapidly across the Earth's surface. Signals are typically received by the mobile at very high elevation angles, so that only environmental features which are very close to the mobile contribute significantly to the propagation process (Figure 14.1). Atmospheric effects, similar to those described in Chapter

Figure 14.1: Megacell propagation geometry

Antennas and Propagation for Wireless Communication Systems Second Edition   Simon R. Saunders and
Alejandro Aragón-Zavala
© 2007 John Wiley & Sons, Ltd

7, may also be significant in systems operated at SHF and EHF. Megacell propagation prediction techniques must also combine predictions of fast (multipath) fading and of shadowing effects, as these tend to occur on similar distance scales and cannot therefore be easily separated. The predictions tend to be highly statistical in nature, as coverage across very wide areas must be included, while still accounting for the large variations due to the local environment.

Mobile satellite systems are usually classified according to their orbit type: Low earth orbits (LEOs) involve satellites at an altitude of 500–2000 km and require a relatively large number of satellites to provide coverage of the whole of the Earth (e.g. 66 at 780 km in the case of the Iridium system). Medium earth orbits (MEOs) involve altitudes of around 5000–12 000 km and involve fewer, slower moving satellites for whole-earth coverage (e.g. 12 at 10 370 km in the case of the Odyssey system). Geostationary satellites (GEO), at the special height of around 36 000 km, have the benefit of requiring only three satellites for whole-earth coverage and requiring almost no tracking of satellite direction. The large altitudes in GEO systems lead to a very large free space loss component. For example, the free space loss for a GEO at 1.5 GHz would be around 186 dB at zenith. The transmit powers needed at both the satellite and mobile to overcome this loss are excessive, so LEO and MEO systems are far more attractive for mobile communications, whereas GEO systems are more usually applied to fixed satellite links (Chapter 7). For further details, see [Evans, 99] and [Pattan, 98].

In all orbits other than GEO, the satellite position changes relative to a point on the Earth, so the free space loss for a particular mobile position becomes a function of time. In the extreme case of an LEO system, an overhead pass might last only for few minutes, so the path loss will change rapidly between its maximum and minimum values (Figure 14.2). Similarly, the motion of the satellite relative to the location of the mobile user will create significant Doppler shift, which will change rapidly from positive to negative as the satellite passes overhead. In Figure 14.2 the maximum Doppler shift would be around ±37 kHz. This *shift* can be compensated by retuning either the transmitter or receiver; it should not be confused with the Doppler *spread* which arises from motion of the mobile relative to sources of multipath as described in Chapter 10; although this is of a much smaller size, it cannot be compensated.

The main local sources of propagation impairments in mobile satellite propagation are

- trees
- buildings
- terrain.

These interact with wave propagation via the following mechanisms:

- reflection
- scattering
- diffraction
- multipath.

14.2 SHADOWING AND FAST FADING

14.2.1 Introduction

In mobile satellite systems, the elevation angle from the mobile to the satellite is much larger than for terrestrial systems, with minimum elevation angles in the range 8–25°. Shadowing effects therefore tend to result mainly from the clutter in the immediate vicinity of the mobile. For example, when the mobile is operated in a built-up area, only the building closest to the

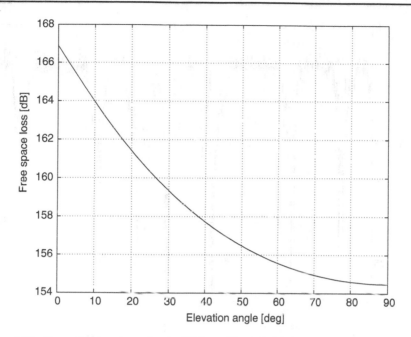

Figure 14.2: Free space loss for a circular LEO satellite orbit at an altitude of 778 km at 1.625 GHz. These values would be encountered over a period of around 7.5 min

mobile in the direction of the satellite is usually significant. As the mobile moves along the street, the building contributing to this process changes rapidly, so the shadowing attenuation may also change at a relatively high rate. By contrast, terrestrial macrocellular systems involve elevation angles of order 1° or less, so a large number of buildings along the path are significant (Chapter 8).

As a consequence of this effect, there may be rapid and frequent transitions between line-of-sight and non-line-of-sight situations in mobile satellite systems, causing a variation in the statistics of fast fading which is closely associated with the shadowing process. Figure 14.3 shows a typical measured example of the variation of signal level with location in a suburban area; note that the fade depths are much greater during the obstructed periods.

It is therefore most convenient to treat the shadowing and narrowband fast fading for mobile satellite systems as a single, closely coupled process, in which the parameters of the fading (such as the Rice k factor and local mean signal power) are time-varying. In subsequent sections we will examine the basic mechanisms of channel variation in mobile satellite systems and then describe a number of models which can be used to predict these effects.

14.2.2 Local Shadowing Effects

Roadside buildings are essentially total absorbers at mobile satellite frequencies, so they can be regarded as diffracting knife-edges. They can be considered to block the signal significantly when at least $0.6 \times$ the first Fresnel zone around the direct ray from the satellite to the mobile is blocked (Figure 14.4). Thus, shadowing may actually be less at higher frequencies due to the

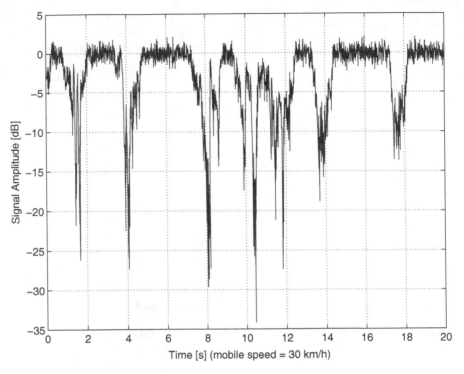

Figure 14.3: Example channel variations measured in a suburban area

narrower Fresnel zones for a given configuration. Once shadowing has occurred, building attenuation may be estimated via the single knife-edge diffraction formula in Section 3.5.2.

Tree shadowing also occurs primarily when the tree is contained within $0.6 \times$ the first Fresnel zone. In this case, however, the tree is not a complete absorber, so propagation occurs through the tree as well as around it. The attenuation varies strongly with frequency and path length. The simplest approach is to find the path length through the tree and calculate the attenuation based on an exponential attenuation coefficient in [dB m^{-1}], according to the values given in Chapter 6. Single-tree attenuation coefficients have been measured at 869 MHz, and the largest values were found to vary between 1 and 2.3 dB m^{-1}, with a mean value of 1.7 dB m^{-1} [Vogel, 86].

14.2.3 Local Multipath Effects

As well as the existence of a direct path from the satellite to the mobile, reflection and scattering processes lead to other viable wave paths. These multiple paths interfere with each other, leading to rapid fading effects as the mobile's position varies. The multipath may result from adjacent buildings, trees or the ground. The level of the multipath is usually rather lower than that of the direct path, but it may still lead to significant fading effects, similar to those described in Chapter 10. Note that multiple scattering paths, such as via points x and y in Figure 14.5, are attenuated by two reflection coefficients and are therefore unlikely to be significant.

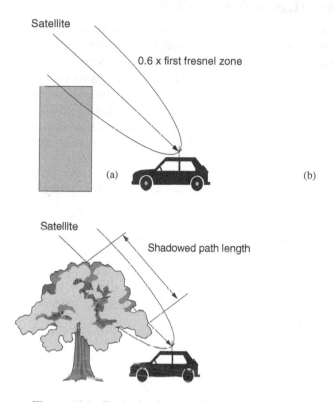

Figure 14.4: Shadowing by (a) buildings and (b) trees

The multipath can also lead to wideband fading if the differential path lengths are sufficiently large. In practical systems, however, the dominant source of wideband effects may be the use of multiple satellites to provide path diversity (Section 14.8).

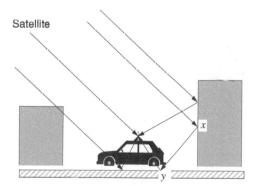

Figure 14.5: Multipath propagation

14.3 EMPIRICAL NARROWBAND MODELS

Models for narrowband propagation in mobile satellite systems are different from those used in terrestrial systems in two key ways. First, they include the excess path loss and shadowing effects as dynamic processes, along with the fast fading. Second, they rarely use direct deterministic calculation of physical effects, as this is not practical for predicting satellite coverage of areas exceeding tens of thousands of square kilometres. Instead they use statistical methods, although these may be based on either empirical or physical descriptions of the channel.

Empirical models, particularly the empirical roadside shadowing (ERS) model, have been constructed for mobile satellite systems operated in areas characterised mainly by roadside trees. The ERS model is expressed as follows [ITU, 681]:

$$
\begin{aligned}
L(P, \theta) = &-(3.44 + 0.0975\theta - 0.002\theta^2)\ln P \\
&+ (-0.443\theta + 34.76)
\end{aligned}
\tag{14.1}
$$

where $L(P, \theta)$ is the fade depth [dB] exceeded for P percent of the distance travelled, at an elevation angle θ (degrees) to the satellite, where for elevation angles from $7°-20°$, Eq. (14.1) is used with $\theta = 20°$. If the vehicle travels at constant speed, P is also the percentage of the time for which the fade exceeds L. This model applies only to propagation at L band 1.5 GHz, elevation angles from $20°-60°$ and at fade exceedance percentages from 1 to 20%. The result is shown in Figure 14.6.

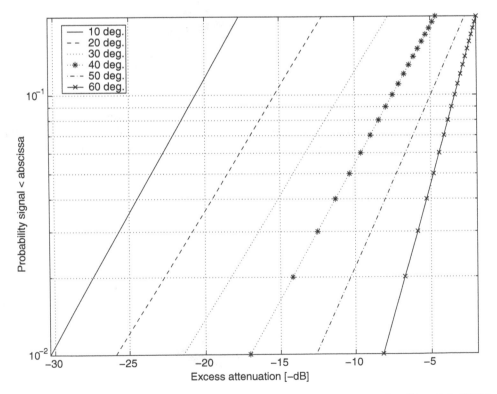

Figure 14.6: Predictions from ERS model: elevation angle increases from left ($10°$) to right ($60°$) in steps of $10°$

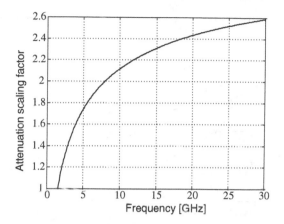

Figure 14.7: ERS model frequency-scaling function

The ERS model can then be extended up to 20 GHz using the following frequency-scaling function:

$$L(f_2) = L(f_1) \exp\left[1.5\left(\frac{1}{\sqrt{f_1}} - \frac{1}{\sqrt{f_2}}\right)\right] \tag{14.2}$$

where $L(f_2)$ and $L(f_1)$ are the attenuations in decibels at frequencies $f_1 = 1.5\,\text{GHz}$ and f_2, with $0.8\,\text{GHz} \leq f_2 \leq 20\,\text{GHz}$. This function is shown in Figure 14.7.

The ERS model has also been extended to larger time percentages from the original distributions at L band. The model extension is given by

$$L(P, \theta) = \frac{L(20\%, \theta)}{\ln 4} \times \left(\frac{80}{P}\right) \tag{14.3}$$

for $80\% \geq P > 20\%$. Predictions from the extended model at 20 GHz for elevation angles from $20°$ to $60°$ are shown in Figure 14.8.

Foliage effects can also be modelled empirically using a model which was developed from mobile measurements taken in Austin, Texas, in 1995 [Goldhirsh, 92]. In February the trees had no leaves and in May they were in full foliage. A least-squares fit to the equal probability levels of the attenuations, yielded the following relationship:

$$L_{\text{foliage}} = a + bL_{\text{no foliage}}^c \tag{14.4}$$

where the constants are $a = 0.351$, $b = 6.8253$ and $c = 0.5776$ and where the model applies in the range $1 \leq L_{\text{no foliage}} \leq 15\,\text{dB}$ and $8 \leq L_{\text{foliage}} \leq 32\,\text{dB}$.

14.4 STATISTICAL MODELS

Statistical models give an explicit representation of the channel statistics in terms of parametric distributions which are a mixture of Rice, Rayleigh and log-normal components. These models use statistical theory to derive a reasonable analytical form for the distribution of the narrowband fading signal and then use measurements to find appropriate values of the

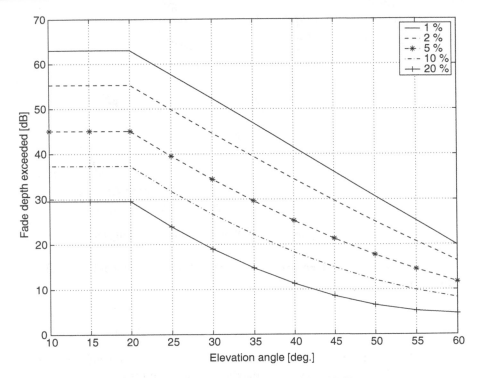

Figure 14.8: Extended ERS model at 20 GHz

parameters in the distribution. They all have in common an assumption that the total narrowband fading signal in mobile satellite environments can be decomposed into two parts: a coherent part, usually associated with the direct path between the satellite and the mobile, and a diffuse part arising from a large number of multipath components of differing phases. The magnitude of the diffuse part is assumed to have a Rayleigh distribution. Thus, the multiplicative complex channel α corresponding to all such models can be expressed as

$$\alpha = A_c s_c e^{j\phi} + r s_d e^{j(\theta+\phi)} \tag{14.5}$$

where A_c is the coherent part, s_c and s_d are the shadowing components associated with the coherent and diffuse parts, respectively, and r has a complex Gaussian distribution (i.e. its magnitude is Rayleigh distributed).

The simplest model of this form is the Rice distribution introduced in Chapter 10 [Rice, 48] which assumes that both components of the signal have constant mean power. More recent studies such as [Loo, 85], [Corazza, 94] and [Hwang, 97] have generalised this model to account for the rapidly changing conditions associated with attenuation and shadowing of both the coherent and diffuse components which arise from mobile motion. The models are summarised in Table 14.1. Note that, as with terrestrial shadowing, the distribution of the mean power arising from s_c and s_d is widely assumed to be log-normal. These models can all be implemented within the structure shown in Figure 14.9.

Table 14.1: Statistical satellite mobile channel models

Model	Coherent part	Correlation[a]	Diffuse part
Rice [Rice, 48]	Constant	Zero	Rayleigh
Loo [Loo, 85]	Log-normal	Variable	Rayleigh
Corazza [Corazza, 94]	Log-normal	Unity	Log-normal–Rayleigh
Hwang [Hwang, 97]	Log-normal	Zero	Log-normal–Rayleigh

[a]Between coherent and diffuse part.

If the parameters of these models are appropriately chosen, they can provide a good fit to measured distributions over a wide range of environment and operating conditions, although the Loo model is really applicable to only moderate rural situations. The Hwang model has been shown [Hwang, 97] to include the Rice, Loo and Corazza models as special cases. Note however that Table 14.1 reveals how a further generalisation of the Hwang model to allow correlation in the range (0,1) would allow still further generality.

14.4.1 Loo Model

This model [Loo, 85] is specifically designed to account for shadowing due to roadside trees. It is assumed that the total signal is composed of two parts; a line-of-sight component which is log-normally distributed due to the tree attenuation, plus a multipath

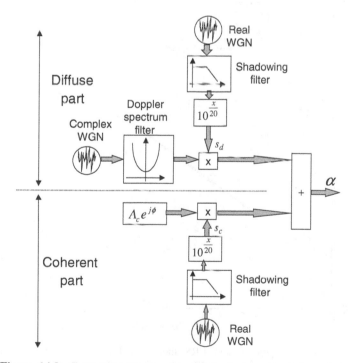

Figure 14.9: Generative structure for statistical narrowband channel models

component which has a Rayleigh distribution. Thus the total complex fading signal α is given by

$$\alpha = d e^{j\phi_0} + s e^{j\phi} \tag{14.6}$$

where d is the log-normally distributed line-of-sight amplitude, s is the Rayleigh distributed multipath amplitude and ϕ_0 and ϕ are uniformly distributed phases.

The p.d.f. of the fading amplitude, $r = |\alpha|$, is complicated to evaluate analytically, but it can be approximated by the Rayleigh distribution for small values and by the log-normal distribution for large values:

$$p(r) \approx \begin{cases} \dfrac{r}{\sigma_m^2} \exp\left[-\dfrac{r^2}{2\sigma_m^2}\right] & \text{for } r \ll \sigma_m \\[4mm] \dfrac{1}{20 \log r \sqrt{2\pi}\sigma_0} \exp\left[-\dfrac{(20 \log r - \mu)^2}{2\sigma_0}\right] & \text{for } r \gg \sigma_m \end{cases} \tag{14.7}$$

where σ_m is the standard deviation of either the real or imaginary component of the multipath part, σ_0 is the standard deviation of $20 \log d$ [dB] and μ is the mean of $20 \log d$ [dB]. An example prediction is shown in Figure 14.10, where the parameters of the Loo model have been chosen to fit the results from the ERS model at an elevation angle of 45°. Expressions for the level-crossing rate and average fade duration are also given in [Loo, 85] and it is found that

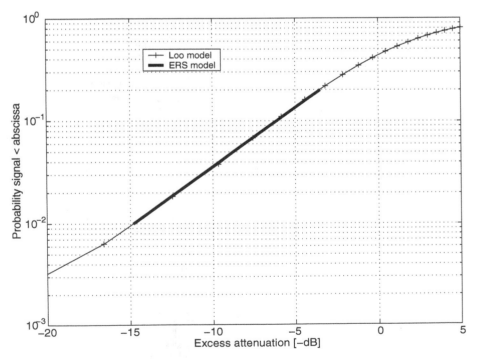

Figure 14.10: Loo model and ERS model at 1.5 GHz and 45° elevation: parameters are $\sigma_m = 0.3, \sigma_0 = 5$ dB and $\mu = 0.1$ dB

they depend upon the correlation between d and s, with the highest crossing rates occurring for low values of the correlation.

14.4.2 Corazza Model

This model [Corazza, 94] may be seen as a development of the Loo approach, where both the direct path and multipath are log-normally shadowed, so that the channel amplitude is given by

$$\alpha = S(e^{j\phi_0} + se^{j\phi}) \tag{14.8}$$

where S is the log-normal shadowing and other parameters follow the definitions in (14.6). The parameters of the model are then the Rice factor k and the log-normal mean μ [dB] and standard deviation σ_L [dB]. A wide range of environments can be modelled by appropriate choice of these parameters. The following empirical formulations for the parameters were extracted from measurements at L band in a rural tree-shadowed environment:

$$k = 2.731 - 0.1074\theta + 2.774 \times 10^{-3}\theta^2$$
$$\mu = -2.331 + 0.1142\theta - 1.939 \times 10^{-3}\theta^2 + 1.094 \times 10^{-3}\theta^3 \tag{14.9}$$
$$\sigma_L = 4.5 - 0.05\theta$$

14.4.3 Lutz Model

In this model [Lutz, 91], the statistics of line-of-sight and non-line-of-sight states are modelled by two distinct *states*, particularly appropriate for modelling in urban or suburban areas where there is a large difference between shadowed and unshadowed statistics. The parameters associated with each state and the transition probabilities for evolution between states are empirically derived. These models permit time series behaviour to be examined and may be generalised to more states to permit a smoother representation of the transitions between LOS and NLOS conditions [Ahmed, 97], or to characterise multiple satellite propagation [Lutz, 96]. For example, an environment which includes building blockage, tree shadowing and line-of-sight conditions can be modelled via a three-state approach.

In [Lutz, 91], the line-of-sight condition is represented by a 'good' state, and the non-line-of-sight condition by a 'bad' state (Figure 14.11). In the good state, the signal amplitude is assumed to be Rice distributed, with a k factor which depends on the satellite elevation angle and the carrier frequency, so that the p.d.f. of the signal amplitude is given by $p_{good}(r) = p_{rice}(r)$, where $p_{rice} = p_R$ in (10.30). The non-coherent (multipath) contribution has a classical Doppler spectrum.

In the bad state, the fading statistics of the signal amplitude, $r = |\alpha|$, are assumed to be Rayleigh, but with a mean power $S_0 = \sigma^2$ which varies with time, so the p.d.f. of r is specified as the conditional distribution $p_{rayl}(r|S_0)$. S_0 varies slowly with a log-normal distribution $p_{LN}(S_0)$ of mean μ [dB] and standard deviation σ_L [dB], representing the varying effects of shadowing within the non-line-of-sight situation.

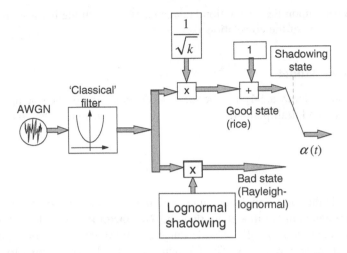

Figure 14.11: Two-state Lutz model of narrowband mobile satellite fading

The overall p.d.f. in the bad state is then found by integrating the Rayleigh distribution over all possible values of S_0, so that

$$p_{\text{bad}}(r) = \int\limits_{0}^{\infty} p_{\text{rayl}}(r|S_0)p_{LN}(S_0)\ dS_0 \qquad (14.10)$$

The proportion of time for which the channel is in the bad state is the *time-share of shadowing*, A, so that the overall p.d.f. of the signal amplitude is

$$p_{\text{r}}(r) = (1-A)p_{\text{good}}(r) + Ap_{\text{bad}}(r) \qquad (14.11)$$

Transitions between states are described by a *first-order Markov chain*. This is a state transition system, in which the transition from one state to another depends only on the current state, rather than on any more distant history of the system. These transitions are represented by the state transition diagram in Figure 14.12, which is characterised by a set of state transition probabilities.

The transition probabilities are

- probability of transition from good state to good state p_{gg}
- probability of transition from good state to bad state p_{gb}

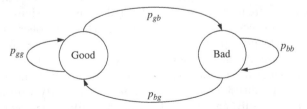

Figure 14.12: Markov model of channel state

- probability of transition from bad state to bad state p_{bb}
- probability of transition from bad state to good state p_{bg}.

For a digital communication system, each state transition is taken to represent the transmission of one symbol. The transition probabilities can then be found in terms of the mean number of symbol durations spent in each state:

$p_{gb} = 1/D_g$ where D_g is the mean number of symbol durations in the good
      state

$p_{bg} = 1/D_b$ where D_g is the mean number of symbol durations in the bad
      state.

The sum of the probabilities leading from any state must sum to 1, so

$$p_{gg} = 1 - p_{gb} \text{ and } p_{bb} = 1 - p_{bg}. \tag{14.12}$$

Finally, the time-share of shadowing (the proportion of symbols in the bad state) is

$$A = \frac{D_b}{D_b + D_g} \tag{14.13}$$

The model can be used to calculate the probability of staying in one state for more than n symbols by combining the relevant transition probabilities. Thus

$$\begin{aligned}
\text{Probability of staying in good state for more than } n \\
\text{symbols} = p_g(>n) = p_{gg}^n \\
\text{Probability of staying in bad state for more than } n \\
\text{symbols} = p_b(>n) = p_{bb}^n
\end{aligned} \tag{14.14}$$

An example of the signal variations produced using the Lutz model is shown in Figure 14.13, where the two states are clearly evident in the signal variations. Typical parameters, taken from measurements at 1.5 GHz, are shown in Figure 14.14 (highway environment) and Figure 14.15 (city environment).

Example 14.1

A mobile satellite system is found to have probability 0.03 of staying in an NLOS situation for more than 10 bits and a probability 0.2 of staying in the bad state for the same number of bits. Calculate the transition probabilities, assuming a two-state Markov model, and calculate the time-share of shadowing.

Solution

Since $p_g(> 10) = p_{gg}^n = 0.03$ we have $p_{gg} = 0.03^{1/10} = 0.7$
Similarly $p_{bb} = 0.2^{1/10} = 0.85$
Hence $p_{gb} = 1 - p_{gg} = 1 - 0.7 = 0.3$
and $p_{bg} = 1 - p_{bb} = 1 - 0.85 = 0.15$

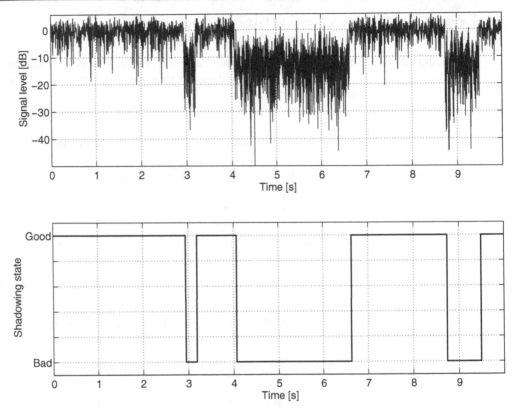

Figure 14.13: Example time series output from the Lutz model. Parameters used are $D_g = 24\,\text{m}$, $D_b = 33\,\text{m}$, $\mu = -10.6\,\text{dB}$, $\sigma_L = 2.6\,\text{dB}$, $k = 6\,\text{dB}$, $f_c = 1500\,\text{MHz}$ and $v = 50\,\text{km}\,\text{h}^{-1}$

The time-share of shadowing is then

$$A = \frac{D_b}{D_b + D_g} = \frac{1/p_{bg}}{1/p_{bg} + 1/p_{gb}} = \frac{p_{gb}}{p_{gb} + p_{bg}} = \frac{0.3}{0.3 + 0.15} = \frac{2}{3}$$

This two-state model of the mobile satellite channel is very useful for analysing and simulating the performance of mobile satellite systems. It is, however, inaccurate in representing the second-order statistics of shadowing, as it assumes the transition from LOS to NLOS situations is instantaneous. In practice this is not true, due to the smooth transitions introduced by diffraction and reflection effects, particularly at lower frequencies. One way of overcoming this is to introduce extra states, which represent intermediate levels of shadowing with smaller Rice k factors than the LOS 'good' state. It has been shown [Ahmed, 97] that at least three states are needed in order to accurately model these transitions, even at 20 GHz, and the expectation is that even more would be needed for accurate modelling at L band. Alternatively, the sharp transitions between states can be smoothed by filtering. In other work [Vucetic, 92], the statistics of the fading process vary according to the environment, with several environment types (urban, open, suburban) distinguished as model states.

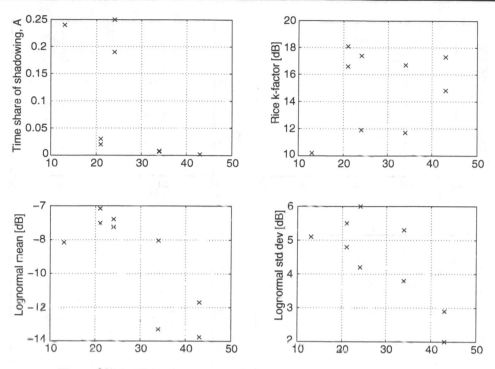

Figure 14.14. Highway parameters for the Lutz model: *x*-axis is elevation angle

14.5 SHADOWING STATISTICS

The model structures shown in Figures 14.9 and 14.11 require parameters of the shadowing statistics to allow proper design of the Doppler and shadowing filters. A simple approach, suggested in [Lutz, 91], is to select an independent value for the shadowing for the duration of each bad state. This does not, however, account for variations in shadowing which may occur during bad states of long duration. Measurements [Taaghol, 97] suggest that satellite shadowing autocorrelation can be modelled by a negative exponential, just as in the terrestrial case (Chapter 9), as shown in Figure 14.16. Measurements at L band produced a correlation distance between 9 and 20 m, depending somewhat on the elevation angle and environment.

14.6 PHYSICAL-STATISTICAL MODELS FOR BUILT-UP AREAS

Deterministic physical models of satellite mobile propagation have been created and have produced good agreement with measurements over limited areas [van Dooren, 93]. For practical predictions, however, an element of statistics has to be introduced. One approach is to devise physical models to examine typical signal variations in environments of various categories (Table 8.2) and then use them to generalise over wider areas. A more appropriate approach is to use *physical-statistical models*, which derive fading distributions directly from distributions of physical parameters using simple electromagnetic theory. This class of models will be described in some detail here, as the modelling approach has been examined only slightly in the open literature.

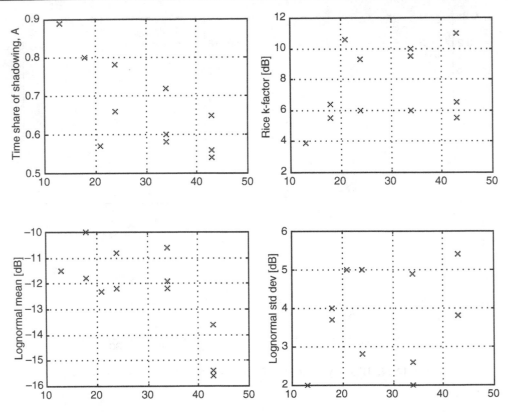

Figure 14.15: City parameters for the Lutz model: *x*-axis is elevation angle

In modelling any propagation parameter, the aims of modelling are broadly similar. The key point is to predict a particular parameter with maximum accuracy, consistent with minimum cost in terms of the quantity and expense of the input data and in terms of the computational effort required to produce the prediction.

For empirical models, the input knowledge consists almost entirely of previous measurements which have been made in environments judged to be representative of practical systems. An approximation to this data, usually consisting of a curve-fit to the measurements, is used for predictions. The input data is then fairly simple, consisting primarily of operating frequency, elevation angle, range and a qualitative description of the environment (e.g. rural, urban). Such models are simple to compute and have good accuracy within the parameter ranges spanned by the original measurements. However, as the models lack a physical basis, they are usually very poor at extrapolating outside these parameter ranges. There is a classification problem involved in describing the environment, as an environment judged to be urban in some countries may be little more than a small town elsewhere. Additionally, the use of a curve-fitting approach implies that the real data will generally be considerably scattered around the predicted values and this represents a lower limit on the prediction accuracy. For example, predictions of loss are subject to an error resulting from the effects of shadowing and fading.

The input knowledge used in physical models, by contrast, consists of electromagnetic theory combined with engineering expertise which is used to make reasonable assumptions

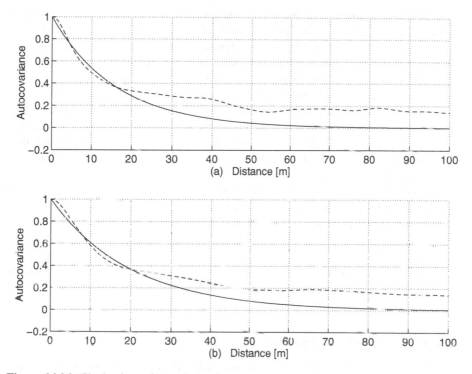

Figure 14.16: Shadowing autocorrelation from measurements in a suburban environment, L-band, 60° and 80° elevation: (---) measured, (---) negative exponential model

about which propagation modes are significant in a given situation. Provided that the correct modes are identified, the theoretical approach is capable of making very accurate predictions of a wide range of parameters in a deterministic manner. The output which can be given is point by point rather than an average value, so the model can apply to very wide ranges of system and environment parameters, certainly well beyond the range within which measurements have been made. In order to make such predictions, however, the models may require very precise and detailed input data concerning the geometrical and electrical properties of the environment. This may be expensive or even impossible to obtain with sufficient accuracy. Also, the computations required for a full theoretical calculation may be very complex, so extra assumptions often have to be made for simplification, leading to compromised accuracy. Physical-statistical modelling is a hybrid approach, which builds on the advantages of both empirical and physical models while avoiding many of their disadvantages. As in physical models, the input knowledge consists of electromagnetic theory and sound physical understanding. However, this knowledge is then used to analyse a *statistical* input data set, yielding a *distribution* of the output predictions. The outputs can still effectively be point by point, although the predictions are no longer linked to specific locations. For example, a physical-statistical model can predict the distribution of shadowing, avoiding the errors inherent in the empirical approach, although it does not predict what the shadowing value will be at a particular location. This information is usually adequate for the system designer. Physical-statistical models therefore require only simple input data such as input distribution parameters (e.g. mean building height, building height variance). The environment description is entirely objective, avoiding problems of subjective classification, and capable of high statistical accuracy. The models are based on sound physical principles, so they are applicable over very wide parameter ranges. Finally, by precalculating the effect of specific input distributions, the required computational effort can be very small.

One example of a physical-statistical model has been used to predict the attenuation statistics of roadside tree shadowing, using only physical parameters as input [Barts, 92]. This modelled the trees as consisting of a uniform slab whose height and width were uniformly distributed random variables. For a given direction from mobile to satellite, the mean and standard deviation of the path length through the block was calculated and used with a version of the modified exponential decay model described in Section 6.9 to calculate the mean and standard deviation of the tree attenuation. These values are then taken as the mean and standard deviation of the log-normal distribution in the Lutz structure (Figure 14.11). Another example uses a UTD analysis of diffraction and scattering from buildings as a physical basis, and then applies the statistics of building height to compute attenuation probabilities [Oestges, 99].

Sections 14.6.2 and 14.6.3 describe two physical-statistical models for megacells operated in built-up areas [Saunders, 96]; [Tzaras, 98a]. First, the basic physical parameters used in both these models are introduced.

14.6.1 Building Height Distribution

The geometry of the situation to be analysed is illustrated in Figure 14.17. It describes a situation where a mobile is situated on a long straight street with the direct ray from the satellite impinging on the mobile from an arbitrary direction. The street is lined on both sides with buildings whose height varies randomly.

In the models to be presented, the statistics of building height in typical built-up areas will be used as input data. A suitable form was sought by comparing them with geographical data for the City of Westminster and for the City of Guildford, UK (Figure 14.18). The probability density functions that were selected to fit the data are the log-normal and Rayleigh

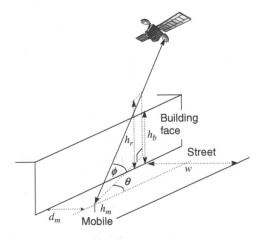

Figure 14.17: Geometry for mobile satellite propagation in built-up areas

distributions with parameters the mean value μ and the standard deviation σ_b. The p.d.f. for the log-normal distribution is

$$p_b(h_b) = \frac{1}{h_b\sqrt{2\pi}\sigma_b}\exp(-(1/2\sigma_b^2)\ln^2(h_b/\mu)) \tag{14.15}$$

and the p.d.f. for the Rayleigh distribution is

$$p_b(h_b) = \frac{h_b}{\sigma_b^2}\exp(-h_b^2/2\sigma_b^2) \tag{14.16}$$

The best-fit parameters for the distributions are shown in Table 14.2. The log-normal distribution is clearly a better fit to the data, but the Rayleigh distribution has the benefit of greater analytical simplicity.

Table 14.2: Best fit parameters for the theoretical p.d.f.s

| City | Log-normal p.d.f. | | Rayleigh p.d.f. |
	Mean μ	Standard deviation σ_b	Standard deviation σ_b
Westminster	20.6	0.44	17.6
Guildford	7.1	0.27	6.4

14.6.2 Time-Share of Shadowing

This model [Saunders, 96] estimates the time-share of shadowing (A in (14.11)) for the Lutz model [Lutz, 91] using physical-statistical principles. The direct ray is judged to be shadowed when the building height h_b exceeds some threshold height h_T relative to the direct ray height h_s at that point. Parameter A can then be expressed in terms of the probability density function of the building height $p_b(h_b)$:

$$P_s = \Pr(h_b > h_T) = \int_{h_T}^{\infty} p_b(h_b)\ dh_b \tag{14.17}$$

Assuming that the building heights follow the Rayleigh distribution (14.16) this yields

$$A = \int_{h_T}^{\infty} \frac{h_b}{\sigma_b^2}\exp\left(-\frac{h_b^2}{2\sigma_b^2}\right) dh_b = \exp\left(-\frac{h_T^2}{2\sigma_b^2}\right) \tag{14.18}$$

The simplest definition of h_T is obtained by considering shadowing to occur exactly when the direct ray is geometrically blocked by the building face (a more sophisticated approach would

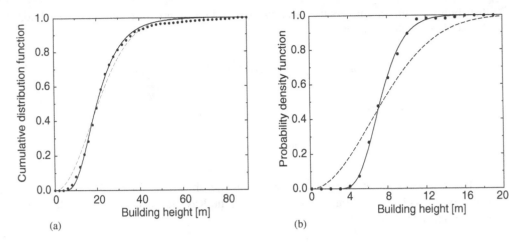

Figure 14.18: Building height distribution: (a) Westminster (b) Guildford

account for the size of the Fresnel zone at that point). Simple trigonometry applied to this yields the following expression for h_T:

$$h_T = h_r = \begin{cases} h_m + \dfrac{d_m \tan \phi}{\sin \theta} & for \quad 0 < \theta \le \pi \\[3mm] h_m + \dfrac{(w - d_m) \tan \phi}{\sin \theta} & for \quad -\pi < \theta \le 0 \end{cases} \qquad (14.19)$$

Figure 14.19 compares this model with measurements of A versus elevation angle in city and suburban environments at L band, taken from [Parks, 93], [Lutz, 91] and [Jahn, 1996]. The model parameters are $\sigma_b = 15$, $w = 35$, $d_m = w/2$, $h_m = 1.5$ and $\theta = 90°$.

14.6.3 Time Series Model

Although the approach in the last section allowed one of the parameters in the Lutz model to be predicted using physical-statistical methods, the rest of the parameters still have to be determined empirically. All of the statistical models described earlier assumed that the log-normal distribution was valid for predicting the shadowing distribution. However, Chapter 9 showed that this distribution comes as a result of a large number of individual effects acting together on the signal. In the case being treated here, only a single building is involved, so the log-normal approximation is questionable. The model described here avoids this assumption and directly predicts the statistics of attenuation [Tzaras, 98a]. The power received in this case is predicted as a continuous quantity, avoiding the unrealistic discretisation of state-based models such as [Lutz, 91].

The total received power for a mobile in a built-up area consists of the direct diffracted field associated with the diffraction of the direct path around a series of roadside buildings, plus a multipath component whose power is set by computing reflections from the

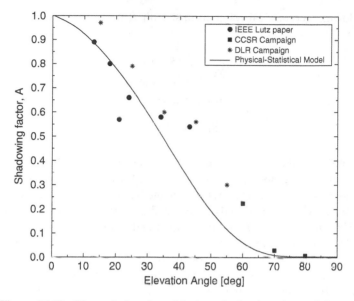

Figure 14.19: Theoretical and empirical results for time-share of shadowing

buildings on the opposite sides of the street and from the ground. The direct field is given by

$$u_0(P) = u_0(P) \left\{ 1 - \frac{j}{2} \sum_{m=1}^{N} \left[\left(\int_{-\infty}^{v_{x2}} e^{-(j\pi/2)v_x^2} dv_x \right) \cdot \left(\int_{v_{y1m}}^{v_{y2}} e^{-(j\pi/2)v_y^2} dv_y \right) \right] \right\} \quad (14.20)$$

where

$$v_{x1} = -\infty \quad v_{x2} = \sqrt{\frac{2(d_1 + d_2)}{\lambda d_1 d_2}} (x_{22} - x_m)$$

$$v_{yi} = \sqrt{\frac{k(d_1 + d_2)}{\pi d_1 d_2}} (y_{2i} - y_m) \quad i = 1, 2$$

$$(14.21)$$

where x_{22} defines the building height, y_{21} defines the position of the left building edge, y_{22} defines the position of the right building edge, d is the distance from the satellite to the building and d_2 is the distance from the building to the mobile. This result is similar to the Fresnel integral formulation of single knife-edge diffraction used in Chapter 3 but accounts for diffraction from around the vertical edges of buildings as well as over the rooftops. For the direct-diffracted field, $d_2 \equiv d_m$ and $x_3 = h_m$. For the wall-reflected path, Eq. (14.20) is again used but it is applied to the image of the source, so $d_2 = 2w - d_m$. For the ground-reflected field $x_3 = -h_m$. The satellite elevation angle and azimuth angles ϕ and θ are introduced using the following geometrical relationships, with R being the direct distance from mobile to satellite:

$$y_{\text{sat}} = \frac{R\cos\phi}{\sqrt{1 + \tan^2\theta}}$$
$$x_{\text{sat}} = R\sin\phi + h_m$$
$$d_1 = \sqrt{R^2\cos^2\phi - y_{\text{sat}}^2} - d_2$$

(14.22)

The formulation above is then used with a series of roadside buildings, randomly generated according to the log-normal distribution (14.15), with parameters applicable to the environment under study, including gaps between the buildings to represent some open areas. Figure 14.20 illustrates measured and simulated time series data for a suburban environment at 18.6 GHz with the same sampling interval and with 90° azimuth angle and 35° elevation angle. The other model parameters were $w = 16\,\text{m}$, $d_m = 9.5\,\text{m}$, open area $= 35\%$ of total distance, $\mu = 7.3\,\text{m}$, $\sigma_b = 0.26$. For the building and ground reflections, the conductivity was set to 0.2 and 1.7 S m^{-1}, and the relative permittivity was set to 4.1 and 12, respectively. Qualitatively, the characteristics of the signal variation are similar, although the statistical nature of the prediction implies that the model should not be expected to match the predictions at any particular locations.

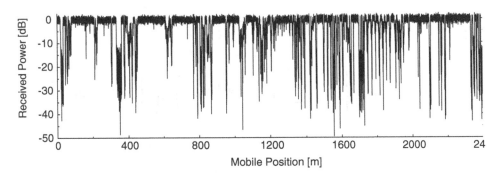

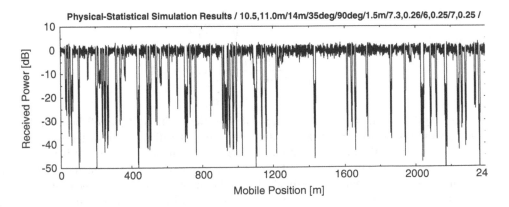

Figure 14.20: Theoretical output and measurement data

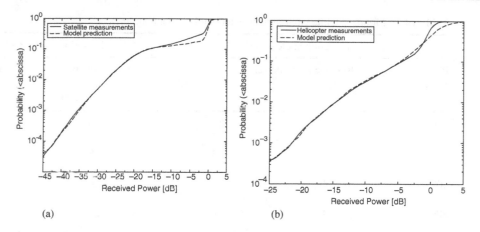

Figure 14.21: First-order statistics and real outdoor measurements: (a) satellite (b) helicopter

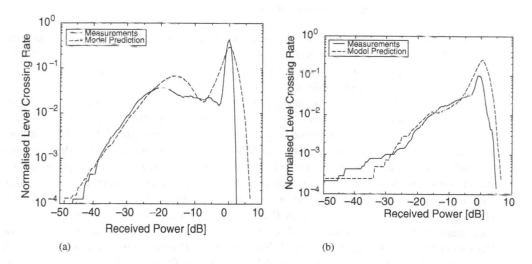

Figure 14.22: Second-order statistics from prediction and measurement: (a) satellite (b) helicopter

Figures 14.21 and 14.22 compare results for the first-order and second-order statistics for the same satellite measurements as in Figure 14.20 and also for helicopter measurements [Ahmed, 97]. The frequency for the helicopter measurements was 1.5 GHz, with 90° azimuth angle and 60° elevation angle.

14.7 WIDEBAND MODELS

Wideband measurement and modelling have received relatively little attention for mobile satellite systems, partly because the delay spreads encountered are far smaller than in most terrestrial

Table 14.3: Mobile satellite environment: wideband channel parameters[a]

	Elevation angle $< 45°$			Elevation angle $< 45°$	
Relative delay τ [ns]	Relative mean power [dB]	Rice k factor [dB]		Relative mean power[dB]	Rice k factor [dB]
0	0	12		0	16
100	−9.7	3		−15	6
200	−19.2	0		−20	0
300	−22.8	0		−26	0
400	—	0		−28	0
500	—	0		−30	0

[a]Doppler spectrum for multipath component of all taps is classical.

systems, rendering the channel essentially narrowband for most first-generation mobile satellite systems. However, future systems will offer multimedia services requiring very large channel bandwidths and will use spread spectrum techniques to provide high reliability and capacity, increasing the significance of wideband effects.

Shadowing effects in mobile satellite systems are dominated by the clutter in the immediate vicinity of the mobile. This is also true when considering wideband scattering, where the significant scatterers tend to contribute only relatively small excess path delays. The resulting wideband channel has been found to have an RMS delay spread of around 200 ns. The wideband channel model is then very similar to the terrestrial case except that the tap gain processes are each represented as instances of narrowband models such as those described in Sections 14.3, 14.4 and 14.6. Additionally, the first arriving tap will usually have a significant coherent component and hence a relatively high Rice k factor compared to the later echoes. Several measurements have been conducted [Jahn, 1996]; [Parks, 96]. Parks' work has led to results for the variation of delay spread at L and S bands in five different environmental categories at elevation angles from 15° to 80°. A tapped delay line model (Table 14.3) has been created based on the analysis of a representative fraction of the L band data [Parks, 97].

A more sophisticated structure has also been proposed [Jahn, 96] in which the earliest arriving echoes, which arise from scatterers within 200 m excess delay 600 ns of the mobile, are assumed to have exponentially decreasing power with increasing delay. In this region the number of echoes is assumed to be Poisson distributed and the delays are exponentially distributed. Occasional echoes also occur with longer delays, and the delays are uniformly distributed up to a maximum of around 15 µs.

14.8 MULTI-SATELLITE CORRELATIONS

When considering the effects of multiple-satellite diversity (or for more accuracy in the single-satellite case) it is crucial to consider the effects of the correlation between the fading

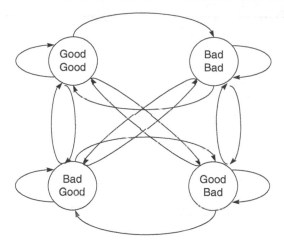

Figure 14.23: Four-state model of satellite diversity

encountered for satellites at different elevation *and* azimuth angles. Two satellites at the same elevation angle may exhibit very different outage probabilities when, for example, one is viewed perpendicular to the direction of a building-lined street but the other is viewed down the street and is therefore much more likely to be unshadowed.

One method of accounting for these effects is to generalise the Lutz model (Section 14.4.3) to four states, representing all the good and bad combinations of the two channels between mobile and satellite, as shown in Figure 14.23 [Lutz, 96]. The transition probabilities can again be found using information from measurements and models of shadowing time-share; it is also necessary, however, to have knowledge of the correlation between the shadowing states of the two channels.

When two satellites are present, their locations are defined by two pairs of elevation and azimuth angles (θ_1, ϕ_1) and (θ_2, ψ_2). The shadowing state of satellite i is defined as follows:

$$S_i = \begin{cases} 0 & \text{if } h_b \geq h_r \text{ (bad channel state)} \\ 1 & \text{if } h_b < h_r \text{ (good channel state).} \end{cases} \qquad (14.23)$$

where (θ_i, ϕ_i) is used in (14.19) to find the corresponding value for h_r. The correlation between these shadowing states is then defined by

$$\rho = \frac{E[(S_1 - \bar{S}_1)(S_2 - \bar{S}_2)]}{\sigma_1 \sigma_2} = \frac{1}{\sigma_1 \sigma_2}[E(S_1 S_2) - \bar{S}_1 \bar{S}_2] \qquad (14.24)$$

where $E[S] = \bar{S}$ is the mean shadowing state. If ρ is close to unity, the satellites suffer simultaneous shadowing always and satellite diversity does not increase system availability significantly. If ρ is close to zero, however, the satellites suffer simultaneous shadowing only rarely, and the system is available far more of the time than would be the case with a single satellite. Still better is the situation where ρ is close to -1, and the availability of the satellites complement each other perfectly.

Measurements of the correlation are available [Bischl, 96] which have used the results of circular flight-path measurements to derive a correlation function. The correlation function is

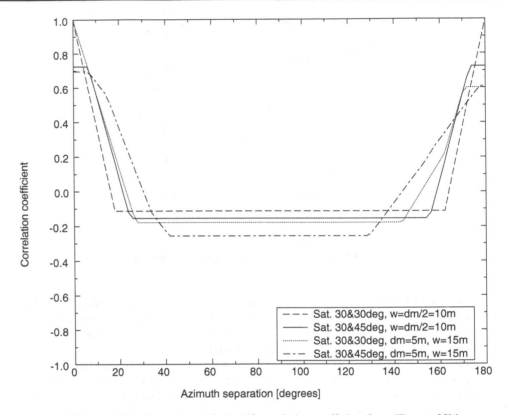

Figure 14.24: Example prediction of correlation coefficient from [Tzaras, 98b]

then used to derive appropriate transition probabilities for the model. Other approaches are to derive the correlations using fisheye lens photographs [Meenan, 1998] or using physical expressions [Tzaras, 98b]. In all cases the correlation encountered in built-up areas diminishes rapidly with increasing azimuth angle between the satellites, being sufficiently small for appreciable diversity gain above around 30° difference (Figure 14.24). Negative correlations are also possible when the environment has a particular geometrical structure. Parameters such as the time-share of shadowing can also be extracted from fisheye lens photographs and a very close correspondence between measurements and predicted parameters has been found [Lin, 98].

14.9 OVERALL MOBILE SATELLITE CHANNEL MODEL

The overall mobile satellite channel model is shown in Figure 14.25. There are two parts to the model: a *satellite process* which includes effects between the satellite and the Earth's surface and a *terrestrial process* which accounts for all the effects in the vicinity of the mobile.

The satellite process includes a delay τ_{sg} which arises from the total propagation path length; a total path loss (excluding shadowing) A, which includes free space and atmospheric loss components; and a Doppler shift through a frequency of $f_{sg} = \omega_{sg}/2\pi$, which arises from the relative speed of the satellite and a point on the ground adjacent to the mobile.

The terrestrial process is modelled as a time-variant transversal filter (tapped delay line) representation of the wideband channel, with tap-gain processes $r_1(t)$, $r_2(t), \ldots, r_n(t)$. Each

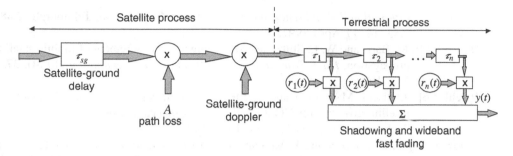

Figure 14.25: Complete mobile satellite channel model

of these processes may be modelled using any of the narrowband models described earlier in this chapter, so essentially the same parametric representation applies. Note that there are two quite distinct sources of Doppler in the megacell channel: one is a Doppler *shift* arising from satellite motion relative to the ground; the other is a Doppler *spread* arising from motion of the mobile relative to the scatterers in the immediate vicinity. It is assumed in this structure that all waves arriving at the mobile are subject to the same Doppler shift, which is a good approximation for scatterers in the near-vicinity of the mobile.

14.10 CONCLUSION

The megacell channel is a relatively new field of study, as non-GEO satellite systems have only recently been deployed. Nevertheless, the same basic physical principles apply as in terrestrial systems, with the differences in the resulting behaviour arising mainly due to the increased elevation angle from the mobile to the satellite (compared with a terrestrial base station) and due to the increased free space path length. Here are the key differences:

- The distance between the Earth and the satellite is very large, from a few thousand kilometres in the case of LEOs to around 36 000 km for geostationary orbits. This leads to a very large free space loss component.
- The elevation angle is much larger than for terrestrial systems, with minimum elevation angles in the range 8–25°. There is usually a direct path between mobile and satellite, so that little shadowing loss is experienced and relatively little fade margin needs to be included.
- When the elevation angle does become small, the shadowing is dominated by trees and buildings in the near-vicinity of the mobile. Empirical and theoretical models of this shadowing process enable the distribution of shadow fades to be calculated. Note that this distribution is no longer log-normal.
- The long path through the atmosphere is perturbed by effects such as scintillation and absorption by atmospheric gases. Rain attenuation becomes significant at higher frequencies.

REFERENCES

[Ahmed, 97] B. Ahmed, S. Buonomo, I. E. Otung, M. H. Aziz, S. R. Saunders and B. G. Evans, Simulation of 20 GHz narrowband mobile propagation data using *N*-state Markov channel modelling Approach, in *Proceedings of the 10th*

International Conference on Antennas and Propagation, Edinburgh, 2.48–2.53, 14–17 April 1997.

[Barts, 92] R. M. Barts and W. L. Stutzman, Modelling and measurement of mobile satellite propagation, *IEEE Transactions on Antennas and Propagation*, 40 (4), 375–382, 1992.

[Bischl, 96] H. Bischl, M. Werner and E. Lutz, Elevation-dependent channel model and satellite diversity for NGSO S-PCNs, in *Proceedings on the IEEE Vehicular Technology Conference*, 1038–1042, 1996.

[Corazza, 94] G. E. Corazza and F. Vatalaro, A statistical-model for land mobile satellite channels and its application to nongeostationary orbit systems, *IEEE Transactions on Vehicular Technology*, 43 (3), 738–742, 1994.

[van Dooren, 93] G. A. J. van Dooren, M. H. A. J. Herben, G. Brussard, M. Sforza and J. P. V. Poiares Baptista, Electromagnetic field strength prediction in an urban environment: a useful tool for the planning of LMSS, in *Proceedings on International Mobile Satellite Conference*, 343–348, 1993.

[Evans, 99] B. G. Evans (ed.), *Satellite communication systems*, 3rd edn., IEE, London, ISBN 0 85296 899 X, 1999.

[Goldhirsh, 92] J. Goldhirsh and W. J. Vogel, Propagation effects for land mobile satellite systems: Overview of experimental and modelling results, *NASA Reference Publication 1274*, February 1992.

[Hwang, 97] Seung-Hoon Hwang, Ki-Jun Kim, Jae-young Ahn and Keum-Chan Whang, A channel model for nongeostationary orbiting satellite system, in *Proceedings on IEEE International Vehicular Technology Conference*, Phoenix AZ, 5–7 May 1997.

[ITU, 681] International Telecommunication Union, *ITU-R Recommendation P.681-3: Propagation data required for the design of earth-space land mobile telecommunication systems*, Geneva, 1997.

[Jahn, 96] A. Jahn, H. Bischl and G. Heiss, Channel characterization for spread-spectrum satellite-communications, in *Proceedings on IEEE Fourth International Symposium on Spread Spectrum Techniques and Applications*, 1221–1226, 1996.

[Lin, 98] H. P. Lin, R. Akturan and W. J. Vogel, Photogrammetric prediction of mobile satellite fading in roadside tree-shadowed environment, *Electronics Letters*, 34 (15), 1524–1525, 1998.

[Loo, 85] C. Loo, A statistical model for a land mobile satellite link, *IEEE Transactions on Vehicular Technology*, 34 (3), 122–127, 1985.

[Lutz, 91] E. Lutz, D. Cygan, M. Dippold, F. Dolainsky and W. Papke, The land mobile satellite communication channel-recording, statistics and channel model, *IEEE Transactions on Vehicular Technology*, 40 (2), 375–375, 1991.

[Lutz, 96] E. Lutz, A Markov model for correlated land mobile satellite channels, *International Journal of Satellite Communications*, 14, 333–339, 1996.

[Meenan, 98] C. Meenan, M. Parks, R. Tafazolli and B. Evans, Availability of 1st generation satellite personal communication network service in urban environments, in *Proceedings on IEEE International Vehicular Technology Conference*, 1471–75, 1998.

[Oestges, 99] C. Oestges, S. R. Saunders and D. Vanhoenacker-Janvier, Physical-statistical modelling of the land mobile satellite channel based on ray-tracing, *IEE Proceedings on Microwaves, Antennas and Propagation*, 146 (1), 45–49, 1999.

[Parks, 93] M. A. N. Parks, B. G. Evans and G. Butt, High elevation angle propagation results applied to a statistical model and an enhanced empirical model, *Electronics Letters*, 29 (19), 1723–1725, 1993.

[Park, 96] M. A. N. Parks, B. G. Evans, G. Butt and S. Buonomo, Simultaneous wideband propagation measurements for mobile satellite communications systems at L- and S-bands, in *Proceedings on 16th International Communications Systems Conference*, Washington DC, 929–936, 1996.

[Parks, 97] M. A. N. Parks, S. R. Saunders and B. G. Evans, Wideband characterisation and modelling of the mobile satellite propagation channel at L and S bands, in *Proceedings on International Conference on Antennas and Propagation*, pp. 2.39–2.43, Edinburgh 1997.

[Pattan, 98] B. Pattan, *Satellite-Based Cellular Communications*, McGraw-Hill, New York, ISBN 0-07-0494177, 1998.

[Rice, 48] S. O. Rice, Statistical properties of a sine wave plus random noise, *Bell System Technical Journal*, 27, 109–157, 1948.

[Saunders, 96] S. R. Saunders and B. G. Evans, A physical model of shadowing probability for land mobile satellite propagation, *Electronics Letters*, 32 (17), 1548–1589, 1996.

[Taaghol, 97] P. Taaghol and R. Tafazolli, Correlation model for shadow fading in land-mobile satellite systems, *Electronics Letters*, 33 (15), 1287–1289, 1997.

[Tzaras, 98a] C. Tzaras, B. G. Evans and S. R. Saunders, A physical-statistical analysis of the land mobile satellite channel, *Electronics Letters*, 34 (13), 1355–1357, 1998.

[Tzaras, 98b] C. Tzaras, S. R. Saunders and B. G. Evans, A physical-statistical propagation model for diversity in mobile satellite PCN, in *Proceedings on IEEE International Vehicular Technology Conference*, Ottawa, Canada, 525–29, May 1998.

[Vogel, 86] W. J. Vogel and J. Goldhirsh, Tree attenuation at 869 MHz derived from remotely piloted aircraft measurements, *IEEE Transactions on Antennas and Propagation*, 34 (12), 1460–1464, 1986.

[Vucetic, 92] B. Vucetic and J. Du, Channel model and simulation in satellite mobile communication systems, *IEEE Transactions on Vehicular Technology*, 10, 1209–1218, 1992.

PROBLEMS

14.1 Use the ERS model to calculate the availability of a mobile satellite system with an elevation angle of 70°, carrier frequency of 10 GHz, and a fade margin of 10 dB.

14.2 Describe two ways in which a high-directivity terminal antenna in a mobile satellite system will reduce the impact of multipath fading.

15 Antennas for Mobile Systems

'This new form of communication could have some utility'.
Guglielmo Marconi

15.1 INTRODUCTION

The mobile systems described in Chapters 8, 12, 13 and 14 all require carefully designed antennas at both the base station and the user terminal (mobile) end for efficient operation. It is essential to carefully control radiation patterns to target coverage to desired coverage areas while minimising interference outside. The antenna structure should minimise interactions with its surroundings, such as supporting structures and the human body, while maximising the efficiency for radiation and reception. The requirements of antennas at both the mobile terminal (Section 15.2) and the base station (Section 15.3) are examined in this chapter, together with an outline of the main structures which are suitable and the key design issues. The chapter draws on the fundamentals described in Chapter 4. For further detail, the reader is referred to books such as [Stutzmann, 98] and [Vaughan, 03] for detailed design theory and to [Fujimoto, 00] and [Balanis, 07] for practical information specific to mobile systems.

15.2 MOBILE TERMINAL ANTENNAS

15.2.1 Performance Requirements

Mobile terminal antennas include those used in cellular phones, walkie-talkies for private and emergency service applications and data terminals such as laptops and personal digital assistants. Such antennas are subjected to a wide range of variations in the environment which they encounter. The propagation conditions vary from very wide multipath arrival angles to a strong line-of-sight component. The orientation of the terminal is often random, particularly when a phone is in a standby mode. They must be able to operate in the close proximity of the user's head and hand. They must also be suitable for manufacturing in very large volumes at an acceptable cost. Increasingly, also, users prefer that the antenna be fully integrated with the casing of the terminal rather than being separately identifiable. In general, these challenging requirements may be summarised as follows:

- *Radiation pattern*: Approximately omnidirectional in azimuth and wide beamwidth in the vertical direction, although the precise pattern is usually uncritical and given the random orientation, the large degree of multipath and the pattern disturbance which is inevitable and given the close proximity of the user.
- *Input impedance*: Should be stable and well-matched to the source impedance over the whole bandwidth of operation, even in the presence of detuning from the proximity of the

Antennas and Propagation for Wireless Communication Systems Second Edition   Simon R. Saunders and
Alejandro Aragón-Zavala
© 2007 John Wiley & Sons, Ltd

user and other objects. Many user terminals now operate over a wide variety of standards, so multi-band, multi-mode operation via several resonances is increasingly a requirement.

- *Efficiency*: Given the low gain of the antenna, it is important to achieve a high translation of input RF power into radiation over the whole range of conditions of use.
- *Manufacturability*: It should be possible to manufacture the antenna in large volumes efficiently, without the need for tuning of individual elements, while being robust enough against mechanical and environmental hazards encountered while moving.
- *Size*: Generally as small as possible, consistent with meeting the performance requirements. Increasingly, the ability to adapt the shape to fit into casing acceptable to a consumer product is important.

This section outlines the most important issues to consider in analysing this performance and discusses a selection of the various antenna types available and emerging.

15.2.2 Small Antenna Fundamentals

Mobile antennas can often be classified as electrically small antennas, defined as those whose radiating structure can be contained within a sphere of radius r such that $kr < 1$, where k is the wave number (i.e. $r \leq \frac{\lambda}{6}$ approximately). There is a body of literature which highlights the performance trade-offs inherent in such antennas. In particular, there is a basic trade-off between size, bandwidth and directivity. Small antennas, with limited aperture size, cannot achieve high directivity. Similarly, the bandwidth and directivity cannot both be increased if the antenna is to be kept small [McLean, 96]. As the size decreases, the radiation resistance decreases relative to ohmic losses, thus decreasing efficiency. Increasing the bandwidth tends to decrease efficiency, although dielectric or ferrite loading can decrease the minimum size. This is best illustrated by examining the relationship between the quality factor, Q of the antenna and the size. The unloaded Q of an antenna is defined as

$$Q = \frac{f_0}{\Delta f_{3dB}} \qquad (15.1)$$

Note that the loaded bandwidth is twice this value. The smallest Q (and hence highest fractional bandwidth) within a given size of enclosing sphere is obtained from a dipole-type field, operating at a fundamental mode of the antenna. Higher-order modes should be avoided as they have intrinsically higher Q. Assuming a resistive matched load, for a first-order mode, whether electrical or magnetic, the Q is related to the electrical size of the antenna by [McLean, 96]

$$Q_{01} = \frac{1}{(kr)^3} + \frac{1}{kr} \qquad (15.2)$$

If both the electrical and magnetic first-order modes are excited together, then the combined Q is somewhat lower,

$$Q_{E,M01} = \frac{1}{2(kr)^3} + \frac{1}{kr} \qquad (15.3)$$

These values are illustrated in Figure 15.1. Practical antennas will always exceed these values. A spherical antenna, such as a spherical helix, approaches the limits most closely but is unlikely to be practical for mobile terminal applications.

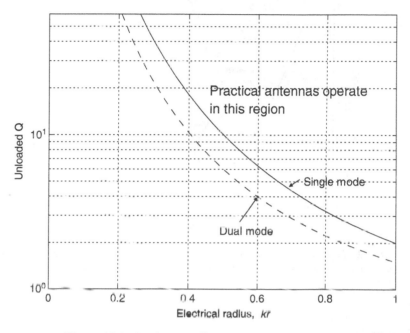

Figure 15.1: Fundamental limits on small antenna efficiency

This analysis assumes a resistive matched load. The bandwidth can be enhanced using a matching network, such as a bandpass filter, which provides a varying load impedance to the antenna. An even better approach is to make the matching network part of the structure, so that the associated currents radiate. This can be achieved via multiple resonant portions to the antenna, such as dipoles of various sizes connected in parallel or stacked patches of varying sizes.

To illustrate this near resonance, an antenna can be represented by an equivalent series *RLC* tuned circuit with an impedance given by

$$
\begin{aligned}
Z &= R + j\left(\omega L - \frac{1}{\omega C}\right) \\
&= R\left(1 + jQ\left(\frac{f}{f_0} - \frac{f_0}{f}\right)\right)
\end{aligned}
\tag{15.4}
$$

From this we can calculate the VSWR via Eqs. (4.13) and (4.14). For a simple half-wave dipole with $R = 72\,\Omega$, we obtain the simple dipole characteristics shown in Figure 15.2. By feeding two dipoles with different resonant frequencies at the same input terminals, the impedances appear as a parallel combination and the result has a much wider bandwidth (mutual coupling interactions are neglected here).

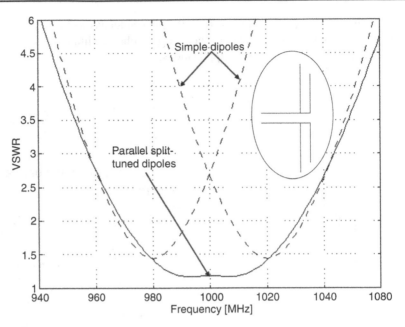

Figure 15.2: Bandwidth enhancement via parallel dipoles. The dipoles are combined in parallel and have resonant frequencies of 980 MHz and 1020 MHz. The source resistance is assumed to be 50 Ω

15.2.3 Dipoles

The most basic starting point for a mobile terminal antenna is a half-wave dipole as described in Section 4.4. Such an antenna has a bandwidth which increases with the diameter of the wire elements in comparison with the overall length. For a length:diameter ratio of 2500, the VSWR = 2 bandwidth is around 8% of the resonant frequency, increasing to 16% for a ratio of 50. A common practical realisation of such an antenna is the folded dipole, Figure 15.3(a), which increases the impedance four-fold, providing a good match to 300 Ω balanced transmission line and improving the bandwidth. Similarly, Figure 15.3(b) shows a configuration which provides a coaxial feed. Such configurations are in common practical use as receive antennas for home FM/VHF radio or as the driven element in Yagi-Uda antennas (Section 4.5.4). The bandwidth may be further increased by broadening or end-loading one or both of the dipole elements, which can produce antennas with bandwidths of many octaves, suitable for ultra-wideband systems (UWB) (Figure 15.4).

Although the half-wave dipole is a versatile and useful antenna in some applications, it is likely to be too large for convenient operation on portable mobiles (e.g. 16.5 cm at 900 MHz). In principle, the lower arm of the dipole could be made the conducting case of the mobile terminal, but then the impedance and the radiation pattern will be severely affected by the interaction between the currents on the case and the user's hand, leading to poor performance.

A quarter-wave monopole (or *quarter-wave whip*), operating over a ground plane, would seem to reduce the antenna size and permit a coaxial feed. However, since the dominant radiation direction of a monopole is along the ground plane, the ground plane size needs to be

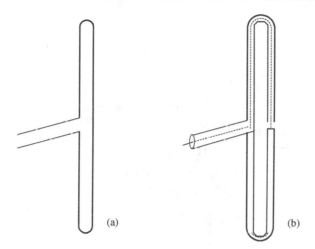

Figure 15.3: Folded half-wave dipoles: (a) balanced feed; (b) coaxial feed

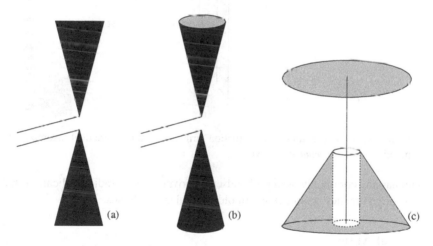

Figure 15.4: Wideband dipole variants: (a) bow-tie antenna; (b) bi-conical dipole; (c) top-loaded monopole on conical groundplane or 'discone' with coaxial feed

several wavelengths to produce stable input impedance, rendering the antenna impractical. Nevertheless, such antennas are sometimes used when performance is uncritical.

A common solution to this is to use a *sleeve dipole* as shown in Figure 15.5 [Fujimoto, 00]. Here the radiating element of the antenna is fed coaxially, with the outer conductor of the coaxial line surrounded by a metal sleeve. The metal sleeve is filled with a cylindrical dielectric insert. If the sleeve is made large in diameter, and the sleeve length and the dielectric constant are chosen appropriately, the sleeve can act as a resonant 'choke', which allows RF currents to flow in the sleeve outer but minimises currents in the terminal case. The antenna is thus fairly robust against variations due to the user's hand on the case and it acts as an asymmetrically-fed half-wave dipole with a dipole-like radiation pattern. At frequencies more than around 5% away from the choke resonant frequency, the vertical radiation pattern tends to be multi-lobed. Sleeve

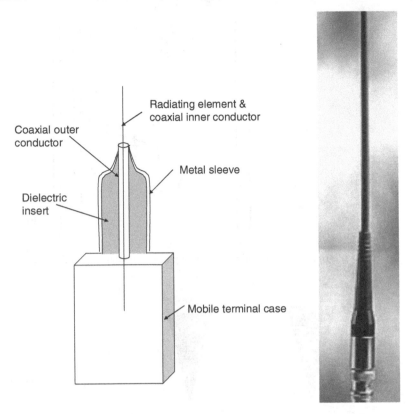

Figure 15.5: Sleeve dipole construction principles and a practical example (Reproduced by permission of *Jaybeam Wireless*)

antennas are commonly used for handsets in private mobile radio applications (such as for taxi firms or emergency services) and in older cellular telephones.

15.2.4 Helical Antennas

Given the large size of even a quarter-wave whip at VHF frequencies for hand-held operation, it is common to reduce the physical size of the radiating element by using a helical antenna radiating in normal mode as described in Section 4.8. However, shortening the element in this way increases the losses, so it is usual to make the element physical axial length no smaller than approximately $\lambda/12$ (while maintaining its electrical length at $\lambda/4$). Making the element too small also increases the shadowing of the antenna by the user's head. This configuration relies on the use of the case as the balancing element, so it is again strongly influenced by the position of the user's hand on the case. Combining a helical element with a sleeve is one approach, but the sleeve will then dominate the overall size.

15.2.5 Inverted-F Antennas

Given the large ground-plane required for efficient operation with a monopole, one solution is to deploy an antenna which produces its maximum radiation normal to the ground plane. The ground plane can then be one side of the terminal case. One such antenna is the inverted-F antenna (Figure 15.6). If the image of this antenna reflected in the ground plane is considered, the antenna appears as a two-wire balanced transmission line with a short circuit at one end. Analysis in this way allows the dimensions to be set to provide resonance and an appropriate impedance match.

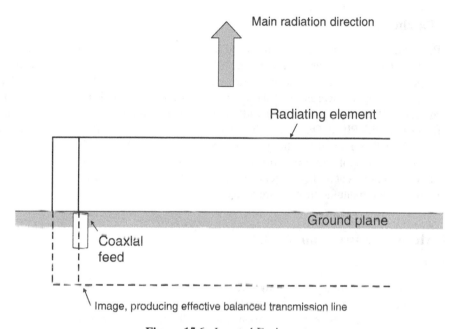

Figure 15.6: Inverted-F wire antenna

A popular development of this antenna type is the *planar inverted-F antenna* (*PIFA*), otherwise referred to as an open microstrip antenna, as shown in Figure 15.7 [James, 89]. The wires in the inverted-F antenna are replaced by metal plates, yielding a parallel-plate waveguide between the ground plane and patch, which is often dielectric loaded to reduce the size. The fields set up in this waveguide are sinusoidal along the length and uniform across the open width, so the radiation pattern is similar to a uniform current dipole in the space above the ground plane. Although the overall dimensions of the antenna are not

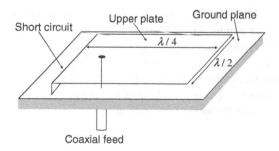

Figure 15.7: Planar inverted-F antenna

particularly small, the low profile makes it compact and unobtrusive for cellular telephone applications. The bandwidth is reasonably large, increasing with the height up to around 14%. The bandwidth can be increased still further by adding parasitic conductor layers connected to the upper plate. Its major disadvantage is the awkwardness in manufacture of the short circuit plate.

15.2.6 Patches

Patch antennas, such as those introduced in Section 4.9, are also a common form of portable mobile antenna. They are commonly a half-wavelength square, which is not especially small (depending on the dielectric constant) overall, but is compact vertically and avoids the manufacturing awkwardness of the vertical short-circuit in the PIFA. The fields are approximately uniform across the patch width and sinusoidal along the length, resulting in similar patterns to the PIFA. The directivity reduces with increases in the dielectric constant, from around 10 dB for an air substrate to about 5.5 dB for $\varepsilon_r = 10$. The Q is inversely proportional to the patch height and proportional to $\sqrt{\varepsilon_r}$. As with the PIFA, additional layers, producing offset resonances at higher frequencies, can be added to increase the bandwidth, although this increases the manufacturing complexity.

15.2.7 Mean Effective Gain (MEG)

Introduction

The performance of a practical mobile antenna in its realistic operating environment may be very different than would be expected from measurements of the gain of the antenna in isolation. This arises because the mobile is usually operated surrounded by scattering objects which spread the signal over a wide range of angles around the mobile. The detailed consequences of this on the fading signal were examined in Chapter 10. The question arises, given this complexity, as to what value of mobile antenna gain should be adopted when performing link budget calculations. The concept of a *mean effective gain* which combines the radiation performance of the antenna itself with the propagation characteristics of the surrounding environment was introduced by [Taga, 1990] to address this and is described in some detail here.

Formulation

Consider a mobile antenna which receives power from a base station after scattering has occurred through a combination of buildings, trees and other clutter in the environment. The total average power incident on the mobile is composed of both horizontally and vertically polarised components, P_H and P_V, respectively. All powers are considered as averages, taken after the mobile has moved along a route of several wavelengths. The mean effective gain (MEG) of the antenna, G_e, is then defined as the ratio between the power which the mobile actually receives and the total which is available,

$$G_e = \frac{P_{rec}}{P_V + P_H} \tag{15.5}$$

The received power at the antenna can then be expressed in spherical coordinates taking into account the three-dimensional spread of incident angles as follows:

$$P_{rec} = \int\limits_{0}^{2\pi} \int\limits_{0}^{\pi} [P_1 G_\theta(\theta, \phi) P_\theta(\theta, \phi) + P_2 G_\phi(\theta, \phi) P_\phi(\theta, \phi)] \sin\theta d\theta d\phi \qquad (15.6)$$

where P_1 and P_2 are the mean powers which would be received by ideally θ (elevation) and ϕ (azimuth) polarised isotropic antennas, respectively, G_θ and G_ϕ are the corresponding radiation patterns of the mobile antenna and P_θ and P_ϕ represent the angular distributions of the incoming waves. The following conditions must be satisfied to ensure that the functions are properly defined:

$$\int\limits_{0}^{2\pi} \int\limits_{0}^{\pi} [G_\theta(\theta, \phi) + G_\phi(\theta, \phi)] \sin\theta d\theta d\psi = 4\pi \qquad (15.7)$$

$$\int\limits_{0}^{2\pi} \int\limits_{0}^{\pi} P_\theta(\theta, \phi) \sin\theta d\theta d\phi = 1 \qquad (15.8)$$

$$\int\limits_{0}^{2\pi} \int\limits_{0}^{\pi} P_\phi(\theta, \phi) \sin\theta d\theta d\phi = 1 \qquad (15.9)$$

The angular distribution of the waves may be modelled by, for example, Gaussian distributions in elevation and uniform in azimuth as follows:

$$P_\theta(\theta, \phi) = A_\theta \exp\left\{ -\left[\theta - \left(\frac{\pi}{2} - m_V\right)\right]^2 \times \frac{1}{2\sigma_V^2} \right\} \qquad (15.10)$$

$$P_\phi(\theta, \phi) = A_\phi \exp\left\{ -\left[\phi - \left(\frac{\pi}{2} - m_H\right)\right]^2 \times \frac{1}{2\sigma_H^2} \right\} \qquad (15.11)$$

where m_V and m_H are the mean elevation angles of the vertically and horizontally polarised components, respectively, σ_v and σ_H are the corresponding standard deviations and A_θ and A_ϕ are chosen to satisfy Eqs. (15.8) and (15.9). The precise shape of the angular distribution is far less important than its mean and its standard deviation.

Example MEG Calculation

Since the arrival angle has been assumed uniform in azimuth, any variations from omnidirectional in the radiation pattern will have no impact upon the MEG. Although this assumption is likely to be valid in the long term as the mobile user's position changes, there may be short-term cases where this is not so, and the power arrives from a dominant direction. This may particularly be the case in a rural setting where a line-of-sight, or near-line-of-sight path exists. Figure 15.8 shows the radiation pattern obtained from a pair of dipoles arranged $\lambda/4$ apart and fed with equal phase and amplitude. This could, for example, represent an attempt to

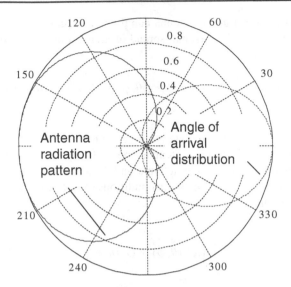

Figure 15.8: Azimuth radiation pattern and angle-of-arrival distribution

reduce radiation into the human head by placing a null in the appropriate direction. This results in a gain pattern of the form

$$G_\theta = \cos^2\left(\frac{\pi}{4} + \frac{\pi}{4}\cos\phi\right) \tag{15.12}$$

Also shown is the arrival angle distribution, assumed Gaussian in azimuth and shown with a standard deviation of 50° and a mean of 0°. Calculation of Eq. (15.6) in comparison with the assumption of uniform arrival angle in the azimuth plane yields the results shown in Figure 15.9. This is performed with the centre of the arriving waves both within the pattern null and directly opposite and is shown as a function of the standard deviation of the spread relative to the mean. Considerable gain reduction is evident in both cases, particularly when the angular spread is small.

15.2.8 Human Body Interactions and Specific Absorption Rate (SAR)

In the frequency range of interest for practical radio communications, electromagnetic radiation is referred to as 'non-ionising radiation' as distinct from the ionising radiation produced by radioactive sources. The energy associated with the quantum packets or *photons* at these frequencies is insufficient to dissociate electrons from atoms, whatever be the power density, so the main source of interactions between non-ionising radiation and surrounding human tissue is simple heating. Nevertheless, given that we are all continually exposed to EM radiation from a variety of sources, including mobile phone systems, common sources of radiowaves include radio and television broadcasts, there is understandable concern to ensure that human health is not adversely affected.

The potential health impact of EM fields has been studied for many years by both civil and military organisations, as well as the effects and interactions of handheld antennas with the human head (e.g. [Suvannapattana, 99]). A number of bodies have commissioned research

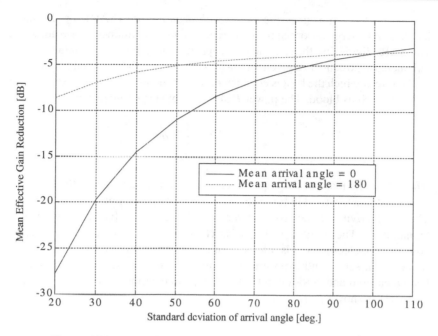

Figure 15.9: Reduction in mean effective gain due to angular spreading

into such effects and the World Health Organisation has produced guidelines to ensure that this research is conducted according to appropriate standards [Repacholi, 97]. The conclusions from these investigations have been used to set regulatory limits on exposure which reflect a precautionary principle based on the current state of knowledge. Many administrations require equipment manufacturers to ensure that the fields absorbed are below given limits and to quote the values produced by individual equipment under suitable reference conditions. Therefore, it is essential at this stage to establish procedures and metrics to assess the impact of antennas on absorption within the body.

Formulation

The evaluation of human exposure in the near field of RF sources, like portable mobile phones, can be performed by measuring the electric field (E-field) *inside* the body [Fujimoto, 00]. Given a current density vector **J** and an electric field **E** the power absorbed per unit volume of human tissue with conductivity $\sigma[\Omega^{-1}\,m^{-1}]$ is

$$P_V = \frac{1}{2}\mathbf{J}\cdot\mathbf{E}^* = \frac{1}{2}\sigma|\mathbf{E}|^2 \quad (\mathrm{W\,m^{-3}}) \tag{15.13}$$

By introducing the density of the tissue ρ in $[\mathrm{kg\,m^{-3}}]$, then the absorption per unit mass is obtained as follows:

$$P_g = \frac{1}{2}\frac{\sigma}{\rho}|E|^2 \quad [\mathrm{W\,kg^{-1}}] \tag{15.14}$$

The term P_g is known as the *specific absorption rate* (SAR), or the rate of change of incremental energy absorbed by an incremental mass contained in a volume of given density. The SAR (divided by the specific heat capacity) indicates the instantaneous rate of temperature increase possible in a given region of tissue, although the actual temperature rise depends on the rate at which the heat is conducted away from the region, both directly and via the flow of fluids such as blood. The power P absorbed by a specific organ is given by

$$P = \int_M P_g dm \tag{15.15}$$

where M is the mass of the object in consideration, which can be the entire human body.

The parameters used in Eqs. (15.13) and (15.14) are complex and depend on many factors. The conductivity of the tissue σ varies with frequency [Gabriel, 96] and increases with temperature. The tissue density ρ also changes with tissue and is a function of the water content. The electric field is often determined by the dielectric and physical properties of the tissue, polarisation and exposure environment. The conductivity at various frequencies has been measured and reported in [Gabriel, 96]. An example of some tissue dielectric values at 900 MHz is shown in Table 15.1.

SAR Measurements

A popular method to perform SAR measurements is by logging E-field data in an artificial shape acting as a representation of the human body in normal use, from which SAR distribution and peak averaged SAR can be computed. To do so repeatably, it is important to use an appropriate body shape and dielectric material. Appropriate 'phantoms' for the human head and other body parts are standardised by relevant committees [IEEE, 00]; [CENELEC, 00], along with the associated positioning of phones and other devices to be tested. Examples of some commercial SAR measurement systems are shown in Figure 15.10. Here a robotic arm is used to hold the E-field probe and scan the whole exposed volume of the phantom, in order to evaluate the 3D field distribution. The use of the robot allows high

Table 15.1: Tissue dielectric properties at 900 MHz, from [Gabriel, 96]

Tissue	Permittivity (ε)	Conductivity (σ)
Bladder	18.93	0.38
Fat (Mean)	11.33	0.10
Heart	59.89	1.23
Kidney	58.68	1.39
Skin (dry)	41.40	0.87
Skin (wet)	46.08	0.84
Muscle (parallel fibre)	56.88	0.99
Muscle (transverse fibre)	55.03	0.94
Cerebellum	49.44	1.26
Breast fat	5.42	0.04
Average brain	45.80	0.77
Average skull	16.62	0.24
Average muscle	55.95	0.97

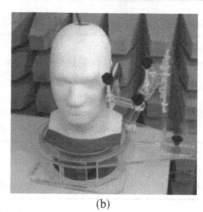

(b)

(a)

Figure 15.10: SAR measurement systems; (a) CENELEC compliant system, (b) Phantom used in SAR tests, with mobile phone head mounting bracket. (Reproduced by permission of IndexSAR Ltd.)

repeatability and very high position accuracy. A similar procedure can also be used for testing the SAR arising from indoor antennas to establish whether they are 'touch safe'.

At the early design stages of an antenna, it is also possible to make an assessment of the likely SAR via analytical or numerical calculations. The near-field nature of the problem means that the geometrical optical approaches discussed in earlier chapters are not appropriate and full-wave solutions of Maxwell's equations must be applied [Sewell, 95].

SAR Regulations

After characterising the RF exposure produced through SAR measurements, it is necessary to assess whether this exposure falls beyond acceptable limits. ICNIRP (International Committee on Non-ionising Radiation Protection) is an independent non-governmental scientific organisation, set up by the World Health Organisation and the International Labour Office, responsible for providing guidance and advice on the health hazards of non-ionising radiation exposure [ICNIRP, 98], and levels based on its recommendations are widely adopted in Europe. On the contrary, the IEEE C95.1-1999 Standard for Safety Levels with Respect to Human Exposure to Radio Frequency Electromagnetic Fields [IEEE, 91] has been adopted in America as a reference, which includes frequencies between 3 kHz and 300 GHz. In 1999, the European Committee for Electrotechnical Standardisation [CENELEC, 99] endorsed the guidelines set by ICNIRP on exposure reference levels, and recommended that these should form the basis of the European standard .

Table 15.2 shows the basic SAR limits for both ICNIRP and IEEE standards. Both standards make a clear distinction between general public (uncontrolled environment) and occupational (controlled environment). For the former, people with no knowledge of or no control over their exposure are included, and hence the exposure limits need to be tighter.

Table 15.2 SAR exposure limits $[\text{W kg}^{-1}]$ [Fujimoto, 00]

Standard	Condition	Frequency	Whole body	Local SAR (head and trunk)	Local SAR (limbs)
ICNIRP	Occupational	100 kHz–10 GHz	0.4	10	20
	General public	100 kHz–10 GHz	0.08	2	4
IEEE	Controlled	100 kHz–6 GHz	0.4	8	20
	Uncontrolled	100 kHz–6 GHz	0.08	1.6	4

Table 15.3: ICNIRP reference field strength levels $[\text{V m}^{-1}]$

Standard	Condition	>10 MHz <400 MHz	900 MHz	1.8 GHz	>2 GHz <300 GHz
ICNIRP	General public	28	41.25	58.3	61
	Occupational	61	90	127.3	137

Table 15.4: IEEE reference E-field and power density levels

Standard	Condition	E-Field $[\text{V m}^{-1}]$		Power density $[\text{mW cm}^{-2}]$	
		>30 MHz <300 MHz	900 MHz	1.8 GHz	>15 GHz <300 GHz
IEEE	Uncontrolled	27.5	0.6	1.2	10
	Controlled	61.4	3	6	10

However, the general public values are often regarded as representing best practice, whoever be the affected parties.

The ICNIRP and IEEE standards also establish field strength and power density limits for far-field exposure as shown in Tables 15.3 and 15.4. Notice the variations in maximum E-field exposure with frequency. These are the levels typically adopted when testing and predicting the fields around macrocellular base station antennas.

15.2.9 Mobile Satellite Antennas

The key requirements for the antenna on a mobile handset for non-geostationary satellite systems, such as those described in Chapter 14, can be summarised as follows:

- *Omnidirectional, near-hemispherical radiation pattern*. This allows the handset to communicate with satellites received from any elevation and azimuth angle, without any special cooperation by the user. The elevation pattern should extend down to at least the minimum elevation angle of the satellite, but should not provide too much illumination of angles below the horizon, since this would lead to pick-up of radiated noise from the ground and degradation of the receiver noise figure. The pattern need not necessarily be uniform within the beamwidth; indeed, it may be an advantage in some systems to emphasise lower elevation angles at the expense of higher ones in order to overcome the extra free space and shadowing losses associated with lower satellites [Agius, 97].

- *Circular polarisation, with axial ratio close to unity.* This limits the polarisation mismatch. A typical specification is that the axial ratio should be no more than 5 dB at elevation angles down to the minimum elevation angle of the satellite constellation.

Quadrifilar Helix Antenna (QHA)

A commonly used structure which can be used to meet these requirements is the *quadrifilar helix antenna* (QHA). This was invented by [Kilgus, 68], [Kilgus, 69] and a typical example is shown in Figure 15.11, mounted on top of a conducting box to represent the case of a mobile phone. The four elements of the QHA are placed at 90° to each other around a circle and are fed consecutively at 90° phase difference (0°, 90°, 180°, 270°). The QHA is resonant when the length of each element is an integer number of quarter-wavelengths. When the elements are each one-quarter of a wavelength, for example, the QHA can be considered as a pair of crossed half-wave dipoles, but with the elements folded to save space and to modify the radiation pattern. Adjustment of the number of turns and the axial length allows the radiation pattern to be varied over a wide range, according to the statistics of the satellite constellation and the local environment around the user, while maintaining circular polarisation with a small axial ratio from zenith down to low elevation angles [Agius, 98]. The QHA is relatively unaffected by the presence of the user's head and hand, since the antenna has a fundamentally balanced configuration, leading to relatively little current flowing in the hand-held case, in contrast to terrestrial approaches which use a monopole as the radiating element and rely on the currents in the case to provide a ground plane. The QHA can also be made resonant at multiple frequencies and reduced in size via various means (e.g. [Chew, 02].

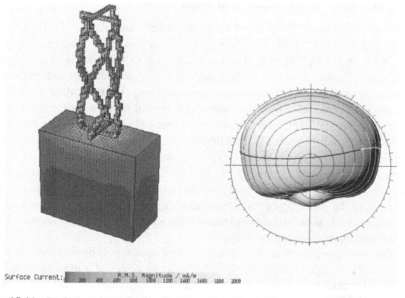

Surface Current: R.M.S. Magnitude / mA/m

Figure 15.11: Surface current distribution of a quadrifilar helix antenna mounted on a conducting box, plus corresponding (left-hand circular polarised) radiation pattern

Although the QHA is dominant for application to hand-held terminals in frequency bands up to around 3 GHz, other antennas are used in other cases. In terminals which are mounted on a vehicle rather than being hand-held, it may be more appropriate to use a circularly polarised patch antenna. For use at higher frequency bands, it is necessary to provide some antenna gain through directivity in order to improve the available fading margin. This necessitates the use of either electrical or mechanical steering, in order to ensure the antenna is properly pointed.

An adaptive, intelligent variant of the QHA, the I-QHA, was invented by [Agius, 99], [Agius, 00] and described in detail in [Leach, 00a]. The purpose of this configuration is to adapt the antenna to changes in the incoming signal imposed by the environment, the system and the user handling of the terminal. The I-QHA has been demonstrated to achieve a large potential diversity gain of up to 14 dB [Leach, 00b]. The MEG of the I-QHA has been evaluated in detail by [Brown, 03].

Patch Antennas

Patch antennas are also used for satellite mobile terminals, and become very attractive due to their low cost and easy manufacturing, as well as the reduction in size as a result of the technology used in their construction. Circular polarisation may be achieved by dual feeding a square or circular patch at right angles with quadrature phasing or by perturbing the patch shape (e.g. cutting off one corner) so as to create anti-phase currents which produce the same result. A good example of the use of such antennas is in the global positioning system (GPS), where patch antennas are the most common configuration, although the QHA is still often applied for high-performance requirements.

15.3 BASE STATION ANTENNAS

15.3.1 Performance Requirements in Macrocells

The basic function of a macrocell base station antenna is to provide uniform coverage in the azimuth plane, but to provide directivity in the vertical plane, making the best possible use of the input power by directing it at the ground rather than the sky. If fully omnidirectional coverage is required, vertical directivity is usually provided by creating a vertical array of dipoles, phased to give an appropriate pattern. This is usually called a *collinear* antenna and has the appearance of a simple monopole. A typical radiation pattern for such an antenna is shown in Figure 15.12. Such antennas are commonly used for private mobile radio systems.

More commonly in cellular mobile systems, however, some limited azimuth directivity is required in order to divide the coverage area into sectors, as described in Chapter 1. A typical example is shown in Figure 15.13. The choice of the azimuth beamwidth is a trade-off between allowing sufficient overlap between sectors, permitting smooth handovers, and controlling the interference reduction between co-channel sites, which is the main point of sectorisation. Typical half-power beamwidths are 85–90° for 120° sectors. Figure 15.13 also shows some typical patterns.

The elevation pattern of the antenna has also to be carefully designed, as it allows the edge of the cell coverage to be well defined. This can be achieved using either mechanical downtilt (where the antenna is simply pointed slightly downwards) or electrical downtilt (where the phase weighting of the individual elements within the panel is chosen to produce a downward-pointing pattern) with the antenna axis maintained vertical. Electrical downtilt is usually

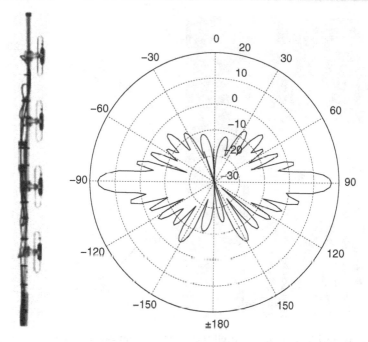

Figure 15.12: Elevation pattern for Omnidirectional collinear antenna (Reproduced by permission of *Jaybeam Wireless*) and typical elevation radiation pattern: radial axis is gain in [dBi]. This consists of a set of vertical folded dipoles, co-phased by a set of coaxial cables of appropriate lengths to produce the correct phase relationships to synthesise the desired radiation pattern from a single feed

preferred as it can produce relatively even coverage in the azimuth plane. Example elevation patterns produced using electrical downtilt are shown in Figure 15.14.

The combination of a downtilted radiation pattern with the macrocell path loss models described earlier in this chapter can have the effect of increasing the effective path loss exponent as shown in Figure 15.15. This causes the power received from the base station to fall off more abruptly at the edge of the cell, reducing the impact of interference and permitting the available spectrum to be reused more efficiently.

15.3.2 Macrocell Antenna Design

Modern macrocell antennas usually achieve the desired radiation pattern by creating an array of individual elements in the horizontal and vertical directions, built together into a single panel antenna. The array is typically divided into several subarrays. The array elements are fed via a feed network, which divides power from the feed to excite the elements with differing amplitudes and phases to produce the desired pattern.

The overall elevation pattern, $G(\theta)$ of such an array is given by

$$G(\theta) = g_0(\theta) \sum_{n=1}^{N/M} \sum_{m=1}^{M} I_{nm} \times \exp(j\phi_{nm}) \times \exp(jkd_{nm} \sin\theta) \times \exp(-jkd\phi_r) \qquad (15.16)$$

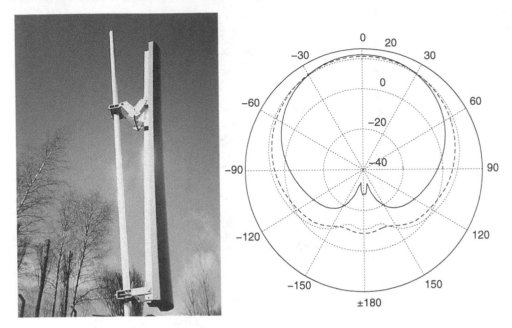

Figure 15.13: Typical macrocell sector antenna and example radiation patterns in azimuth; 3 dB beamwidths are 60° (———), 85° (- - - -) and 120° (. . .), (Reproduced by permission of Jaybeam Wireless)

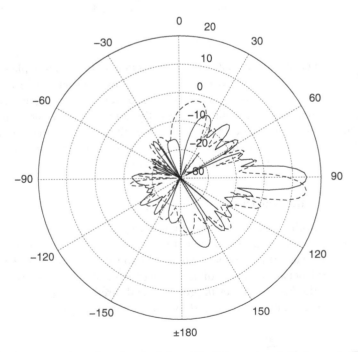

Figure 15.14: Effect of varying electrical downtilt: 0 (———) and 6° (- - - -), (Reproduced by permission of Jaybeam Wireless)

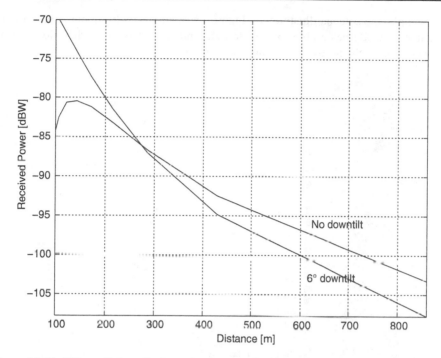

Figure 15.15: Effect of downtilt from the antennas in Figure 15.14 on received signal power: transmitter power 10 W, base station antenna height 15 m, mobile antenna height 1.5 m. Calculated assuming the Okumura–Hata path loss model from [Okumura, 68]

where $g_o(\theta)$ is the radiation pattern of an individual array element (or subarray), I_{nm} and ϕ_{nm} are the amplitude and phase of the excitation of the array element numbered $m + (n-1)M$, $d_{nm} = d(m + (n-1)M)$[m] is the distance along the array of the n, m element, M is the number of element rows in a subarray, m is the row number, N is the total number of rows, n is the subarray number and

$$\phi_r = m \sin \theta_{sub} + (n-1)M \sin \theta_{tilt} \qquad (15.17)$$

Here θ_{sub} is the wave front down tilt associated with a subarray and θ_{tilt} is the desired electrical down tilt of the whole antenna.

The excitation phases and amplitudes are chosen to maximise gain in the desired direction and to minimise sidelobes away from the mainlobe, particularly, in directions which will produce interference to neighbouring cells. The selection of the excitation coefficients follows the array principles which will be described in Chapter 18. In the vertical plane, it is important to minimise sidelobes above the mainlobe and to ensure that the first null below the mainlobe is filled to avoid the presence of a coverage hole close to the base station. The patterns in Figure 15.14 are good examples.

Since, as illustrated in the preceding section, downtilt is such a critical parameter for optimising the coverage area of a network, it is often desirable to vary the down tilt as the network evolves from providing wide-area coverage to providing high capacity over a limited area. This can be achieved by varying the element feed phases in an appropriate network. For example, the antenna illustrated in Figure 15.16 allows multiple operators to share the same antenna, while being able to provide independent variable electrical downtilt per operator according to the needs of each operator's network.

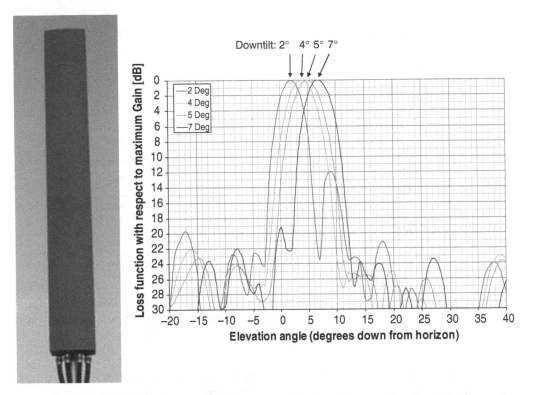

Figure 15.16: A multi-operator antenna with independently-selectable electrical down tilt (*Reproduced by permission of Quintel Technology Ltd.*)

Array elements can be composed of dipoles operating over a corner reflectors over a corner reflector (as described in Section 4.5.7). It is more common in modern antennas, however, to use patch antennas to reduce the antenna panel thickness. These are often created with metal plates suspended over a ground plane rather than printed on a dielectric to maximise efficiency and to avoid arcing arising from the high RF voltages which can be developed in high-power macrocells. Another important practical consideration in the antenna design is to minimise the creation of passive intermodulation products (PIMs). These arise from non-linearities in the antenna structure which create spurious transmission products at frequencies which may be far removed from the input frequency. They occur due to rectification of voltages at junctions between dissimilar metals or at locations where metal corrosion has occurred, so the choice of metals and the bonding arrangements at junctions must be carefully considered to minimise such issues.

15.3.3 Macrocell Antenna Diversity

Figure 15.17 shows three typical contemporary macrocell antenna installations. The first is a three-sector system, where each sector consists of two panels arranged to provide spatial

Figure 15.17: Typical macrocell installations (Reproduced by permission of O$_2$ (UK) Ltd.)

diversity to overcoming narrowband fading, with a spacing following principles which will be described in Section 16.3.3. A more compact arrangement is produced by using polarisation diversity, where each panel provides two orthogonally polarised outputs as will be described in Section 16.4. This is typically achieved by dual orthogonal feeds to patch antenna elements. The third case shows how several panel antennas can be mounted onto a building for increased height and reduced visual impact. Figure 15.18 shows some more visionary examples of how macrocell masts might be designed to be a more integral part of the built environment.

15.3.4 Microcell Antennas

The large number of individual rays which can contribute to microcell propagation, as discussed in Chapter 12, make it clear that the cell shape is not determined directly by the radiation pattern of the base station antennas, since each of the rays will emerge with a different power. Nevertheless, it is still important to ensure that power is radiated in generally the right directions so as to excite desirable multipath modes (usually with lobes pointing along streets) and to avoid wasting power in the vertical direction.

In determining the practical antenna pattern, the interactions between the antenna and its immediate surroundings are also very important. These objects may include walls and signs, which may often be within the near field of the antenna, so that accurate prediction and analysis of these effects requires detailed electromagnetic analysis using techniques such as the finite-difference time-domain method [Yee, 66].

Some typical examples of practical microcell antennas are shown in Figure 15.19. All are designed to be mounted on building walls and to radiate in both directions along the streets they are serving. An alternative approach is to use a directional antenna such as a Yagi-Uda antenna (Figure 15.20, as described in Section 4.5.4), which may help with minimising interference to other cells. Directional antennas are also useful for containing cell coverage along roads which are not lined with buildings in a sufficiently regular pattern.

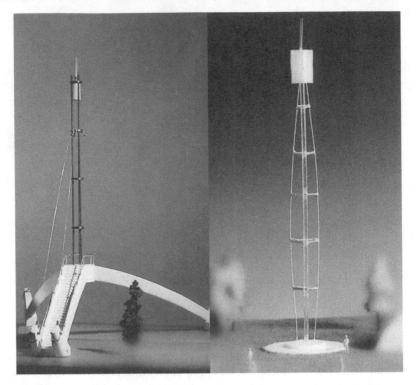

Figure 15.18: Conceptual mast designs for macrocells (Reproduced by permission of O$_2$ (UK) Ltd.)

Figure 15.19: Examples of typical microcell antennas (Reproduced by permission of O$_2$ (UK) Ltd. and Jaybeam Wireless)

It is common practice to use microcell antennas for either outdoor or in-building environments, and often antenna manufacturers do not distinguish between these applications. However, although some microcell antennas can be used in-building, there are other types which due to their size and construction would be aesthetically inappropriate for indoor use. Yagi-Uda, shrouded omnidirectional, ceiling-mount and monopoles are often employed.

15.3.5 Picocell Antennas

A practical ceiling-mounted antenna for indoor coverage at 900 MHz is shown in Figure 15.21. Particular requirements of indoor antennas are very wide beamwidth, consistent

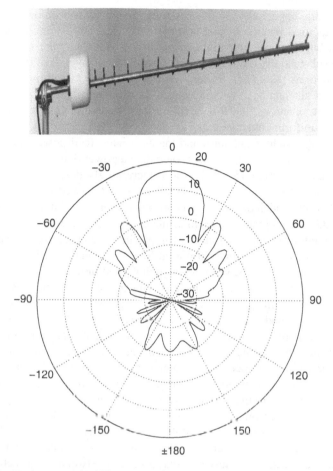

Figure 15.20: Yagi antenna for 900 MHz and its radiation pattern (radial scale in dBi) (Reproduced by permission of Jaybeam Wireless)

Figure 15.21: Typical ceiling-mounted indoor antenna (Reproduced by permission of Jaybeam Wireless)

with a discrete appearance, so this particular antenna has been designed to look similar to a smoke detector. Linear polarisation is currently used almost universally for indoor communications, but there are potential benefits in the use of circular polarisation. This has been shown to substantially reduce fade depth and RMS delay spread due to the rejection of odd-order reflections [Kajiwara, 95] as well as reducing polarisation mismatch loss. Similarly, reduction in antenna beamwidth has been shown to substantially reduce the delay spread in line-of-sight situations, but this effect must be traded against the difficulty of providing a reasonably uniform coverage area.

Increasingly, indoor antennas and the associated feed powers also typically have to be compliant with specific requirements on radiated power density and specific absorption rates as described in Section 15.2.8. It is also increasingly desirable to achieve a high level of integration between the systems and technologies deployed by different operators so wideband and multiband indoor antennas are increasingly of interest, providing, for example, WLAN, 2G and 3G technologies in a single antenna housing.

Printed antennas, including microstrip patches, are attractive for indoor antenna designs, with wire antennas being more useful at lower frequencies. Biconical antennas have been proposed for millimetre-wave systems, giving good uniformity of coverage [Smulders, 92].

An issue that has become more important for picocell antennas is that of finding gain values everywhere in space. When performing propagation predictions in outdoor environments, using the models described in Chapter 8, distances between base stations and mobiles are large compared to the base station antenna height, and therefore the signal would be estimated for radiation angles very near to the base station antenna horizontal plane where the manufacturer typically specifies the antenna radiation pattern. However, for picocells, elevation angles typically span the whole range, and hence simple extrapolation methods, such as Eq. (4.16), will lead to unacceptable prediction errors.

A method to overcome such inaccuracies has been proposed by [Gil, 99] and has been found to give better results. The so called *Angular Distance Weight Model* suggests that the generalised gain in any direction P (θ, ϕ), $G(\theta, \phi)$ is obtained from the previous ones by weighting them with the relative angular distances between the direction of interest and the horizontal (θ_2) and the vertical $(\theta_1, \phi_1, \phi_2)$ planes, i.e. the four points on the sphere closest to the point of interest in Figure 15.22. The basic idea of the model is that the weight by which the value of the gain on a given radiation plane is inversely proportional to the angular distance, so that the closer the direction of interest is to the given radiation plane, the higher the weight. Therefore, continuity of the extrapolation is ensured on each plane.

The final formulation for the angular distance weight model is given by

$$G(\theta, \phi) = \frac{[\phi_1 \cdot G_{\phi 2} + \phi_2 \cdot G_{\phi 1}] \cdot \dfrac{\theta_1 \cdot \theta_2}{(\theta_1 + \theta_2)^2} + [\theta_1 \cdot G_{\theta 2} + \theta_2 \cdot G_{\theta 1}] \cdot \dfrac{\phi_1 \cdot \phi_2}{(\phi_1 + \phi_2)^2}}{[\phi_1 + \phi_2] \cdot \dfrac{\theta_1 \cdot \theta_2}{(\theta_1 + \theta_2)^2} + [\theta_1 + \theta_2] \cdot \dfrac{\phi_1 \cdot \phi_2}{(\phi_1 + \phi_2)^2}} \qquad (15.18)$$

where all the angles and gain values are defined in Figure 16.13. This method is recommended as an alternative for pattern extrapolation for indoor environments, since as reported in [Aragon, 03], when compared with anechoic chamber measurements taken at various planes, it shows an improvement in accuracy of around 2.5 dB of standard deviation of error when compared to the method stated in Eq. (4.16).

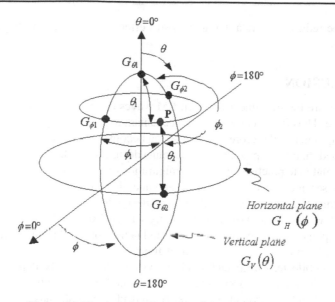

Figure 15.22: Angular distance weight model, from [Gil, 99]

15.3.6 Antennas for Wireless Lan

The vast majority of wireless local area networks can be found operating at two frequency bands: the 2.4 GHz ISM band (IEEE 802.11b and 802.11g standards) and the 5.4 GHz band (802.11a standard), with maximum data rates from 11 Mbps (802.11b) up through to 54 Mbps (802.11g/a) and up to over 100 Mbps (IEEE 802.11n, operating in either frequency band).

Spatial diversity is often employed in WLAN access points to overcome multipath fading effects and combine the various replicas of the received signal coherently, achieving substantial spatial diversity gain. Indeed, in the 802.11n standard multiple antennas are an absolute requirement to achieve high data rates (see Chapter 18). Omnidirectional antennas are preferred for some applications, but this depends on whether uniform coverage is required; i.e. if the access point and antennas are located in the middle of a room.

Some WLAN access points have integrated antennas, which are often microstrip elements, designed to provide coverage underneath the access point, in an 'umbrella' fashion. Floor penetration is sometimes difficult to achieve, especially at the relatively low transmit powers used in access points (50–200 mW EIRP depending on the regulatory regime in the country of use).

When coverage enhancement is required, especially for corridors, tunnels or to connect two buildings, directional antennas with narrow beamwidth are employed. In this case, parabolic reflectors, Yagi-Uda antennas and phased-array panels are often used as shown in Figure 16.25. As the number of channels which can be used is very limited (only three non-overlapping channels in the 2.4 GHz band in the many countries where 11 or 12 channels are available), interference management and sectorisation (also known as zoning for indoor systems) is also important, and hence stringent directional requirements must be enforced to maximise system performance. Such high-gain antennas will usually increase effective transmit power beyond the regulatory limits, so transmit power from the access point should be reduced pro-rata; the gain is still effective in increasing the range at the receiver, however.

15.4 CONCLUSION

Antennas are the transducer between EM waves carried in a transmission line and coupled into space. The effectiveness of such mechanism depends on a great extent in the efficiency of such coupling, and therefore antenna design becomes essential if system performance is to be maximised. In this chapter, the art of antenna design for mobile systems has been revised, and aspects related to practical antenna configurations have been analysed and presented. As new wireless services become a reality, antenna design techniques should encompass such improvements, to guarantee that the system design loop can be closed. Stringent requirements such as low cost, small size and simplicity will still dominate the market requirements and user acceptance for many years to come, and hence formidable and exciting antenna research opportunities are envisaged for the near future.

Both mobile terrestrial and satellite systems increasingly depend on mobile terminal antenna performance to overcome various channel impairments, such as those discussed in Chapters 10 and 11. Diversity (Chapter 16) and adaptive antennas (Chapter 18) are techniques especially designed to maximise system performance, which can be applied not only to the base station but also to the mobile terminal, given appropriate antenna design following the principles in this chapter.

REFERENCES

[Agius, 98] A. A. Agius, S. R. Saunders, Effective statistical G/T (ESGUT): a parameter describing the performance for non-GEO satellite systems, *Electronics Letters*, 34 (19), 1814–1816, 1998.

[Agius, 97] A. A. Agius, S. R. Saunders and B. G. Evans, The design of specifications for satellite PCN handheld antennas, *47nth IEEE International Vehicular Technology Conference*, Phoenix, Arizona, 5–7 May 1997.

[Agius, 99] A. A. Agius and S. R. Saunders, Adaptive multifiliar antenna, UK Patent No. WO9941803, August 1999.

[Agius, 00] A. A. Agius, S. M. Leach, S. Stavrou and S. R. Saunders, Diversity performance of the intelligent quadrifiliar helix antenna in mobile satellite systems, *IEE Proceedings-Microwaves Antennas and Propagation*, 147 (4), 305–310, 2000.

[Balanis, 07] C. A. Balanis (ed.), *Modern Antenna Handbook*, John Wiley & Sons, Inc., New York, 2007, ISBN 0470036346.

[Brown, 03] T. W. C. Brown, K. C. D. Chew and S. R. Saunders, Analysis of the diversity potential of an intelligent quadrifiliar helix antenna, *Proceedings of Twelfth Internatinal Conference on Antennas and Propagation, IEEICAP2003*, Vol. 1, 2003, pp. 194–198.

[CENELEC, 99] European Council recommendation on the limitation of exposure of the general public to electromagnetic fields 0 Hz–300 GHz, *European Journal*, 197, 30 July 1999.

[CENELEC, 00] TC211 WGMBS European Committee for Electrotechnical Standardisation, *Basic Standard for the Measurement of Specific Absorption Rate Related* to *Human Exposure to Electromagnetic Fields from Mobile Phones* (300 MHz–3 GHz), 2000.

[Chew, 02] D. K. C. Chew and S. R. Saunders, Meander line technique for size reduction of quadrifilar helix antenna, *IEEE Antennas and Wireless Propagation Letters*, 1 (5), 109–111, 2002.

[Fujimoto, 00] K. Fujimoto and J. R. James (eds.), *Mobile Antenna Systems* Handbook, 2nd edn, Artech House, London, ISBN 1-58053-007-9, 2000.

[Gil, 99] F. Gil, A. R. Claro, J. M. Ferreira, C. Pardelinha, L. M. Correia, A 3-D extrapolation model for base station antennas' radiation patterns, IEEE Vehicular Technology Conference, 3, 1341–1345, September 1999.

[Gabriel, 96] S. Gabriel, R. W. Lau and C. Gabriel, The dielectric properties of biological tissues: measurement in the frequency range 10 Hz to 20 GHz, *Physics in Medicine and Biology*, 41 (11), 2251–2269, 1996.

[ICNIRP, 98] International Commission on Non-Ionizing Radiation Protection, Guidelines for limiting exposure to time-varying electric, magnetic, and electromagnetic fields (up to 300 GHz), *Health Physics*, 75 (4), 494–522, 1998.

[IEEE, 91] IEEE C95.1-1991, *IEEE Standard for Safety Levels with Respect to Human Exposure to Radio Frequency Electromagnetic Fields, 3 kHz to 300 GHz*, 1991.

[IEEE, 00] IEEE SCC34/SC2 Institute of Electrical and Electronics Engineers, *IEEE Recommended Practice for Determining the Spatial-Peak Specific Absorption Rate (SAR) in the Human Body Due to Wireless Communications Devices: Experimental Techniques*, 2000.

[James, 89] J. R. James and P. S. Hall, *Handbook of Microstrip Antennas*, Peter Peregrinus, London, ISBN 0 86341 150 9, 1989.

[Kajiwara, 95] A Kajiwara, Line-of-sight indoor radio communication using circular polarized waves, *IEEE Transactions on Vehicular Technology*, 44 (3), 487–493, 1995.

[Kilgus, 68] C. C. Kilgus, Multielement, fractional turn helices, *IEEE Transactions on Antennas and Propagation*, 16, 499–500, 1968.

[Kilgus, 69] C. C. Kilgus, Resonant quadrifilar helix, *IEEE Transactions on Antennas and Propagation*, 17, 349–351, 1969.

[King, 60] R. King, C. W. Harrison & D. H. Denton, Transmission-line missile antennas, *IRE Transactions on Antennas and Propagation*, 88–90, 1960.

[Leach, 00a] S. M. Leach, A. A. Agius and S. R. Saunders, Intelligent quadrifilar helix antenna, *IEE Proceedings-Microwaves Antennas and Propagation*, 147 (3), 219–223, 2000.

[Leach, 00b] S. M. Leach, A. A. Agius, S. Stavrou and S. R. Saunders, Diversity performance of the intelligent quadrifilar helix antenna in mobile satellite systems, *IEE Proceedings on Microwaves, Antennas and Propagation*, 147 (4), 305–310, August 2000.

[McLean, 96] J. S. McLean, A Re-examination of the fundamental limits on the radiation Q of electrically small antennas, IEEE *Transactions on Antennas and Propagation*, 44 (5), 1996.

[Okumura, 68] Y. Okumura, E. Ohmori, T. Kawano and K. Fukuda, Field strength and its variability in VHF and UHF land mobile radio service, *Review of the Electrical Communications Laboratories*, 16, 825–873, 1968.

[Repacholi, 97] M. H. Repacholi and E. Cardis, Criteria for EMF health risk assessment, *Radiation Protection Dosimetry* 72, 305–312, 1997.

[Sewell, 95] P. D. Sewell, D. P. Rodohan and S. R. Saunders, Comparison of analytical and parallel FDTD models of antenna-head interactions, *IEE International Conference on Antennas and Propagation, ICAP '95*, pp. 67–71, IEE Conf. Publ. No. 407, 1995.

[Smulders, 92] P. F. M. Smulders and A. G. Wagemans, Millimetre-wave biconical horn antennas for near uniform coverage in indoor picocells, *Electronics Letters*, 28 (7), 679–681, 1992.

[Stutzman, 98] W. L. Stutzman and G. A. Thiele, *Antenna Theory and Design*, 2nd edn, John Wiley & Sons, Inc., New York, ISBN 0-471-025909, 1998.

[Suvannapattana, 99] P. Suvannapattana and S. R. Saunders, Satellite and terrestrial mobile handheld antenna interactions with the human head, *IEE Proceedings-Microwaves Antennas and Propagation*, 146 (5), 305–310, 1999.

[Taga, 90] T. Taga, Analysis for mean effective gain of mobile antennas in land mobile radio environments, *IEEE Transactions on Vehicular Technology*, 39 (2), 117–131, 1990.

[Vaughan, 03] R. Vaughan and J. B. Andersen, *Channels, Propagation and Antennas for Mobile Communications*, Institution of Electrical Engineers, London, ISBN 0-85296-084-0, 2003.

[Yee, 66] K. S. Yee, Numerical solution of initial boundary value problems involving Maxwell's equations in isotropic media, *IEEE Transactions on Antennas and Propagation*, 14, 302–307, 1966.

PROBLEMS

15.1 A new cellular system is to be deployed in a rural area, to provide coverage to a village of around 3000 people. Most of the houses are built from wood, and the town hall often congregates a significant amount of people during special events, for which cellular coverage is desired. This town hall is made of concrete, with wall penetration factors in excess of 12 dB.

    (a) Make suitable recommendations for the types of antennas to be employed inside the town hall as well as within the village.

    (b) Will antenna downtilting benefit the desired coverage? Explain why?

15.2 The authorities of a primary school are concerned with the deployment of a new cellular base station within the vicinity of their building. A health and safety consultancy company has been commissioned to assess whether the school is safe within the local standards. For this purpose, this company, called 'Safety Consultants Ltd.', use an E-field probe to perform electric field measurements in the school. 7000 samples are collected, which follow a Gaussian distribution, with a mean of $5.9\,\mathrm{mV\,m^{-1}}$ and a standard deviation of $1.6\,\mathrm{mV\,m^{-1}}$. Determine if the school complies with the standards if the local standards state that the electric field level should be below $6.2\,\mathrm{mV\,m^{-1}}$ in 90% of the locations within the site under test.

15.3 A quadrifilar helix antenna has an axial ratio of 5 dB at the minimum elevation angle of the satellite system in which it is designed to operate. What is the polarisation mismatch loss, assuming that the antenna receives perfect circularly polarised waves (axial ratio = 0 dB?

15.4 An 802.11g WLAN link is to be established to connect two buildings in a University campus. The geometry of this site indicates that these buildings should be connected and signal leakage to other surrounding buildings is to be avoided. Determine the maximum transmit power for legal operation in your country. Search relevant standards to determine the minimum signal strength required. Create a suitable link budget. Perform an exhaustive search from various antenna manufacturers and recommend the most suitable antenna for this application, justifying your selection. What is the maximum path length then possible?

16 Overcoming Narrowband Fading via Diversity

'Two antennas on a rooftop fell in love and got married.
The wedding was terrible – but the reception was fantastic'.

Anon

16.1 INTRODUCTION

Chapter 10 showed how the narrowband effects of the multipath channel cause very significant impairment of the quality of communication available from a mobile radio channel. Diversity is an important technique for overcoming these impairments and will be examined in this chapter. Chapters 17 and 18 will describe means of overcoming other impairments, respectively, wideband fading and co-channel interference. In some cases these techniques work so successfully that communication quality is improved beyond the level which would be achieved in the absence of the channel distortions.

The basic concept of diversity is that the receiver should have available more than one version of the transmitted signal, where each version is received through a distinct channel, as illustrated in Figure 16.1. In each channel the fading is intended to be mostly independent, so the chance of a deep fade (and hence loss of communication) occurring in all of the channels simultaneously is very much reduced. Each of the channels in Figure 16.1, plus the corresponding receiver circuit, is called a *branch* and the outputs of the channels are processed and routed to the demodulator by the *diversity combiner*

Suppose the probability of experiencing a loss in communications due to a deep fade on one channel is p and this probability is independent on all of N channels. The probability of losing communications on all channels simultaneously is then p^N. Thus, a 10% chance of losing contact for one channel is reduced to $0.1^3 = 0.001 = 0.1\%$ with three independently

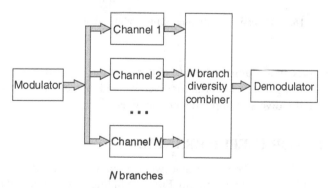

Figure 16.1: Diversity channel model

Antennas and Propagation for Wireless Communication Systems Second Edition   Simon R. Saunders and Alejandro Aragón-Zavala

fading channels. This is illustrated in Figure 16.2, which shows two independent Rayleigh signals. The thick line shows the trajectory of the stronger of the two signals, which clearly experiences significantly fewer deep fades than either of the individual signals. Another motivation for diversity is to improve the system availability for slow-moving or stationary mobiles, which would otherwise sometimes be stuck in a deep fade for a long period, even though the local mean power is sufficient for reliable operation.

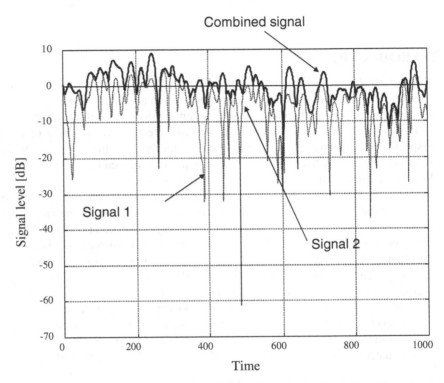

Figure 16.2: Diversity combining of two independent classical Rayleigh-fading signals

This chapter first examines how to obtain multiple branches with appropriate characteristics for obtaining a high potential for diversity, and then examines how the combiner design can optimise the diversity improvement obtained.

16.2 CRITERIA FOR USEFUL BRANCHES

In Section 16.7, it will be shown that two criteria are necessary to obtain a high degree of improvement from a diversity system. First, the fading in individual branches should have low cross-correlation. Second, the mean power available from each branch should be almost equal. If the correlation is too high, then deep fades in the branches will occur simultaneously. If, by contrast, the branches have low correlation but have very different mean powers, then the signal in a weaker branch may not be useful even though it is less faded (below its mean) than the other branches.

Assuming that two branches numbered 1 and 2 can be represented by multiplicative narrowband channels α_1 and α_2, then the correlation between the two branches is expressed by the *correlation coefficient* ρ_{12} defined by

$$\rho_{12} = \frac{E[(\alpha_1 - \mu_1)(\alpha_2 - \mu_2)^*]}{\sigma_1 \sigma_2} \tag{16.1}$$

where $E[\ ]$ is the statistical expectation of the quantity in brackets. If both channels have zero mean (true for Rayleigh, but not for Rice fading), this reduces to

$$\rho_{12} = \frac{E[\alpha_1 \alpha_2^*]}{\sigma_1 \sigma_2} \tag{16.2}$$

The mean power in channel i is defined by

$$P_i = \frac{E\left[|\alpha_i|^2\right]}{2} \tag{16.3}$$

To design a good diversity system, therefore, we need to find methods of obtaining channels with low correlation coefficients and high mean powers.

16.3 SPACE DIVERSITY

16.3.1 General Model

The most fundamental way of obtaining diversity is to use two antennas, separated in space sufficiently that the relative phases of the multipath contributions are significantly different at the two antennas. The required spacing differs considerably at the mobile and the base station in a macrocell environment, as follows.

Figure 16.3 shows two antennas separated by a distance d; both receive waves from the mobile via two scatterers, A and B. The phase differences between the total signals received at

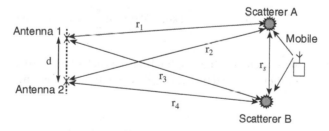

Figure 16.3: Path length differences for spatially separated antennas

each of the antennas are proportional to the differences in the path lengths from the scatterers to each antenna, namely $(r_1 - r_3)$ and $(r_2 - r_4)$. If the distance between the scatterers, r_s, or the distance between the antennas, d, increases then these path length differences also increase. When large phase differences are averaged over a number of mobile positions, they give rise to a low correlation between the signals at the antennas. Hence we expect the correlation to decrease with increases in either d or r_s.

Examining this effect more formally, Figure 16.4 shows the path to a single scatterer at an angle θ to the *broadside* direction (the normal to the line joining the antennas). It is assumed that the distance to the scatterer is very much greater than d, so both antennas view the scatterer from the same direction. The phase difference between the fields incident on the antennas is then

$$\phi = -kd \sin \theta \tag{16.4}$$

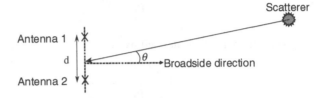

Figure 16.4: Geometry for prediction of space diversity correlation

We can then represent the fields at the two antennas resulting from this scatterer as

$$\alpha_1 = r \quad \text{and} \quad \alpha_2 = re^{j\phi} \tag{16.5}$$

If a large number of scatterers are present, the signals become a summation of the contributions from each of the scatterers:

$$\alpha_1 = \sum_{i=1}^{n_s} r_i \text{ and } \alpha_2 = \sum_{i=1}^{n_s} r_i e^{j\phi_i} \tag{16.6}$$

where r_i are the amplitudes associated with each of the scatterers. The correlation between α_1 and α_2 is then given by

$$\rho_{12} = E \sum_{i=1}^{n_s} \exp(-j\phi_i) = E \sum_{i=1}^{n_s} \exp(jkd \sin \theta_i) \tag{16.7}$$

It has been assumed here that the amplitudes from each of the scatterers are uncorrelated. The expectation may then be found by treating θ as a continuous random variable with a p.d.f. $p(\theta)$, leading to

$$\rho_{12}(d) = \int_{\theta=0}^{2\pi} p(\theta) \exp(jkd \sin \theta) \, d\theta \tag{16.8}$$

Equation (16.8) can be used in a wide range of situations, provided reasonable distributions for $p(\theta)$ can be found. Note that (16.8) is essentially a Fourier transform relationship between $p(\theta)$ and $\rho(d)$. There is therefore an inverse relationship between the widths of the two functions. As a result, a narrow angular distribution will produce a slow decrease in the correlation with antenna spacing, which will limit the usefulness of space diversity, whereas environments with significant scatterers widely spread around the antenna will produce good space diversity for modest antenna spacings. It also implies that if the mobile is situated close to a line through antennas 1 and 2 (the *endfire* direction), the effective value of d will become close to zero and the correlation will be higher.

16.3.2 Mobile Station Space Diversity

As described in Chapter 10, the mobile station is surrounded by scatterers, so the angular distribution of scatterers can be described by a uniform p.d.f over $[0, 2\pi]$. Solution of (0.8) with $p(\theta) = 1/2\pi$ yields

$$\rho(d) = J_0\left(\frac{2\pi d}{\lambda}\right) \tag{16.9}$$

where $J_0()$ is the Bessel function of the first kind and zeroth order; see Appendix B. This is exactly the result which was produced in Chapter 10 Eq. (10.48) for the autocorrelation function of the fading signal at a single antenna between two moments in time when the vehicle is in motion, except with the time delay introduced by the vehicle motion reinterpreted as a horizontal antenna spacing by putting $\tau = d/v$. If the arrival angles at the mobile are restricted to a limited range, as in Figure 10.28, then the correlation for a given spacing is increased. Figure 16.5 shows the result for various angles.

Although restricted angles may indeed be encountered over certain mobile routes, a reasonable compromise for horizontal antenna spacing is around 0.5 wavelengths. Note,

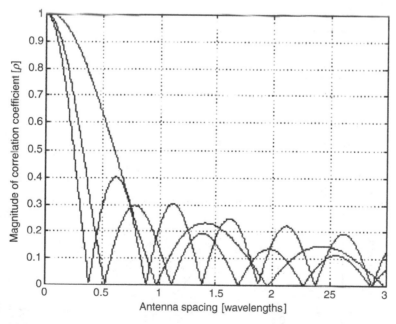

Figure 16.5: Correlation for horizontal space diversity at a mobile with restricted arrival angles over a range $\beta = 60°$, $120°$ and $180°$. The $180°$ case is described by Eq. 16.9 and β is defined by Figure 10.28

however, that the results in Figure 16.5 show only the correlation between the fields incident on the antennas. The actual voltages at the antenna terminals are additionally affected by the mutual coupling between the antennas, which may be particularly significant for antennas spaced with their main lobes aligned with each other, as would typically be the case with horizontal spacing of vertically polarised antennas. Perhaps counterintuitively, it has been shown by both theory and experiment that this mutual coupling tends to reduce the correlation coefficient, so that acceptable diversity can be obtained with horizontal spacings as small as 0.1 λ [Fujimoto, 94].

If a more compact antenna structure is required, then vertical space diversity becomes attractive, as the two antennas can then be packaged together into a single vertical structure. However, the previous assumption that all waves arrive in the horizontal plane, would lead to $p(\theta) = \delta(\theta)$ and the signals would be perfectly correlated for all separations. In a more realistic assumption [Parsons, 91], waves arrive moderately spread relative to the horizontal, according to this p.d.f.:

$$p(\theta) = \begin{cases} \dfrac{\pi}{4|\theta_m|}\cos\dfrac{\pi\theta}{2\theta_m} & |\theta| \le |\theta_m| \le \dfrac{\pi}{2} \\ 0 & \text{elsewhere} \end{cases} \qquad (16.10)$$

where θ_m is half of the vertical angular spread. Substituting (16.10) into (16.8) yields the results shown in Figure 16.6. It is clear that considerably larger spacing is required for vertical separation than for horizontal separation. Nevertheless it may be more convenient to separate elements vertically within a single structure than to have two horizontally spaced elements.

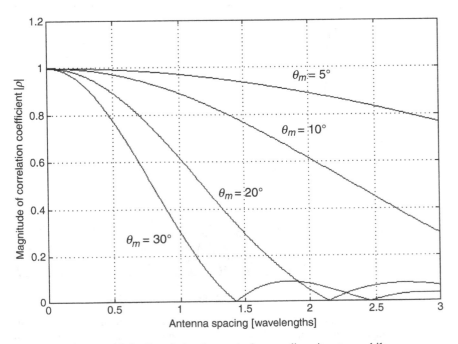

Figure 16.6: Correlation for vertical space diversity at a mobile

16.3.3 Handset Diversity Antennas

Practical handset diversity antennas are of increasing interest to increase reliability when users are slow moving, to provide an increase in downlink signal-to-noise ratios to increase download data rates, and to increase overall channel capacity via MIMO techniques (see Chapter 18).

This chapter has established the basic theory for mobile handset diversity. For practical implementation, the fundamental challenge is the antenna size. One approach is to combine an external antenna such as a loaded whip or sleeve dipole with a compact internal antenna such as a PIFA or patch (see Chapter 15). Although the internal antenna is likely to provide lower MEG, this still allows it to overcome the majority of fading nulls encountered at the 'main' element. With a 3 dB reduction in MEG for one antenna, a two-branch diversity system loses only around 1 dB of diversity gain at 1% fade probability in a Rayleigh environment. Additionally, the dissimilar patterns and polarisation states mean that the fading correlation coefficient is usually much smaller than might be expected from simple theory based on the element spacing alone.

Likewise, the use of two similar antennas spaced apart by a small fraction of wavelength is far more effective than might be expected. The mutual coupling between the elements interacts with the spatial field patterns to produce low cross-correlation even with a spacing of only 0.05–0.10 wavelengths [Vaughan, 93].

16.3.4 Base Station Space Diversity

The angular distribution of scattering at the base station antenna may be very different from the mobile case, particularly for macrocells. Figure 16.7 illustrates a typical geometry for a macrocell in a built-up area. If it is assumed that the main scatterers contributing to the signal

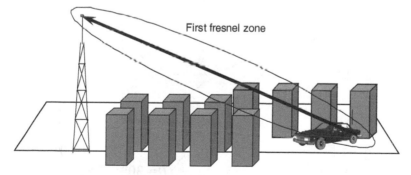

First fresnel zone

Figure 16.7: Effect of base antenna height on scatterer distribution in macrocells

are those within the first Fresnel zone around the direct ray from the mobile to the base, then the separation of the scatterers will clearly reduce as the base height is increased. The main scatterers are likely to be located close to the mobile, so the angular distribution at the base station may be rather narrow.

It is commonly assumed that the scatterers lie on a ring centred on the mobile location, as illustrated in plan view in Figure 16.8. In practice there will be scatterers within the ring, but this serves as a useful approximation for analytical purposes.

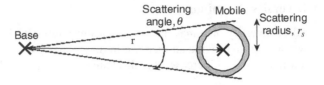

Figure 16.8: Ring of scatterers for calculating macrocell space diversity performance

In the simple case of all scattering occurring within the plane of horizontally spaced base station antennas, the result is as follows [Jakes, 94]:

$$\rho(d) = J_0\left(\frac{2\pi d}{\lambda}\frac{r_s}{r}\cos\theta\right)J_0\left(\frac{\pi d}{\lambda}\left(\frac{r_s}{r}\right)^2\sqrt{1-\frac{3}{4}\sin^2\theta}\right) \qquad (16.11)$$

Example calculations are given in Figure 16.9, with $r_s/r = 0.006$. It is clear that the spacings required are very much greater than in the mobile case, particularly when the mobile is not incident from the broadside direction. These calculations are rather pessimistic as they include no vertical spreading of angle-of-arrival. This is essential when calculating the effect of vertical spacing, which nevertheless requires even larger spacings than the horizontal case [Turkmani, 91].

Despite the large required spacings, horizontal space diversity is very commonly applied in cellular base stations to allow compensation for the low transmit power possible from hand-portables compared to the base stations. Vertical spacing is rarely used; this is because of the large spacings required to obtain low cross-correlation and because the different heights

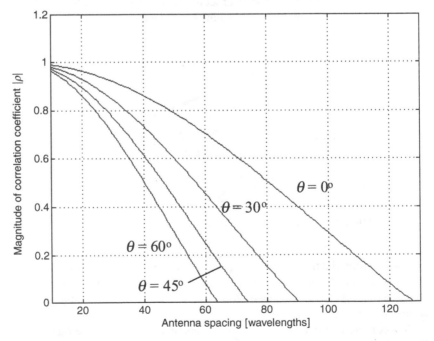

Figure 16.9: Correlation for horizontally spaced macrocell base station antennas with scattering radius equal to 0.6% of base-mobile range

of the antennas can lead to significant differences in the path loss for each antenna, which degrades the diversity effect.

In microcell environments, both the base and mobile antennas are submerged among scatterers, so the angular spread of scatterers is very high. However, there is a high probability of encountering a strong line-of-sight component, so the p.d.f. of the angles may be strongly non-uniform. The usefulness of space diversity will depend on the particular geometry of the scatterers in the cell.

In picocells the angles of arrival will be distributed even more widely in three dimensions, particularly when propagation takes place between floors. Space diversity with spacings comparable to those calculated in Section 16.3.2 for horizontal spacing will then yield low correlations, even at the base station.

16.4 POLARISATION DIVERSITY

Chapter 3 showed that both reflection and diffraction processes are polarisation sensitive and can produce a rotation of the polarisation of the scattered wave compared to the incident wave. The compound effect of multiple instances of these processes in the propagation path depolarises a vertically polarised transmission, producing a significant horizontally polarised component at the receiver. This allows *polarisation diversity* when two collocated but differently polarised antennas are used as the branches of a diversity receiver. Collocation is attractive to reduce the aesthetic impact of base station antennas and to allow a very compact solution in the hand-held case. Base station polarisation diversity also helps to reduce the polarisation mismatch which may be produced by hand-held users who tend to hold their hand-held with an average angle of around 45° to the vertical, although this is mainly significant in line-of-sight cases.

16.4.1 Base Station Polarisation Diversity

It has been found experimentally [Kozono, 84] that the horizontal and vertical field components are almost uncorrelated at the base station, so a pair of cross-polarised antennas can provide diversity with no spacing between them. The disadvantage is that the cross-polarised component typically has considerably lower power than the co-polar component, which tends to reduce the diversity gain.

The cross-polar ratio, Γ, is defined as the ratio between the mean powers from the horizontally and vertically polarised electric fields:

$$\Gamma = E[|E_v|^2]/E[|E_h|^2] \tag{16.12}$$

The mean value of Γ has been measured as around 6 dB in macrocell environments [Kozono, 84] and around 7.4 dB in microcells in the 900 MHz band. The mean correlation coefficients are around 0.1 in both cases, and they were found to increase somewhat with range in the microcells. Thus, polarisation diversity has the potential for significant gains, although the mean power received by the two branches may be significantly different when Γ is high. Many practical experiments, e.g., [Turkmani, 95], [Eggers, 93], have shown that the diversity gains obtained with polarisation diversity in areas with reasonable scattering are almost as high as those obtained with space diversity alone. In consequence, many new installations of cellular base stations use polarisation diversity to reduce wind loading and the visual impact of multiple antennas for space diversity. In open areas, however, the scattering is often insufficient for polarisation diversity, then space diversity provides a more reliable alternative. Under the assumption that the vertical and horizontal components of the field are independently Rayleigh distributed, the correlation coefficient can be calculated [Kozono, 84] as follows:

$$\rho = \frac{\tan^2 \alpha \cos^2 \beta - \Gamma}{\tan^2 \alpha \cos^2 \beta + \Gamma} \tag{16.13}$$

where the fields are received by antennas inclined at an angle α to the vertical and the mobile is situated at an angle β to the antenna boresight.

One method to improve the performance of polarisation diversity is with a mixed scheme where antennas are both spatially separated and differently polarised. In this case it has been shown [Vaughan, 90]; [Eggers, 93] that the correlation coefficients are approximately multiplicative, thus

$$\rho \approx p_\times(\alpha)\rho(r, h) \tag{16.14}$$

where $\rho_\times(\alpha)$ is the correlation which would be obtained with pure polarisation diversity with collocated antennas polarised at an angle α to the vertical and $\rho(r, h)$ is the correlation which would be obtained with co-polarised antennas having a horizontal spacing r and a vertical spacing h. The correlation coefficient can therefore be reduced below which would be obtained from polarisation alone, while still keeping a fairly compact structure.

16.4.2 Mobile Station Polarisation Diversity

Polarisation diversity is particularly attractive at the mobile station, given the limited space available. However, analysis of the effects is more complex given the large angular spread and the mutual interactions between antenna elements and the human body. One model has analysed mobile polarisation diversity by considering the geometry illustrated in Figure 16.10 [Brown, 05]. In this figure, four angles are defined: δ, the angle of rotation relative to the vertical field; ϕ, the azimuth angle relative to the normal; θ, the elevation angle relative to the normal; and Ω, the angular spacing between the two antenna branches in the direction of the vertical axis.

In order to calculate the voltages induced on the antennas, two main factors need to be accounted for: the isolation, $[s^{-1}]$, between the antennas and the levels of polarisation impurities for each antenna, m between the cross-polar and co-polar components of the antenna, both defined as voltage ratios. The antennas will exchange power due to coupling, and therefore the induced voltages in the antennas are given by

$$V_1 = [a(1 - s + m) + cs]eE_x + [b(1 - s - m) + ds]fE_y \tag{16.15}$$
$$V_2 = [c(1 - s + m) + as]eE_x + [d(1 - s - m) + bs]fE_y \tag{16.16}$$

where

$$a = \sin\left(\alpha + \frac{\Omega}{2}\right) \qquad c = \sin\left(\alpha - \frac{\Omega}{2}\right) \tag{16.17}$$

$$b = \cos\left(\alpha + \frac{\Omega}{2}\right) \qquad d = \cos\left(\alpha - \frac{\Omega}{2}\right) \tag{16.18}$$

$$e = \cos\phi \qquad f = \cos\theta \tag{16.19}$$

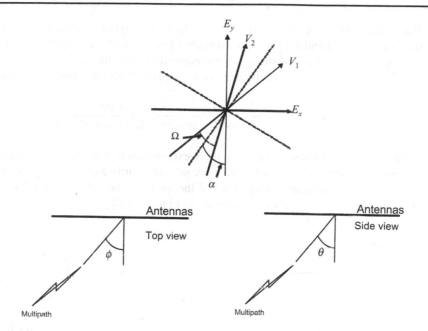

Figure 16.10: Definition of angles for mobile station polarisation diversity antennas

The complex correlation coefficient can be calculated from these terms using

$$\rho_{12} = \frac{\overline{V_1 V_2^*} \quad \overline{V_1}.\overline{V_2^*}}{\sqrt{\overline{V_1 V_1^*} \, \overline{V_2 V_2^*}}}$$ (16.20)

The resultant correlation is then

$$\rho_{12} = \frac{\begin{array}{c}[a(1-s+m)-cs][c(1-s+m)+as]+\\ [b(1-s-m)+ds][d(1-s-m)+hs]\Gamma\end{array}}{\sqrt{\begin{array}{c}([a(1-s+m)+cs]^2+[b(1-s-m)+ds]^2\Gamma)\times\\ ([c(1-s+m)+as]^2+[d(1-s-m)+bs]^2\Gamma)\end{array}}}$$ (16.21)

Calculation shows that with high isolation above 20 dB and low polarisation impurity below -20 dB, Eq. (15.21) can be calculated in terms of the cross-polar ratio, Γ, as

$$\rho_{12} = \frac{\left[\tan\left(\alpha - \tfrac{\Omega}{2}\right)\tan\left(\alpha + \tfrac{\Omega}{2}\right)\cos^2\phi + \Gamma\cos^2\theta\right]}{\sqrt{\left[\tan^2\left(\alpha - \tfrac{\Omega}{2}\right)\cos^2\phi + \Gamma\cos^2\theta\right]\left[\tan^2\left(\alpha + \tfrac{\Omega}{2}\right)\cos^2\phi + \Gamma\cos^2\theta\right]}}$$ (16.22)

The envelope cross-correlation is then $\rho_e \approx |\rho_{12}^2|$. The overall envelope correlation is then found by deriving the average correlation over all significant wave arrival angles:

$$\overline{\rho_e} = \int\limits_{-\pi}^{0} \int\limits_{-\frac{\pi}{2}}^{\frac{3\pi}{2}} \rho_e(\phi,\theta)p(\phi)p(\theta)\cos\theta \, d\phi \, d\theta$$ (16.23)

Taking the distribution suggested in [Taga, 99], then, $p(\phi)$ is a uniform distribution function in the azimuth plane and $p(\theta)$ is a Gaussian distribution function in the vertical plane.

The branch mean signal strength ratio required to estimate the diversity gain can be found by taking the mean square of the induced voltages in both branches as follows:

$$k = \frac{\cos^2\phi + \Gamma\cos^2\theta + \cos(2\alpha + \Omega)(\Gamma\cos^2\theta - \cos^2\phi)}{\cos^2\phi + \Gamma\cos^2\theta + \cos(2\alpha - \Omega)(\Gamma\cos^2\theta - \cos^2\phi)} \qquad (16.24)$$

It emerges from simulations based on these expressions that high polarisation impurities and low isolation between antenna branches are not necessarily a disadvantage to correlation, but just change the optimum arrangement of the antennas branches required to maximise the available diversity gain. This is illustrated in Figure 16.11.

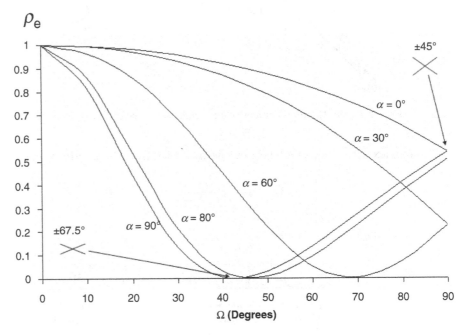

Figure 16.11: Envelope correlation for mobile polarisation diversity antennas

16.5 TIME DIVERSITY

The Bessel function decorrelation of the fading channel (16.9) implies that diversity can be obtained from a single antenna by transmitting the same signal multiple times, spaced apart in time sufficiently that the channel fading is decorrelated i.e. at least around $0.5\,\lambda$ between antenna locations when the repeated signal is received. This is rarely used in practice, however, as the retransmission of information reduces the system capacity and introduces a transmission delay. Nevertheless, the principle is applied to improving efficiency in coded modulation schemes, which apply interleaving to spread errors across fades, allowing better potential for error correction.

16.6 FREQUENCY DIVERSITY

In wideband channels, two frequency components spaced wider than the coherence bandwidth experience uncorrelated fading, providing another means of obtaining diversity. As in time diversity, the simple retransmission of information on two frequencies would be inefficient. Nevertheless, the principle of frequency diversity is implicitly employed in some forms of equaliser, and they will be studied in Chapter 17.

16.7 COMBINING METHODS

Having established how multiple branches with appropriate properties can be created, we now examine means by which the outputs of the branches can be combined to produce a useful signal, resilient against multipath effects.

16.7.1 Selection Combining

In selection combining, the diversity combiner selects the branch which instantaneously has the highest SNR (Figure 16.12).

Thus, if all branches have the same noise power, the amplitude of the output from the combiner is simply the magnitude of the strongest signal:

$$|\alpha_c| = \max(|\alpha_1|, |\alpha_2|, \cdots, |\alpha_N|) \tag{16.25}$$

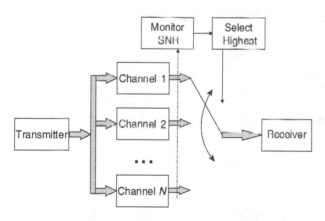

Figure 16.12: Diversity selection combining

Hence the instantaneous SNR is simply

$$\gamma_c = \max(\gamma_1, \gamma_2, \cdots, \gamma_N) \tag{16.26}$$

For N independent branches, the probability of all branches having an SNR less than γ_s is then simply the equivalent probability for a single branch raised to the power N, and for a Rayleigh channel following (10.28) this is given by

$$\Pr(\gamma_1, \gamma_2, \ldots, \gamma_N < \gamma_s) = p_{fade} = (1 - e^{-\gamma_s/\Gamma})^N \tag{16.27}$$

This expression is shown in Figure 16.13. Note that Γ here is the SNR at the input of each branch, assumed the same for all branches. For low instantaneous SNR, i.e. deep fades, Eq. (16.27) can be approximated by

$$p_{\text{fade}} \approx \left(\frac{\gamma_s}{\Gamma}\right)^N \tag{16.28}$$

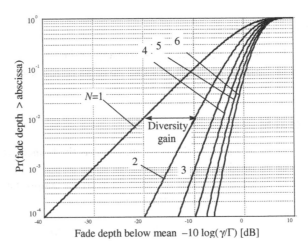

Figure 16.13: Selection combining in a Rayleigh channel

Hence, the mean power increase required to decrease the fade probability by one decade is now only $(10/N)$ decibels, compared with 10 dB in the single-branch Rayleigh case. N is often referred to as the *diversity order*. If some branches do not contribute effectively to the result, due to low mean power or high correlation, it is common to refer to an *effective diversity order* based on the observed fade slope, which may be less than the actual number of branches.

It is often useful to work in terms of a *diversity gain*, defined as the decrease in mean SNR to achieve a given probability of signal exceedance with and without diversity. For example, Figure 16.13 shows that the diversity gain for two branches and 1% availability is approximately 10 dB.

If the power in the two branches is unequal, the diversity gain is reduced. For example, if a two-branch scheme is used with mean SNRs on the branches of Γ_1 and Γ_2, then the probability of a fade below γ_s is

$$\Pr(\gamma_1, \gamma_2 < \gamma_s) = (1 - e^{-\gamma_s/\Gamma_1})(1 - e^{-\gamma_s/\Gamma_2}) \tag{16.29}$$

The result is shown in Figure 16.14 for several values of the ratio between Γ_1 and Γ_2. Although equal branch powers are needed to obtain maximum diversity gain, significant benefit is obtained even for quite large ratios. The weaker branch acts to 'fill in' the deep fades in the strong branch with high probability due to the low correlation between the two. For example, a 3 dB loss in the power of one branch only reduced the diversity gain by about 1 dB at the 1% probability level.

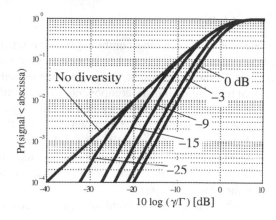

Figure 16.14: Selection combining with unequal branch mean powers

16.7.2 Switched Combining

The disadvantage with selection combining is that the combiner must be able to monitor all N branches simultaneously. This requires N independent receivers, which is expensive and complicated. An alternative is to apply switched combining. Here only one receiver is required, and it is only switched between branches when the SNR on the current branch is lower than some predefined threshold (Figure 16.15). This is a 'switch and stay' combiner.

The performance is less than in selection combining, as unused branches may have SNRs higher than the current branch if the current SNR exceeds the threshold. The threshold therefore has to be carefully set in relation to the mean power on each branch, which must also be estimated with good accuracy. Also the switching rate needs to be limited to avoid excessive switching transients when both signals are close to the threshold.

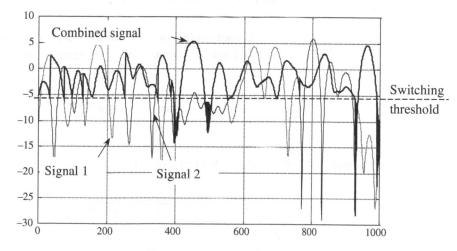

Figure 16.15: Switched combining

16.7.3 Equal-Gain Combining

Although both selection and switching combining receive on only one branch at a given time, the signal energy in the other branches is wasted. One way to improve on this is to add the signals from all the branches. If this were done directly on the complex signals, however, the random real and imaginary components would combine incoherently, resulting in the same fading statistics at the combiner output (although a greater total power). To provide any true diversity, the signals must be co-phased so that they add coherently; the noise on each branch is independent and randomly phased, hence it adds only incoherently. This process is shown in Figure 16.16, where each branch is multiplied by a complex phasor having a phase $-\theta_i$, where θ_i is the phase of the channel associated with branch i. The resultant signals all then have zero phase.

In the case of two-branch equal-gain combining, the received signals are

$$x_1 = s\alpha_1 + n_1$$
$$x_2 = s\alpha_2 + n_2$$

(16.30)

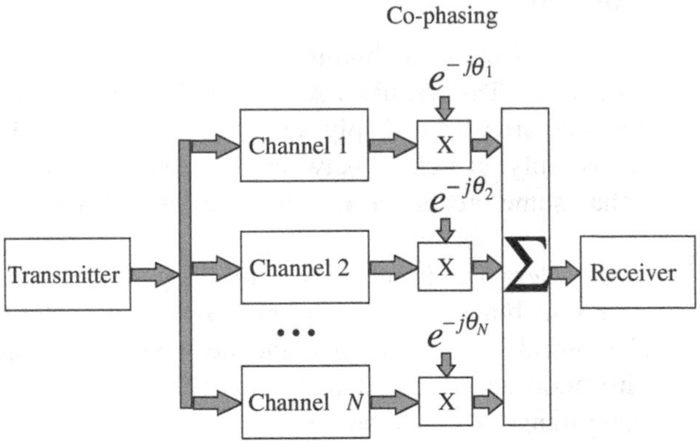

Figure 16.16: Equal-gain combining

where α_i and n_i are the complex channel coefficients and additive noise contributions for branch i, respectively. Following equal-gain combining, the input signal to the receiver is

$$
\begin{aligned}
y &= x_1 e^{-j\theta_1} + x_2 e^{-j\theta_2} \\
&= \left(s r_1 e^{j\theta_1} + n_1\right) e^{-j\theta_1} + \left(s r_2 e^{j\theta_2} + n_2\right) e^{-j\theta_2} \\
&= s(r_1 + r_2) + n_1 e^{-j\theta_1} + n_2 e^{-j\theta_2}
\end{aligned}
$$

(16.31)

Hence the instantaneous SNR at the combiner output γ_c is

$$
\begin{aligned}
\gamma_c &= \frac{(r_1 + r_2)^2}{2} \times \frac{2}{E[|n_1 e^{-j\theta_1} + n_2 e^{-j\theta_2}|^2]} \\
&= \frac{(r_1 + r_2)^2}{4 P_N}
\end{aligned}
$$

(16.32)

where P_N is the noise power on each branch. In terms of the SNRs on the individual branches, this yields

$$\gamma_c = \frac{\gamma_1 + \gamma_2 + 2\sqrt{\gamma_1\gamma_2}}{2} \tag{16.33}$$

In the special case where the channels are of equal (instantaneous) power, $\gamma_c = 2\gamma_1 = 2\gamma_2$ and 3 dB of gain is achieved.

16.7.4 Maximum Ratio Combining

When equal-gain combining is applied, it may sometimes happen that one of the branches has a considerably lower SNR than the others. As it is given equal weighting in the sum, this may occasionally reduce the overall SNR to a low value. A better approach would be to give low-SNR branches a lower weighting. Maximum (or maximal) ratio combining is a method of choosing the branch weights so the SNR at the output is maximised (Figure 16.17).

The optimum weighting for branch i is given by

$$w_i = \frac{\alpha_i^*}{P_N} \tag{16.34}$$

This allows the weighting to reflect changing noise power between the branches, although the noise powers will be equal in most cases. The output signal from the combiner is then

$$y = \sum_{i=1}^{N}(\alpha_i + n_i) \times \frac{\alpha_i^*}{P_N} = \frac{1}{P_N}\left(\sum_{i=1}^{N}|\alpha_i|^2 + \sum_{i=1}^{N}n_i\alpha_i^*\right) \tag{16.35}$$

Then the overall SNR is given by

$$\begin{aligned}\gamma_c &= \frac{1}{2}\left(\sum_{i=1}^{N}r_i^2\right)^2 \times 2 \bigg/ E\left[\left|\sum_{i=1}^{N}n_i\alpha_i^*\right|^2\right] \\ &= \left(\sum_{i=1}^{N}r_i^2\right)^2 \bigg/ P_N\sum_{i=1}^{N}r_i^2 = \sum_{i=1}^{N}\gamma_i\end{aligned} \tag{16.36}$$

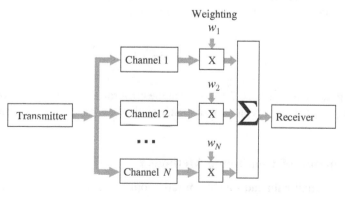

Figure 16.17: Maximum ratio combining

Hence the SNR at the output of a maximum ratio combiner is given by the sum of the SNRs of the individual branches. The best possible use is made of the signal energy and the noise energy is minimised as far as possible.

It has been assumed in the calculations so far that the correlation between branches is zero. Examples of two classical Rayleigh-fading signals with varying correlations are shown in Figure 16.18. If the branches are correlated, then the two-branch fading probability with maximum ratio combining is given [Jakes, 94] as follows:

$$P_2(\gamma_s) = 1 - \frac{1}{2\rho}\left[(1+\rho)\exp\left(-\frac{\gamma_s}{\Gamma(1+\rho)}\right) - (1-\rho)\exp\left(-\frac{\gamma_s}{\Gamma(1-\rho)}\right)\right] \qquad (16.37)$$

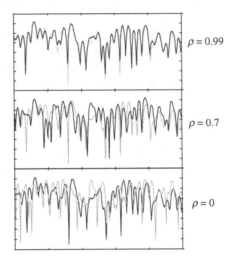

Figure 16.18: Classical Rayleigh time series with varying cross-correlations

and is plotted in Figure 16.19.

It is clear from Figure 16.19 that the diversity gain is significant even when the correlation is quite high. It is commonly assumed that almost the full theoretical diversity gain is obtained when the branch correlation is less than around 0.7. It is also interesting to examine the bit error rate for BPSK when using maximal ratio combining in a Rayleigh channel. The approximate error rate expression [Proakis, 89] is as follows:

$$P_2 \approx_{2N-1} C_N \left(\frac{1}{4\gamma}\right)^N \qquad (16.38)$$

This is plotted in Figure 16.20. Note that the error rate is inversely proportional to the SNR to the Nth power. This is typical of diversity applied to any uncoded modulation scheme in the Rayleigh channel.

16.7.5 Comparison of Combining Methods

Selection, equal-gain and maximum ratio combining are compared in Figure 16.21 for the case of two uncorrelated, equal mean power Rayleigh channels. All three combining

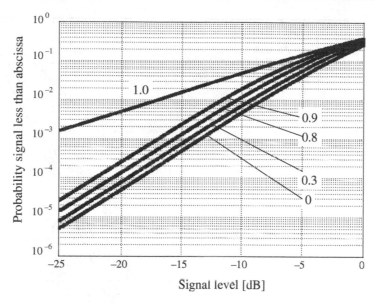

Figure 16.19: The effect of varying branch correlation

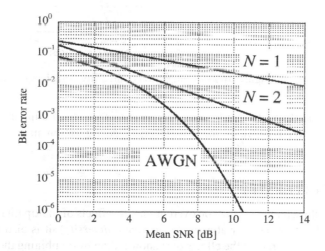

Figure 16.20: Bit error rate reduction via diversity

techniques yield the same gradient of 5 dB per decade, but are shifted slightly, with less than 1 dB performance reduction for equal-gain combining compared with maximum ratio. The performance for switched combining would lie between the selection combining and single-branch curves, with the exact value depending on the switch strategy used.

16.8 DIVERSITY FOR MICROWAVE LINKS

Chapter 6 described how anomalous tropospheric refractivity conditions could give rise to multipath propagation in terrestrial fixed links. This produces fading, just as for mobile

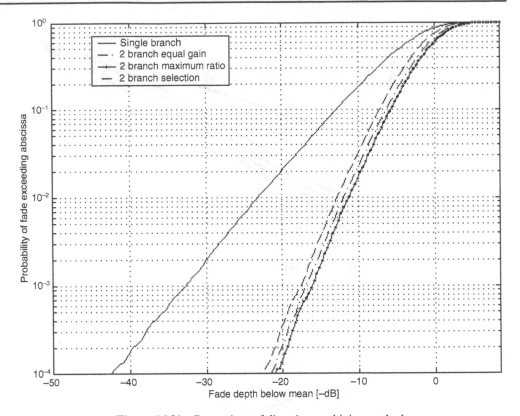

Figure 16.21: Comparison of diversity-combining methods

systems, and diversity can help to overcome this. As the dominant direction of refractive index variation is vertical, the best diversity configuration in this case is vertical space diversity.

16.9 MACRODIVERSITY

Although this chapter focuses on overcoming the effects of multipath fading using techniques associated with a receiver at a single site (*microdiversity*), it is also possible to use similar approaches to overcome the effects of shadow fading by combining the signals from receivers at several sites. This can produce significant improvements in the percentage of locations served by a given number of base stations. The requirement for good results is that the cross-correlation of shadowing between the antennas must be low, and this can be predicted using the method described in Chapter 9.

16.10 TRANSMIT DIVERSITY

All of the diversity schemes described in this chapter can, in principle, be applied at the transmitter as well as the receiver. This involves selecting the best antenna to transmit from at a given moment, or choosing the amplitudes and phases of the signals transmitted so that they combine in-phase at the receiver. The difficulty is that the information available concerning the channel is always less accurate in this direction, due to differences of time and frequency between the incoming and outgoing channels and also due to the different RF characteristics of the transmit and receiver paths, which must be estimated and compensated for. If the transmit and receive signals are separated in frequency by more than the channel coherence bandwidth, then transmit diversity is essentially useless. Nevertheless, the technique is potentially very powerful, and is particularly appropriate for application in systems employing time-division duplex, if the transmit and receive time slots are separated by less than the channel coherence time.

16.11 CONCLUSION

Diversity is an extremely powerful technique for improving the quality of communication systems and it is easy to achieve gains equivalent to power savings in excess of 10 dB. These gains are achieved at the expense of extra hardware, particularly in terms of extra antennas and receivers, which must be balanced against the benefits. The key requirements for achieving the maximum benefit are that the multiple branches of the system should encounter substantially equal mean powers and near-zero cross-correlation of the fading signals. The optimum approach depends on the mechanism and geometry of the multipath scattering which produces the fading.

REFERENCES

[Brown, 05] T. W. C. Brown, S. R. Saunders and B. G. Evans, Analysis of mobile terminal diversity antennas, *IEE Proceedings, Microwaves, Antennas and Propagation*, 152 (1), 1–6, 2005.

[Eggers, 93] P. C. F. Eggers, J. Toftgård and A. M. Oprea, Antenna systems for base station diversity in urban small and micro cells, *IEEE Journal on Selected Areas in Communications*, 11 (7), 1046–1057, 1993.

[Fujimoto, 94] K. Fujimoto and J. R. James (eds.), *Mobile Antenna Systems Handbook*, Artech House, London, ISBN 089006539X, 1994.

[Jakes, 94] W. C. Jakes (ed.), *Microwave Mobile Communications*, IEEE Press, New York, ISBN 0-7803-1069-1, 1994.

[Kozono, 84] S. Kozono, T. Tsuruhara and M. Sakamoto, Base station polarisation diversity reception for mobile radio, *IEEE Transactions on Vehicular Techology*, 33 (4), 301–306, 1984.

[Parsons, 91] J. D. Parsons and A. M. D. Turkmani, Characterisation of mobile radio signals, *IEE Proceedings I*, 138 (6), 549–556, 1991.

[Proakis, 89] J. G. Proakis, *Digital Communications*, 2nd edn, McGraw-Hill, New York, ISBN 0-07-100269-3, 1989.

[Taga, 99] T. Taga, Analysis for mean effective gain of mobile antennas in land mobile radio environments, *IEE Transactions on Vehicular Technology*, 39 (2), 117–131, 1999.

[Turkmani, 91] A. M. D. Turkmani and J. D. Parsons, Characterisation of mobile radio signals: base station crosscorrelation, *IEE Proceedings I*, 138 (6), 557–565, 1991.

[Turkmani, 95] A. M. D. Turkmani, A. A. Arowojolu, P. A. Jefford and C. J. Kellett, An experimental evaluation of the performance of two-branch space and polarisation diversity schemes at 1800 MHz, *IEEE Transactions on Vehicular Technology*, 44 (2), 318–326, 1995.

[Vaughan, 90] R. G. Vaughan, Polarisation diversity in mobile communications, *IEEE Transactions on Vehicular Technology*, 39 (3), 177–186, 1990.

[Vaughan, 93] R. G. Vaughan and N. L. Scott, Closely spaced terminated monopoles, *Radio Science*, 28 (6), 1259–1266, 1993.

PROBLEMS

16.1 Calculate the correlation coefficient for a polarisation diversity system with two antennas polarised at $\pm 45°$ to the vertical, with a cross-polar discrimination from the environment of 6 dB when the mobile is at 30° to the boresight direction.

16.2 Calculate the time separation required for two signals to achieve a high degree of time diversity in a classical Rayleigh channel at 900 MHz with a mobile speed of 10 km h^{-1}.

16.3 Given a two-branch selection-combining system operated with independent Rayleigh fading, estimate then calculate the diversity gain for a fade probability of 10^{-6}.

16.4 Devise an algorithm for a switched combiner which reduces the amount of chatter shown in Figure 16.15.

16.5 Derive an expression for the instantaneous SNR at the output of an equal gain combiner with three branches in terms of the instantaneous SNR at the inputs to each of the branches.

16.6 Explain why an equal gain combiner must co-phase signals before combining them. If two branches are independently Rayleigh distributed with equal mean power P, what is the distribution of the combined signal *without* co-phasing? What is the diversity gain in this case? What is the effective diversity order?

16.7 Discuss the advantages of using polarisation diversity over other diversity techniques.

17

Overcoming Wideband Fading

'Time is the great equalizer in the field of morals'.
H.L. Mencken

17.1 INTRODUCTION

Chapter 16 showed how diversity could be used to produce significant reductions in channel impairments due to multipath fading. The methods it described were aimed mostly at narrowband fading channels. Although the use of antenna diversity does yield benefits in the wideband channel, more advanced techniques are needed to fully exploit the frequency diversity potential of the wideband channel. We can divide these techniques into three broad types depending on the modulation and multiple access scheme employed. See Section 1.10 for defnitions of these schemes.

For TDMA systems and single-user systems employing conventional forms of continuous serial modulation, it is necessary to apply an equaliser in the receive path in order to reduce the impact of inter-symbol interference and, where possible, to benefit from quality improvements resulting from frequency diversity. These are described in Sections 17.3–17.5. In CDMA systems, equalisers are replaced by Rake receivers. These are explained in Section 17.6. OFDM systems (including OFDMA and COFDM) overcome wideband fading partly as an intrinsic feature of the modulation and access technique and partly via their receiver structure. Both are described in Section 17.7.

In order to establish a consistent terminology for analysis of all of these systems, this chapter begins with a section on system modelling.

17.2 SYSTEM MODELLING

17.2.1 Continuous-Time System Model

Figure 17.1 shows a system model which defines the terms that will be used throughout this chapter. The data source produces a sequence of m binary bits, $\mathbf{b} = [b_0, b_1, \ldots, b_{m-1}]$, where each bit can take values $b_i = \pm 1$. In this chapter and the next, lower case quantities in bold sans serif fonts are vectors (e.g. $\mathbf{x}$ or $\mathbf{u}$), while upper case ones are matrices (e.g. $\mathbf{R}$ or $\mathbf{M}$). When handwritten, vectors should be indicated with an underscore or an arrow above (e.g. $\underline{x}$ or $\vec{u}$), while matrices should be indicated by a double underscore (e.g. $\underline{\underline{R}}$ or $\underline{\underline{M}}$).

These values of the bits are assumed random, equally likely, and independent of each other. Practical schemes ensure this independence via scrambling and interleaving processes. The modulator then maps the bit sequence to a waveform, $u(t, \mathbf{b})$, suitable for transmission over the channel. The channel is represented by a linear transversal filter with impulse response $h(\tau)$ which introduces distortion to the signal in the form of inter-symbol interference. The channel is described by the methods of Chapter 11, except it is assumed here to be time-invariant over the time occupied by the m bits.

Antennas and Propagation for Wireless Communication Systems Second Edition   Simon R. Saunders and Alejandro Aragón-Zavala

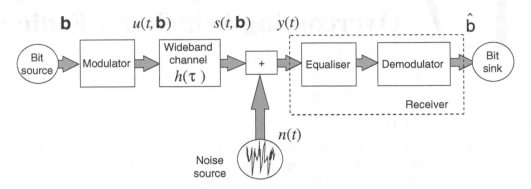

Figure 17.1: System model for equaliser-based receiver

The resulting signal at the output of the channel is then given by the convolution of the modulated signal with the channel impulse response,

$$s(t, \mathbf{b}) = u(t, \mathbf{b}) * h(\tau) \qquad (17.1)$$

The signal is further disturbed by additive white Gaussian noise, so that the waveform finally available at the receiver is

$$y(t) = s(t, \mathbf{b}) + n(t) \qquad (17.2)$$

The equaliser and the receiver then produce an estimate of the transmitted bit sequence, denoted $\hat{\mathbf{b}}$. The task is to design them so that the estimate is as close as possible to the actual sequence, $\mathbf{b}$, in the presence of the full range of channel variations and noise levels likely to be encountered.

17.2.2 Discrete-Time System Model

In almost all real implementations, the continuous waveform $y(t)$ is sampled at the receiver using an analogue-to-digital converter at intervals of the symbol interval T, or higher. Assuming one sample per symbol, the equaliser and the rest of the receiver have to work using a sequence of discrete values $\{y_k\}$ given by

$$y_k = y(t_0 + kT) \qquad (17.3)$$

where t_0 is the sampling instant, chosen to overcome the delay effects of the channel. Hence

$$y_k = s_k + n_k \qquad (17.4)$$

and

$$s_k = \sum_{j=-D}^{D} h_j u_{k-j} \qquad (17.5)$$

where the discretised channel has $(2D + 1)$ taps. The received signal can then be expressed as the sum of three separate terms,

$$y_k = u_k h_0 + \sum_{\substack{-D \le j \le D \\ j \ne 0}} h_j u_{k-j} + n_k$$

$$= \text{Desired Signal} + \text{ISI} + \text{Noise}$$

(17.6)

where the first term is the desired signal, the second is the inter-symbol interference (ISI) resulting from the delay spread of the channel, and the third is the noise. The equaliser will provide best performance if it can maximise the ratio between the desired signal power and the power in the other terms, collectively known as the *disturbance power*.

17.2.3 First Nyquist Criterion

The inter-symbol interference is zero if the channel consists of only a single tap, i.e. the narrowband case, since h_0 is the only non-zero tap. There are other cases, however, in which the ISI may also be zero, if the channel happens to have zeros at all the sample points $t = t_0 + kT$. In all such cases, the channel is said to obey the *first Nyquist criterion*. The transfer function of the channel then has a special shape, calculated as follows.

The spectrum of the discrete sequence following sampling is periodic in frequency, with *aliased* components centred on frequencies $f = k/T$ (see Figure 17.2). If the spectrum of the received waveform $y(t)$ is given by $Y(f)$, then the spectrum of the sampled version of this is $Y_a(f)$, given by

$$Y_a(f) = \sum_{n=-\infty}^{\infty} Y\left(f + \frac{n}{T}\right)$$

(17.7)

where n is an integer. The received signal samples are the inverse Fourier transform of $Y_a(f)$, given by

$$y_k = y(kT + t_0) = \int_{-W/2}^{W/2} Y_a(f) e^{j2\pi fkT} df$$

(17.8)

where the signal bandwidth is W.

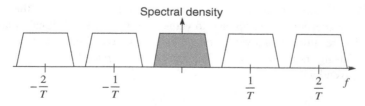

Figure 17.2: Shaded section indicates spectrum of continuous signal; the other segments are aliased components. The 'folded' spectrum is the total of all components

For zero ISI we require that both the following conditions hold simultaneously

$$y_0 = \int_{-W/2}^{W/2} Y_a(f)df = 1 \tag{17.9}$$

$$y_k = y(kT + t_0) = \int_{-W/2}^{W/2} Y_a(f)e^{j2\pi fkT}df = 0 \quad \text{for all } k \neq 0 \tag{17.10}$$

These are only satisfied if Y_a is a constant with frequency, i.e. $Y_a(f) = Y_a$ for all f. This can only happen if the aliased components in Figure 17.2 'fill in the gaps' at the edge of the spectrum of the continuous signal. The signal spectrum must therefore possess odd symmetry around $f = 1/2T$, as illustrated in Figure 17.3. This is the *first Nyquist criterion*; it is usually achieved by distributing the filters at the transmitter and the receiver so that the product of their transfer functions satisfies the criterion. A common example of such a filter is a *root-raised cosine*, specified in several common wireless standards (see Problem 17.1 and [Proakis, 89]).

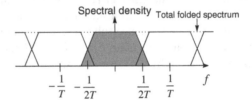

Figure 17.3: Shaded section indicates spectrum of continuous signal; other segments are aliased components. The 'folded' spectrum is the total of all components

17.3 LINEAR EQUALISERS

17.3.1 Linear Equaliser Structure

If the channel has significant delay spread compared with the channel symbol duration, the first Nyquist criterion is no longer satisfied and the system will exhibit an error floor as described in Section 11.1. In order to reduce or remove this error floor in a conventional modulation scheme, an equaliser is required at the receiver. In the simplest case, the equaliser is a linear equaliser, which usually consist of a transverse filter as shown in Figure 17.4. There are $(2K + 1)$ coefficients c_i, chosen to best overcome the effects of the channel. In the case shown, the coefficients are applied to versions of the received signal delayed by the symbol interval T, resulting in a *symbol-spaced equaliser*. Even better performance can usually be obtained using shorter delays, yielding a *fractionally spaced equaliser*. The output of the filter is then given by

$$\hat{u}_k = \sum_{i=-M}^{M} c_i y_{k-i} \tag{17.11}$$

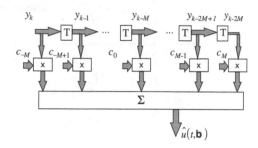

Figure 17.4: Structure of a linear equaliser

This can then be combined with Eq. (17.6) to give

$$\hat{u}_k = u_k c_0 h_0 + \sum_{\substack{-M \le i \le M \\ i \ne 0}} \sum_{\substack{D \le j \le D \\ j \ne 0}} h_j c_i u_{k-j} + \sum_{i=-M}^{M} c_i n_{k-i} \qquad (17.12)$$

This result again has the form of 'desired signal plus ISI plus noise', except that the filter through which the signal and ISI are received consists of the cascade of the channel and the equaliser.

17.3.2 Zero-Forcing Equaliser

If the spectrum of the received signal obeys the first Nyquist criterion, then zero ISI is produced. In the presence of a random channel, this condition is unlikely to be obeyed, so one option is to choose the coefficients of a linear equaliser such that the combination of the channel and the equaliser response obeys the first Nyquist criterion.

In order to set the ISI term in Eq. (17.12) to zero, the folded spectrum of the channel, $H_a(f)$, and the equaliser, $C(f)$, combine to produce a constant,

$$C(f)H_a(f) = \begin{cases} T & |f| \le \frac{1}{2T} \\ 0 & |f| > \frac{1}{2T} \end{cases} \qquad (17.13)$$

Thus the equaliser response has to be a scaled version of the inverse of the folded frequency response of the channel. The combined impulse response of the channel and the equaliser will now have zeroes at all T-spaced instants except at $t = 0$ (Figure 17.5). In practice, meeting this criterion requires an infinite number of taps, even if the channel is only moderately dispersive, so the tap weights are chosen to be a reasonable approximation to Eq. (17.13).

This result would appear to be ideal for removing distortions in the signal. However, when noise is present in the signal, the noise becomes enhanced at frequencies where the channel frequency response is low as illustrated in Figure 17.6. This *noise enhancement* can lead to large performance degradations, even for simple channels. For example, a two-tap channel with equal tap amplitudes will have a zero somewhere in its frequency spectrum. At this frequency the gain of the zero-forcing equaliser will be enormous, leading to an unacceptably low signal-to-noise ratio at the equaliser output.

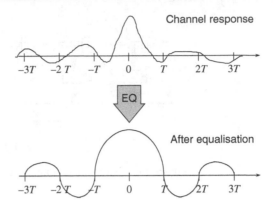

Figure 17.5: Impulse response behaviour of zero-forcing equaliser

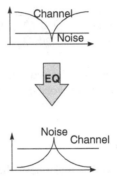

Figure 17.6: Action of zero-forcing equaliser on signal and noise spectra

17.3.3 Least Mean Square Equaliser

The noise enhancement problem of the zero-forcing equaliser arises because it is designed with regard only to the ISI components and not to the noise. The least mean square (LMS) equaliser minimises the total *disturbance* in the received signal, consisting of the sum of all the ISI terms plus the noise. This is equivalent to maximising the signal-to-noise ratio in the received signal. It avoids the noise enhancement problem of the zero-forcing equaliser at the expense of an increase in complexity.

The mean-square error between the desired signal and the estimate at the equaliser output is defined as

$$J = E\left[\left|u_k - \hat{u}_k\right|^2\right] = E\left[\left|u_k - \sum_{i=-M}^{M} c_i y_{k-i}\right|^2\right] \tag{17.14}$$

This equation is solved by choosing the values of c_i which minimise J. The solution is given by the Wiener solution [Haykin, 96]

$$\mathbf{c} = \mathbf{R}_{yy}^{-1}\mathbf{r}_{yu} \tag{17.15}$$

where $\mathbf{R}_{yy}^{-1}$ is the inverse of the *correlation matrix* of the vector of current inputs to the filter,

$$\mathbf{R}_{yy} = E\left[\mathbf{y}(k)\mathbf{y}^H(k)\right] \tag{17.16}$$

where

$$\mathbf{y}(k) = [y_k, y_{k-1}, \cdots, y_{k-2M}]^T \tag{17.17}$$

and

$$\mathbf{r}_{yu} = E\left[\mathbf{y}(k)u_k^*\right] \tag{17.18}$$

Section 18.4 describes another application of the Wiener solution.

17.4 ADAPTIVE EQUALISERS

In practical mobile radio channels, the channel impulse response varies in time according to the maximum Doppler frequency caused by mobile motion or by the motion of scatterers. The optimum coefficients for either the zero-forcing or LMS equalisers then become functions of time. An equaliser in which the coefficients are continually updated in order to approximate the optimum results is an *adaptive equaliser*. Several *convergence algorithms* exist for updating these coefficients, and each represents a different trade-off between the computational complexity required to compute the taps at each step and the speed with which the coefficients come close to the optimum values, i.e. the *convergence speed*.

In all cases, the convergence algorithms require knowledge of the current error at the output of the equaliser, and appropriate means must be provided. It is clear that the correct signals are not known at the receiver in general, as this would require knowledge of the transmitted sequence. The usual approach is illustrated in Figure 17.7. Initially the transmitter sends a *training sequence*, consisting of a standard set of symbols which are known to the receiver. With the switch in position A the errors can be calculated exactly, and the equaliser coefficients are adapted in this *training* or *reference-directed mode* to provide a good initial solution. Subsequently, the transmitter sends the useful data which is unknown to the receiver. The switch is changed to position B and the estimated symbols at the output of the

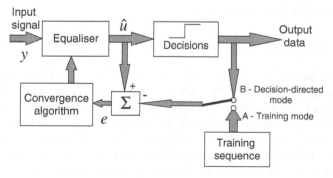

Figure 17.7: Structure of an adaptive equaliser

demodulator are used to estimate the errors in this *decision-directed* mode. Depending on the accuracy of the equaliser coefficients and the noise level, the decisions may sometimes be incorrect. Provided the error rate is reasonably low, however, the error signal is sufficiently accurate to permit the convergence algorithm to maintain the tap coefficients close to their optimum solutions as the channel changes.

The error is defined by

$$e_k = u_k = \hat{u}_k \qquad (17.19)$$

In practical mobile systems, particularly those using TDMA, the data is often sent in bursts. The training sequence can then be sent as a *preamble* at the start of the burst or as a *midamble* in the centre of the burst (Figure 17.8). A midamble has the advantage that the channel changes less during the burst, so the coefficients calculated in training mode stay valid for more of the burst. The disadvantage is that the equaliser must be converged separately in both directions away from the midamble, which involves extra complexity and necessitates storage of the whole burst before demodulation, producing extra delay.

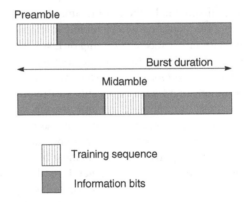

Figure 17.8: Alternative burst structures for equalisation

In some special applications, there may be no training sequence available, so no reference-directed mode is possible. This requires a special class of *blind* algorithms. Blind algorithms have lower performance than conventional algorithms, but they do avoid the loss in capacity due to the redundant training sequence.

17.4.1 Direct Matrix Inversion

The coefficients of both the zero-forcing and LMS equalisers require the solution of a set of linear equations, which may be calculated using matrix techniques. However, the size of the matrix grows rapidly with the length of the equaliser, and the computation time involved in inverting the correlation matrix in Eq. (17.15) grows more rapidly still. Such an approach is therefore often avoided to save complexity. The advantage, however, is that no time is required to allow convergence of the algorithm after this initial calculation. This is particularly helpful if the burst duration is short compared with the channel coherence time, since the equaliser coefficients then remain valid for the whole of the burst.

17.4.2 LMS Algorithm

A popular, low-complexity approach to finding the coefficients of the LMS equaliser is to apply an iterative algorithm known as the LMS algorithm. The algorithm updates the new equaliser weights based on the existing weights and a factor depending on the current input samples and the current estimation error according to the following expression:

$$\mathbf{c}_{k+1} = \mathbf{c}_k + \mu \mathbf{y}_k e_k^* \tag{17.20}$$

where μ is the *step size parameter*, which controls the convergence rate of the algorithm. It can be shown that the mean steady-state value of the coefficients produced by the LMS algorithm is exactly the optimum (Wiener) solution required for the LMS equaliser. However, the coefficients vary around this mean value, resulting in a steady-state error which also depends on the step size parameter. The LMS convergence process is illustrated in Figure 17.9. The initial value of the coefficient at 1 has a large error, and successive

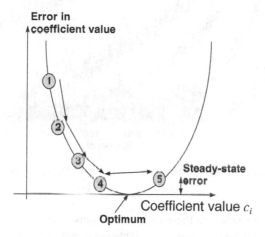

Figure 17.9: Convergence of the LMS algorithm

applications of Eq. (17.20) cause the coefficient to move towards the optimum value via values 2, 3 and 4. After 4 the algorithm causes the coefficient to overshoot the optimum point towards 5. Subsequent iterations cause the value to oscillate between points like 4 and 5, resulting in a mean steady-state error which is non-zero. The steady-state error may be reduced by decreasing μ so that the overshoot is not so large, but this comes at the expense of much slower initial convergence, requiring many more values between 1 and 4. This process is commonly studied using *learning curves* as illustrated in Figure 17.10, which show the error as a function of time, and clearly illustrate the trade-off between convergence time and steady-state error.

17.4.3 Other Convergence Algorithms

Although the LMS algorithm provides a very low-complexity iterative method for computing the optimum equaliser coefficients, the convergence time is too long for many applications. This is particularly the case for fast-changing mobile channels, encountered at high speeds or

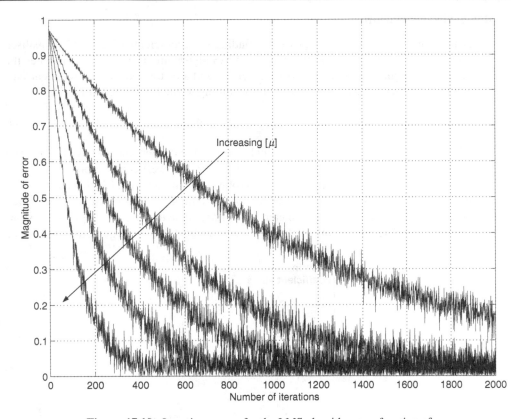

Figure 17.10: Learning curves for the LMS algorithm as a function of μ

high carrier frequencies. In these cases, the channel coherence time may be shorter than the LMS convergence time, so the coefficients never get close to their optimum values.

Other algorithms are available which produce much faster convergence than LMS but require more complex computations. These include the *recursive least squares* (RLS) algorithm. A disadvantage with this is that the results can be very sensitive to the effects of inaccuracy in the computations caused by round-off errors which occur in digital computations. An alternative version of this algorithm, the square root least squares (SRLS) algorithm has greater numerical stability.

When the channel changes particularly rapidly, improved performance may be obtained by constraining (filtering) the variations of the equaliser coefficients according to the known second-order statistics of the channel. The optimum linear structure for this process is the *Kalman filter*.

17.5 NON-LINEAR EQUALISERS

Whichever convergence algorithm is applied, there are fundamental limits to the performance offered by linear equalisers with the structure shown in Figure 17.4. Non-linear structures are therefore often used as an alternative. For example, the large delay spreads encountered in the hilly terrain wideband channels described in Chapter 11 make it absolutely essential to use a non-linear equaliser in the GSM mobile cellular system.

17.5.1 Decision Feedback

The decision feedback equaliser (DFE) is divided into two parts, a *feedforward filter* and a *feedback filter*, as illustrated in Figure 17.11. The feedback filter works to cancel the ISI on the symbol currently being directed which arises from previously detected symbols. The detected symbols are delayed and fed through a filter which replicates the estimated channel impulse response and the modulator, so that the output of the filter represents a noise-free version of the received symbols and can be used to subtract the ISI from the symbols prior to detection. Thus the equaliser output is given by

$$\hat{u}_k = \sum_{i=-M_1}^{0} c_i y_{k-i} + \sum_{i=1}^{M_2} c_i \tilde{u}_{k-i} \tag{17.21}$$

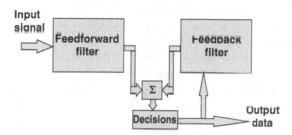

Figure 17.11: Decision feedback equaliser

where the feedforward filter contains $(M_1 + 1)$ taps and the feedback filter contains M_2 taps. So long as the decision process is correct, the symbols fed back contain no noise, and the resultant signal-to-noise ratio at the equaliser output is higher than for a linear equaliser with the same total number of taps.

This process relies on the decision being correct, however; when a detection error is made, the subtraction process may give catastrophically wrong results, which may lead to further detection errors and so on. This *error propagation* phenomenon is a significant disadvantage of the DFE.

17.5.2 Maximum Likelihood Sequence Estimator

The properties of an optimal receiver are defined in this section. Although its complexity is prohibitive in some applications, the optimal receiver serves as a useful reference against which to compare other structures.

The criterion usually adopted to define the optimum receiver is the *maximum likelihood* (ML) criterion. Provided the additive noise is white and Gaussian, the ML receiver is one that finds the estimated data sequence $\hat{\mathbf{b}}$ which minimises the mean squared error between the transmitted and received sequences,

$$\begin{aligned} D^2(\mathbf{b}) &= E\left[|s(t) - u(t, \mathbf{b})|^2 \right] \\ &= E\left[|s(t)|^2 + |u(t, \mathbf{b})|^2 - 2\mathrm{Re}(s(t)u^*(t, \mathbf{b})) \right] \end{aligned} \tag{17.22}$$

The first term in this expression is independent of **b** and can therefore be neglected; the second term depends only on the energy of the sequence, which is usually independent of the data carried. In this case, minimisation of D^2 is equivalent to maximisation of the following *metric*:

$$J(\mathbf{b}) = E[\text{Re}(s(t)u^*(t, \mathbf{b}))] \tag{17.23}$$

If the statistics of the disturbance are constant over time, then the expectation can be replaced with an integration over time, calculated over p symbol durations,

$$J_p(\mathbf{b}) = \int\limits_{t=0}^{pT} \text{Re}(s(t)u^*(t, \mathbf{b}))dt \tag{17.24}$$

This quantity is essentially a correlation of the actual received signal with the signal which would be transmitted if **b** were the sequence to be sent. The optimum receiver is then one which finds the sequence of bits $\hat{\mathbf{b}}$ that gives the smallest possible value of $J_p(\mathbf{b})$ and outputs this sequence as its decision. Another way to think of this is that the most likely transmitted sequence is the one which is most likely to lead to the actual reception of the signal $y(t)$. This maximisation could be achieved by simply calculating Eq. (17.24) for all possible sequences of bits and finding the smallest value. The result is a receiver which implicitly performs equalisation using *maximum likelihood sequence estimation* (MLSE). However, the number of possible sequences is M^p, where M is the number of bits per symbol and this involves far too many computations to be practical in almost all cases.

17.5.3 Viterbi Equalisation

The computations involved in the MLSE receiver described in the previous section can be dramatically simplified if Eq. (17.24) is rewritten as

$$\begin{aligned} J_p(\mathbf{b}) &= \int\limits_{t=0}^{(p-1)T} \text{Re}(s(t)u^*(t, \mathbf{b}))dt + \int\limits_{t=(p-1)T}^{pT} \text{Re}(s(t)u^*(t, \mathbf{b}))dt \\ &= J_{p-1}(\mathbf{b}) + Z_p(\mathbf{b}) \end{aligned} \tag{17.25}$$

where $Z_p(\mathbf{b})$ is known as the *incremental metric*. This represents the correlation of the transmitted signal for a sequence **b** with only the portion of the actual received signal during the pth symbol interval. This is relatively easy to compute, and may be used via Eq. (17.25) to build up the complete metric $J_p(\mathbf{b})$.

As an example, consider a binary modulation scheme in which the bits can take values $+1$ or -1 in any interval T and in which the channel has a delay spread extending over two bit intervals. The ISI caused to a given bit then depends on the values of the previous two bits. Since each of these bits can take two values, the ISI gives rise to four possible states in bit interval b_k, depending on bits b_{k-1} and b_{k-2}. These four states are denoted by S_0 to S_3 as defined in Table 17.1. There are therefore only eight possible waveform segments for $u(t)$ in the interval $[(k-1)T, kT]$ which need to be considered in order to evaluate the incremental branch metrics. These are denoted $Z_{k,0}$ to $Z_{k,7}$ in the *trellis diagram* illustrated in Figure 17.12.

Table 17.1: State table

State	b_{k-2}	b_{k-1}
S_0	-1	-1
S_1	1	-1
S_2	-1	1
S_3	1	1

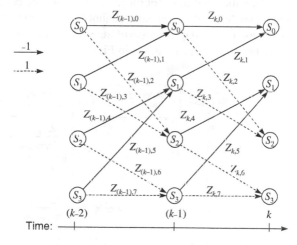

Figure 17.12: Trellis diagram for MLSE example

Note that the number of states rises exponentially with the channel delay spread, so the expected properties of the channel are crucial in determining the complexity of the MLSE receiver.

The trellis diagram provides the means for computing the maximum likelihood sequence as follows. At time instant $(k-2)$, the receiver stores the values of the total metrics J_{k-1} corresponding to paths reaching each of the four states. There are then two possible ways to evolve to each of the four states at time $(k-1)$. The eight incremental metrics describing these are computed and added to the previous values of the total metrics, yielding two possible total metrics at each state corresponding to bit b_{k-1} being either -1, denoted by a solid line, or $+1$, denoted by a dashed line. Only the larger one of these is retained, yielding again one value of the total metric for each of the four states. Note that the discarded transitions may have involved incremental metrics greater than those retained, showing clearly that the retained path will be an overall optimum, rather than depending only on the most likely bit in a single interval T. This process is repeated for each time interval, until a final decision is desired, usually at the end of a transmitted burst. The complete path corresponding to the largest total metric is then the maximum likelihood sequence.

This procedure is known as the *Viterbi algorithm* [Viterbi, 67] and a receiver which utilises it to overcome channel dispersion is said to employ *Viterbi equalisation*. Note that no approximations have been made in this process, so the Viterbi algorithm is simply an efficient implementation of the MLSE equaliser corresponding to maximisation of Eq. (17.24). In practice, it is not usually necessary to retain all surviving paths, and intermediate decisions can periodically be made in order to reduce the storage requirements of the algorithm and to allow a result to be output with reduced delay. It is also common to include in the transmission occasional *pilot symbols*, especially at the beginning and end of a burst, so that the correct state of the trellis is known absolutely at some time instants, allowing all paths which do not pass through this state to be discarded.

Two practical points. First, the Viterbi algorithm described here uses only *T*-spaced samples. To ensure optimum use of the signal energy in this case, the Viterbi algorithm must be preceded by a matched filter [Forney, 72]. Second, the computation of the metrics *J* and *Z* relies on knowledge of the wideband channel, which must be estimated, either using direct matrix inversion or by adaptive algorithms such as LMS or RLS. Thus, the complete structure of a Viterbi equaliser is as shown in Figure 17.13.

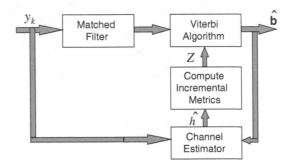

Figure 17.13: Structure of Viterbi equaliser

The resulting algorithm is very widely used in modern mobile systems as well as in terrestrial and satellite fixed links. Although the complexity is relatively high compared to other equaliser types, the performance is potentially so high that the extra complexity is often worthwhile. In particular, since the Viterbi algorithm takes account of the ISI when estimating the ML sequence, constructive use is made of all the energy available from the channel, whereas the other types of algorithm described essentially act to cancel the effect of the ISI and therefore discard potentially useful information. The error rate performance of the Viterbi algorithm with perfect channel estimation is equivalent to maximal ratio diversity combining of the *T*-spaced channel taps [Proakis, 89]. Thus MLSE effectively allows a receiver with a single antenna to exploit the frequency diversity potential of the wideband channel without any need for bandwidth expansion. It is clear that, in these circumstances, wideband fading is actually advantageous for the performance of the system. This type of frequency diversity is usually called *path diversity*. An example is shown in Figure 17.14. The wideband Rayleigh channel used has two delayed taps with relative mean powers of -3 and -5 dB relative to the main tap. Without equalisation, these cause significant inter-symbol interference and hence a large error rate floor at high E_b/N_0 values. With Viterbi equalisation, assuming all three taps can be completely resolved, a substantial reduction in BER is achieved, even relative to the zero-ISI single-tap Rayleigh channel.

Conventional antenna diversity, as in Chapter 15, can also be combined with path diversity to give an increase in the diversity order [Balaban, 91]. One approach is to simply add the incremental metrics corresponding to each symbol as observed at every antenna, which

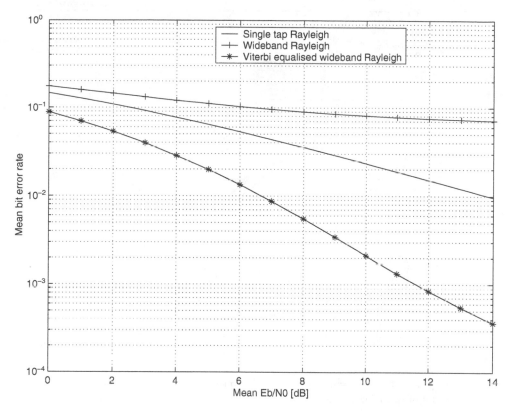

Figure 17.14: Bit error rates for BPSK with and without Viterbi equalisation in a Rayleigh-fading channel

effectively produces a wideband variant of maximal ratio combining. One significant difference between the two cases is that addition of extra antennas increases the total energy available at the receiver, whereas path diversity simply makes best use of the energy available at a single antenna.

17.6 RAKE RECEIVERS

In *direct-sequence spread spectrum* signals, information is transmitted in a wider bandwidth than necessary according to the bit rate, by multiplying the bitstream from the source $b(t)$ by a code signal $c(t)$ at a faster rate, the *chip rate*. This process is called *spreading* and is illustrated in Figure 17.15. If each user in a mobile system transmits in the same bandwidth but uses a different spreading code, and those codes are chosen to have low cross-correlation (ideally they should be *orthogonal*), then the users can be separated at the receiver by multiplying the received signal by a synchronised replica of the spreading code for the desired user. This process is direct-sequence code division multiple access (DS-CDMA) and a basic transmitter and receiver structure suitable for transmission over narrowband channels is shown in Figure 17.16.

When the channel is wideband, however, multiple replicas of the signals from each user are received, and this destroys the orthogonality properties between the codes if the delays are

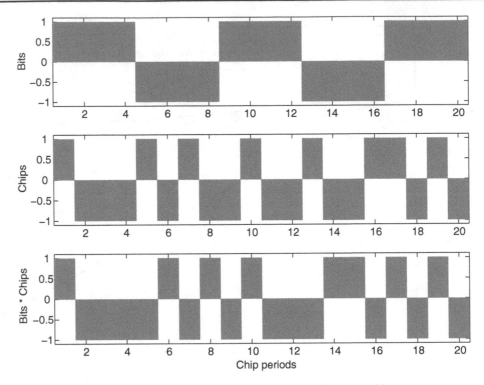

Figure 17.15: Code spreading: initial bit sequence $b(t)$, spreading code $c(t)$, result of spreading, $b(t) \times c(t)$

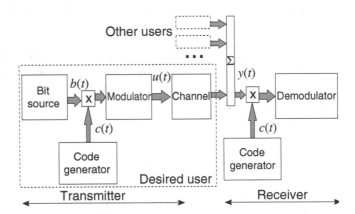

Figure 17.16: DS-CDMA system for narrowband channels

significant compared with a single chip duration. This is highly likely to happen if the chip rate is large compared with the coherence bandwidth of the channel.

In such situations, an alternative receiver structure must be used, known as a *Rake receiver*. The wideband channel is represented using the tapped delay line structure described in Chapter 11, but with delays between taps chosen to be equal to the chip period, T_c (Figure 17.17). The Rake receiver shown in Figure 17.18 multiplies several copies of the received signal by versions

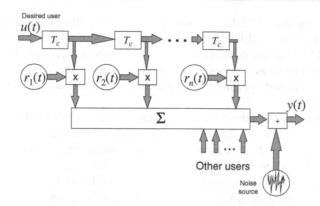

Figure 17.17: Wideband channel representation

of the spreading code, shifted by multiples of T_c. This allows the components of the original (bit rate) signal to be recovered and recombined, but with the time shifts due to the channel removed. Each of these branches, tuned to a different time shift, is known as a *finger* of the receiver. The Rake gets its name by analogy with the action of a garden rake [Price, 58]. Optimally, the combiner produces maximum ratio combining of the fingers, yielding an output signal to the demodulator which has a maximum possible signal-to-noise ratio given the input signals. Thus the performance of the Rake receiver, as with the Viterbi equaliser, can also be calculated from the performance of maximal ratio combining of each of the channel taps. However, greater performance is expected for a given bit rate in this case, because the increased bandwidth allows the receiver to resolve multipath energy which would otherwise appear combined within a single channel tap, thereby increasing the effective diversity order.

The structure shown in Figure 17.18 is only one notional approach to the structure of the Rake receiver. In practice, the signal may be subjected to the time delays rather than the code, which may be more efficient in some implementations. Additionally, the receiver may be limited to only a small number of fingers; these will be used for the time shifts which contain the most energy, as directed by estimates of the channel, so as to yield the maximum possible diversity gain for a given number of fingers.

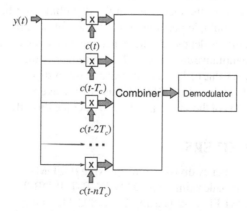

Figure 17.18: Rake receiver for DS-CDMA demodulation in wideband channels

For wideband CDMA systems, the benefits of a Rake receiver can be exploited much more than in narrowband CDMA systems, such as IS-95, due to the nature of this standard. For WCDMA, the chip duration at 3.84 Mcps is only 0.26 μs only. If the time difference between multipath components does not exceed this duration, the WCDMA receiver can separate these components and combine them coherently to obtain multipath diversity, even for small cells [Holma, 00].

The type of combiner used in the Rake receiver for WCDMA systems is a *Maximal Ratio Combiner* (MRC), described in detail in Chapter 16. The receiver corrects the phase rotation caused by a fading channel and then combines the received signals of different paths proportionally to the strength of each path. Since each path undergoes different levels of attenuation, combining them with different weights yield an optimum solution. In a Rayleigh fading channel, the MRC performance is the best achievable.

In [Chyan, 04], two ways of achieving combining are presented: at the chip level and at the symbol level. The difference between them is the location of the descramble/despreading block. For the symbol-level combining, this block is placed *before* the MRC; whereas for chip-level combining, this block is located *after* the MRC.

A *path searcher* finds the exact delay of each path, and the channel estimation finds the fading channel coefficient. For a better multipath diversity, the correlation between the fading paths should be minimal, otherwise performance is deteriorated.

The performance of both combining schemes is essentially the same under perfect channel estimation and path searching, assuming that the fading channel remains constant over a symbol period, as reported in [Chyan, 04]. For symbol-level combining it is easier to perform the channel estimation, and for chip-level combining, this channel is estimated at symbol level and interpolated to the chip level. Therefore, symbol-level channel estimation requires much lower computational effort.

Rake receivers are especially important for WCDMA, as they also account for *soft* and *softer handovers*. During a soft handover, the mobile terminal is in the overlapping region between two sectors of different base stations, and hence the Rake receiver generates two different spreading codes for the downlink direction. In *softer handover*, the two sectors belong to the same base station, and the Rake fingers behave in exactly the same way as for soft handover. More details of soft and softer handover can be found in [Holma, 00].

Finally, it is worth pointing out the effects of multipath in the code orthogonality. At the beginning of this section, it was mentioned that low cross-correlation was a desirable property for codes used to distinguish users in the downlink, and hence high orthogonality was often pursued. When multiple versions of the same symbol arrive at different times to the receiver, this orthogonality is degraded, and hence the efficiency of such codes fade. The o*rthogonality factor* is a quantitative measure of how orthogonal the codes are considered. For maximum orthogonality, a value of 1 is considered; and when strong multipath exists, this factor is reduced to around 0.4 [Holma, 00]. With the use of a Rake receiver, it is then possible to add these multiple versions of the same signal and overcome this effect, providing a higher diversity gain.

17.7 OFDM RECEIVERS

Orthogonal frequency division multiplexing (OFDM) is in common use in systems for digital TV and audio broadcasting (e.g. DAB, DVB-T, ISDB-T etc.) and for wireless LAN and MAN systems (e.g. Wi-Fi systems using IEEE 802.11a and g and WiMax using IEEE 802.16d and e). The basic concept of OFDM is to take a high data rate bitstream of rate R bits per second, subdivide it into N parallel bitstreams, each of rate R/N, and modulate each of the streams onto a subcarrier using waveforms which allow the bit streams to be demodulated independently from each other. As a result, instead of occupying a bandwidth B hertz for the full bit stream, each subcarrier occupies a bandwidth of B/N, so that a delay spread of N times greater can be tolerated before inter-symbol interference occurs.

The block diagram of a generic OFDM system is shown in Figure 17.19. The input bit stream is split into parallel streams, then an inverse fast Fourier transform is performed on the data to produce a complex time waveform corresponding to each of the bitstreams phase rotated at a rate such that each is centred around a frequency of $1/N$ Hz higher than the

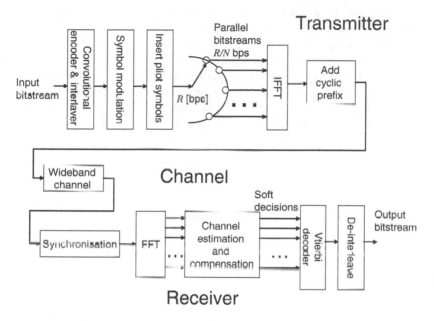

Figure 17.19: Generic OFDM system (reproduced by permission of Nortel plc)

previous one. Thus each bitstream is effectively modulated onto a separate subcarrier, forming a 'comb' of N subcarriers.

If an arbitrary set of waveforms and frequency shifts was adopted, there would be interference between the bitstreams carried on each subcarrier (*inter-channel interference* or ICI). In fact, this is largely avoided due to the choice of orthogonal waveforms for each subcarriers, as will be demonstrated below.

Assume the data set to be transmitted at a given instant of time is

$$U(-N/2), U(-N/2+1), \ldots, U(N/2-1) \tag{17.26}$$

Note that N is assumed to be a power of 2 to facilitate the IFFT process. The continuous-time representation of the signal after IFFT is

$$u(t) = \frac{1}{\sqrt{T_u}} \sum_{k=-N/2}^{N/2-1} U(k)e^{j\pi\Delta fkt} \tag{17.27}$$

Hence the transmitted signal corresponding to the kth data symbol is centred around a frequency $k\Delta f$ relative to the carrier frequency. Due to the finite duration of the symbol at N/R, the spectrum produced by each subcarrier has a shape as follows [Bracewell, 99].:

$$u_{\pm}(f) = \frac{1}{\sqrt{T_u}} \sum_{k=-N/2}^{N/2-1} U(k) \frac{\sin(\pi f - k\Delta f)}{\pi f} \tag{17.28}$$

If we set $\Delta f = 1/T_u$, then the spectrum of each subcarrier is zero at the centre of every other subcarrier as illustrated in Figure 17.20 for a system with $N = 8$. This illustrates the fact that, for this spacing of subcarriers, the individual signals corresponding to each of the subcarriers are *orthogonal*, and can be demodulated without interference between them despite their narrow spacing, without the need for complicated filtering. Note that the total spectrum is reasonably rectangular, and becomes a closer approximation to rectangular as N increases. Thus OFDM occupies a block of spectrum efficiently, with little out-of-band interference.

At the receiver, the symbol on the mth subcarrier is demodulated by multiplying the total received signal by a sinusoid at the subcarrier frequency but with opposite phase rotation and integrating the result over the symbol period T_u, as follows:

$$Y(m) = \frac{1}{\sqrt{T_u}} \int_{t=0}^{T_u} r(t) e^{-j2\pi\Delta fmt} \tag{17.29}$$

Neglecting noise and dispersion, the result is

$$
\begin{aligned}
Y(m) &= \frac{1}{\sqrt{T_u}} \int_{t=0}^{T_u} u(t) e^{-j2\pi\Delta fmt} dt \\
&= \frac{1}{T_u} \int_{t=0}^{T_u} \sum_{k=-N/2}^{N/2-1} U(k) e^{j2\pi\Delta fmt} e^{-j2\pi\Delta fkt} dt \\
&= \frac{1}{T_u} \sum_{k=-N/2}^{N/2-1} U(k) \int_{t=0}^{T_u} e^{j2\pi\Delta f(k-m)} dt \\
&= U(m)
\end{aligned}
\tag{17.30}
$$

because the exponential term integrates to zero for all values of k except $k = m$, where the integral is unity. This proves the orthogonality of the subcarriers. In practice the multiplication by the various subcarrier frequencies is performed via the FFT process, which achieves exactly the same result in a discrete, low-computation fashion [Bracewell, 99].

Although the OFDM scheme reduces the symbol period and hence makes frequency selective fading less likely within a subcarrier, it is still possible for inter-symbol interference (ISI) to occur for strongly dispersive channels. It is common to avoid this by inserting a *cyclic prefix*. Figure 17.21 shows how the signal waveform from the end of each symbol lasting a duration T_g – the guard period – is replicated and inserted at the front of the symbol. The period containing the replicated period is the guard period. If the excess delay spread is τ_{ex}, then ISI only exists during the first portion of the guard period lasting τ_{ex}. If the symbol

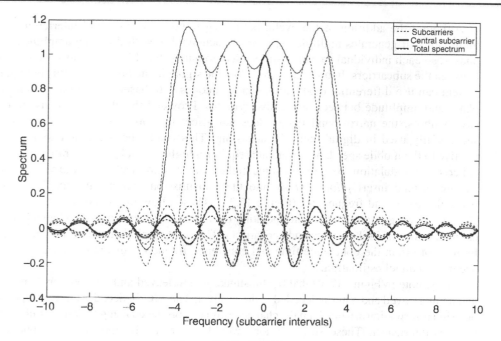

Figure 17.20: OFDM spectrum

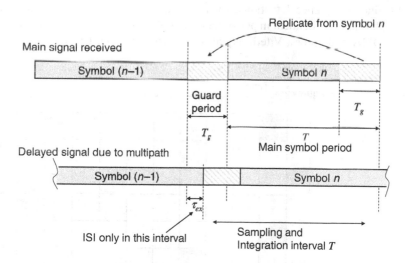

Figure 17.21: Insertion of cyclic prefix

sampling and integration interval T starts at anytime during the remaining $(T_g - \tau_{ex})$ portion of the guard interval, then it gathers all of the required symbol energy, it avoids interference from other channels since it remains orthogonal to them and it suffers no ISI. The guard period must be chosen to avoid ISI in the majority of cases, while not being excessive, since the available channel time and hence capacity is reduced by a factor $\frac{T}{T+T_g}$.

Although the addition of the cyclic prefix can remove inter-symbol interference, the multipath still generates multiple versions of each symbol with differing amplitudes and phases, so each individual carrier is still subject to narrowband fading, whose value varies between the subcarriers. In order to demodulate the signal in the presence of this fading, the system can use differential modulation which encodes data based only on the changes in phase and amplitude between successive symbols. Provided the channel is constant over two symbols, the narrowband fading does not affect this process and this approach is successfully used in digital audio broadcasting. This limits the acceptable symbol rate relative to the mobile speed, however, and requires a higher signal-to-noise ratio relative to coherent demodulation. As a result, some higher-rate systems such as digital video broadcasting instead insert *pilot symbols* into the transmission, which take fixed values at particular times and frequencies (Figure 17.22). These allow the receiver to estimate the channel at these points and by interpolating over time and frequency the channel can be estimated for all symbol times and subcarriers. The pilot symbols need to be spaced apart by no more than the coherence time and coherence frequency of the channel to provide accurate channel estimation.

Lastly, note in Figure 17.19 that the bit stream is interleaved and convolutional encoded, and deinterleaved and Viterbi decoded at the receiver. The Viterbi decoder takes as input soft decision information arising from the channel estimates produced using the pilot symbols and channel interleaver. These soft decisions would, for example, indicate low confidence associated with symbols in subcarriers suffering nulls arising form wideband fading and high confidence associated with strong subcarriers. In this way, it is able to conduct maximum likelihood sequence detection associated with a complete set of subcarriers and symbols. The result is almost identical in performance to Rake reception for the same wideband channel for a CDMA system or a Viterbi-equalised system for a TDMA system.

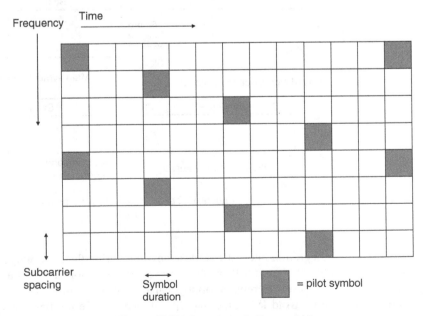

Figure 17.22: Insertion of pilot symbols

Thus an OFDM system (strictly, a coded OFDM or COFDM process) can avoid the negative impacts of wideband fading via the reduction in symbol rates implicit in the multiplexing and guard period insertion processes. It also gains positive path diversity benefit from the use of appropriate coding techniques.

The OFDM concept also lends itself straightforwardly to providing flexible bandwidth and quality characteristics for different users. Each user is assigned to a set of frequency and time intervals appropriate to their needs and this assignment can be changed very rapidly. The result is the OFDMA multiple access scheme incorporated in the WiMax standards.

17.8 CONCLUSION

The equaliser structures described in this chapter allow the negative effects of the wideband channel to be overcome and to be used for improving performance. The appropriate equaliser structure must be chosen as a compromise between computational complexity and performance. For more details, see papers such as [Qureshi, 85] and [Taylor, 98].

More complex and sophisticated Rake receivers are being proposed for W-CDMA systems, where its use, as stated in this chapter, becomes an essential component of such a system. For more information, see [Holma, 00].

OFDM systems achieve similar performance to the others, but can often produce receivers with lower complexity for the same bit rate than TDMA or CDMA systems. On the contrary, CDMA systems are ideally suited to separating signals from multiple users in the same channel. As a result future mobile systems (fourth generation) are likely to use multiple carrier CDMA air interfaces, combining the best qualities of both OFDM and CDMA to provide high data rates and high user capacities at reasonable complexity while benefiting from the diversity inherent in the wideband channel.

REFERENCES

[Balaban, 91] P. Balaban and J. Salz, Dual diversity combining and equalisation in digital cellular mobile radio, *IEEE Transactions on Vehicular Technology*, 40 (2), 342–354, 1991.

[Bracewell, 99] R. Bracewell, *Fourier Transform and Its Applications*, McGraw Hill, ISBN 0071160434, 1999.

[Chyan, 04] G. Kim-Chyan, Maximum ratio combining for a WCDMA rake receiver, *AN2251*, Freescale Semiconductor, Inc., November 2004.

[Forney, 72] G. D. Forney Jr, Maximum likelihood sequence estimation of digital sequences in the presence of intersymbol interference, *IEEE Transactions on Information Theory*, **18**, 367–78, 1972.

[Haykin, 96] S. Haykin, *Adaptive Filter Theory*, 3rd edn, Prentice Hall, Englewood Cliffs, NJ, 1996, ISBN 0-13-322760-X.

[Holma, 00] H. Holma and A. Toskala, *WCDMA for UMTS, Radio Access for Third Generation Mobile Communications*, 1st edn, John Wiley & Sons, Ltd, Chichester, West Sussex, ISBN 0-471-72051-8, 2000.

[Price, 58] R. Price and P. E. Green Jr., A communication technique for multipath channels, *Proceedings of IRE*, 46, 555–570, 1958.

[Proakis, 89] J. G. Proakis, *Digital Communications*, 2nd edn, McGraw-Hill, New York, ISBN 0-07-100269-3, 1989.

[Qureshi, 85] S. U. H. Qureshi, Adaptive equalization, *Proceedings of IEEE*, 73 (9), 1349–1387, 1985.

[Taylor, 98] D. P. Taylor, G. M. Vitetta, B. D. Hart and A. Mammela, Wireless channel equalisation, *European Transactions on Telecommunications*, 9 (2), 117–143, 1998.

[Viterbi, 67] A. J. Viterbi, Error bounds for convolutional codes and an asymptotically optimum decoding algorithm, *IEEE Transactions on Information Theory*, 13, 260–269, 1967.

PROBLEMS

17.1 The output of a transmitter, which transmits symbols of duration T, and the input of a receiver each have a filter with *root-raised cosine* spectrum,

$$S(f) = \begin{cases} \sqrt{T} & 0 \leq |f| \leq \frac{1-\beta}{2T} \\ \sqrt{\frac{T}{2}\left[1 - \frac{1}{\beta}\sin \pi T\left(f - \frac{1}{2T}\right)\right]} & \frac{1-\beta}{2T} \leq |f| \leq \frac{1+\beta}{2T} \end{cases}$$

where $0 \leq \beta \leq 1$ is the *roll-off parameter*. Assuming transmission is over a narrowband channel, prove that the resulting system is ISI-free.

17.2 A wideband channel has impulse response given by $h(\tau) = 1 - 0.1\tau$. Calculate an expression for the coefficients of a T-spaced zero-forcing equaliser which will remove ISI from this channel.

17.3 Given that GSM uses a binary modulation scheme, how many states would a Viterbi equaliser for GSM require to equalise the HT wideband channel illustrated in Figure 11.9?

17.4 What diversity order would you expect from a four-finger Rake receiver in two-way soft handover, given a chip rate of 3.84 Mcps and an RMS delay spread of 0.5 ms? How would this change if the system was in softer handover?

17.5 For an OFDM system with a quaternary modulation scheme, a bit rate over-the-air of 8 Mbps and 256 subcarriers, calculate the minimum period and symbol rate to avoid inter-symbol interference given a channel with an excess delay spread of 1 ms. What would the channel capacity be if the channel were not dispersive?

17.6 Assuming that the system in the previous question operates for mobile receivers at 1.5 GHz, how frequently do pilot symbols need to be inserted in time and frequency to efficiently estimate the channel fading assuming a maximum vehicle speed of $120 \, \text{km h}^{-1}$? What proportion of the channel capacity is sacrificed as a result?

18

Adaptive Antennas

'Every day sees humanity more victorious
in the struggle with space and time'
Marconi

'Space, the final frontier.'
Gene Rodenberry

18.1 INTRODUCTION

The limiting factor on the capacity of a cellular mobile system is interference from co-channel mobiles in neighbouring cells. Adaptive antenna technology can be used to overcome this interference by intelligent combination of the signals at multiple antenna elements at the base station and potentially also at the mobile. In order to perform this combining efficiently and accurately, a thorough knowledge of the propagation channel will be required for every pair of antennas from base to mobile.

This chapter explains the concept of adaptive antennas (often called *smart antennas* in the mobile radio context) and examines the impact of the multiple-branch propagation channel on their operation. An introduction to multiple-input multiple-output (MIMO) systems is also provided which extends adaptive antenna technology to allow a flexible trade-off between enhanced reliability and increased channel capacity.

18.2 BASIC CONCEPTS

In a *phased array*, a set of antenna elements is arranged in space, and the output of each element is multiplied by a complex weight and combined by summing, as shown in Figure 18.1 for a four-element case.

The complete array can be regarded as an antenna in its own right, with a new output y. The radiation patterns of the individual elements are summed with phases and amplitudes depending on both the weights applied and their positions in space; this yields a new combined pattern. Figure 18.2 shows some examples of patterns which may be obtained using a four-element array of isotropic elements. It is clear that a wide range of patterns is possible.

If the weights are allowed to vary in time, the array becomes an *adaptive array*, and it can be exploited to improve the performance of a mobile communication system by choosing the weights so as to optimise some measure of the system performance [Monzingo, 80]. Typically this would be done by estimating the desired weights using a digital signal processor (DSP) and applying them in complex baseband to sampled versions of the signals from each of the elements. The same approach can be used on both transmit and receive due to the reciprocity of the channel and the antenna elements themselves, but there are considerable challenges

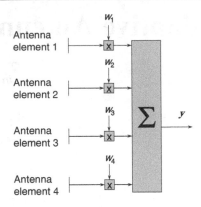

Figure 18.1: A four-element phased receive array

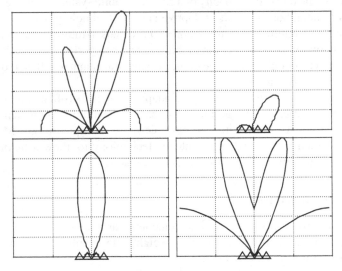

Figure 18.2: Radiation patterns for a four-element phased array with $\lambda/2$ element spacing and various combinations of element weights

associated with assessing the downlink channel state with sufficient accuracy to achieve the full potential of adaptive antennas; see the comments on transmit diversity in Section 16.10.

18.3 ADAPTIVE ANTENNA APPLICATIONS

The basic aim of a mobile adaptive antenna system is to improve the performance of the system in the presence of both noise and interference. This section describes several useful applications for adaptive antennas in broad terms before the detailed theory is introduced in subsequent sections.

18.3.1 Example of Adaptive Antenna Processing

The following example shows how adaptive antennas at a base station can be used to minimise the impact of uplink interference in a simple case. Figure 18.3 shows a desired mobile and an

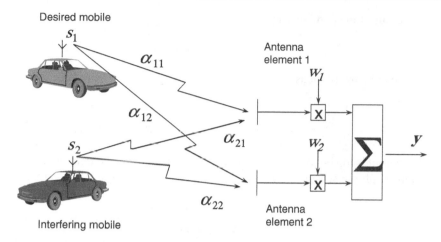

Figure 18.3: Interference reduction using adaptive antennas

interfering mobile transmitting co-channel signals s_1 and s_2, respectively; the signals are received at a base station having two independent antenna elements. In a conventional cellular system, the interferer would be in a distant cell, as the base station would not normally allocate the same channel to two mobiles in the same cell. The channels from mobile i to element j are represented by α_{ij}, including all antenna, path loss, shadowing and fading effects. At the base station, the elements are then multiplied by complex weights w and summed, yielding the output y, which would then be demodulated by the base station in the normal way.

If the base station is able to estimate the channel coefficients α_{ij}, then it would be useful for it to set the weights w_j such that the output y minimises the output due to the interferer while leaving the desired signal unaffected.

The system model for this situation is shown in Figure 18.4. The signals received at the antenna elements are then

$$
\begin{aligned}
x_1 &= s_1\alpha_{11} + s_2\alpha_{21} \quad \text{for element 1} \\
x_2 &= s_1\alpha_{12} + s_2\alpha_{22} \quad \text{for element 2}
\end{aligned}
\tag{18.1}
$$

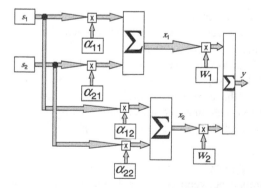

Figure 18.4: System model for the situation shown in Figure 18.3

The output of the combiner is

$$y = x_1 w_1 + x_2 w_2 \tag{18.2}$$

which can be rewritten as

$$
\begin{aligned}
y &= (s_2 \alpha_{11} + s_2 \alpha_{21}) w_1 + (s_1 \alpha_{12} + s_2 \alpha_{22}) w_2 \\
&= s_1 (\alpha_{11} w_1 + \alpha_{12} w_2) + s_2 (\alpha_{21} w_1 + \alpha_{22} w_2)
\end{aligned} \tag{18.3}
$$

If the output is forced to be the same as the signal from the desired mobile, i.e. $y = s_1$, then the term multiplying s_1 should be set to 1, and the term multiplying s_2 should be set to 0, i.e.

$$
\begin{aligned}
\alpha_{11} w_1 + \alpha_{12} w_2 &= 1 \\
\alpha_{21} w_1 + \alpha_{22} w_2 &= 0
\end{aligned} \tag{18.4}
$$

These expressions are then solved simultaneously to give

$$w_1 = \frac{\alpha_{22}}{\alpha_{11}\alpha_{22} - \alpha_{12}\alpha_{21}} \quad w_2 = \frac{-\alpha_{21}}{\alpha_{11}\alpha_{22} - \alpha_{12}\alpha_{21}} \tag{18.5}$$

It can be checked that these do yield the desired result.

This analysis shows that we can use weighting to completely remove interfering co-channel signals. In order to do so, the base station must be able to estimate the channels between each of the mobiles and each of the antenna elements. It also requires that the number of elements is at least equal to the number of mobiles in order to solve the resulting system of simultaneous equations. In practical cases, the optimum weights would have to be estimated in the presence of noise, so an interferer is not removed completely. Instead, the weights are chosen to maximise the signal-to-interference-plus-noise ratio (SINR).

18.3.2 Spatial Filtering for Interference Reduction

If a base station in a cellular system uses an adaptive array to direct its radiation pattern towards the mobile with which it is communicating (Figure 18.5), then several benefits are produced:

- The transmit power for a given signal quality can be reduced in both uplink and downlink directions, or the cell radius can be increased, thereby reducing the number of base stations required to cover a given area.
- As the mobile transmit power is reduced, its battery life can be extended.
- The channel delay spread is reduced because off-axis scatterers are no longer illuminated.
- Depending on the direction of the mobile, the probability of base stations causing interference to co-channel mobiles in surrounding cells is reduced.
- Similarly, the probability of mobiles causing interference to co-channel base stations is reduced.

This last point is known as *spatial filtering for interference reduction* (SFIR). The average level of interference between co-channel cells is reduced, so the reuse distance D can be decreased, thereby increasing the system capacity. Notice that the interference reduction is

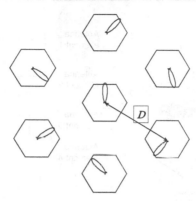

Figure 18.5: SFIR: mobiles are located in the direction of each of the base station antenna main lobes

statistical, in that the level of interference will depend on the positions of the mobiles in each cell at any time. Arrangements therefore have to be made for mobiles to change channels when the interference level is excessive.

SFIR is essentially an extreme form of sectorisation, in which the sectors are much smaller and variable in direction compared with a non-adaptive system. This makes it relatively easy to integrate with existing systems, which already use sectorisation. The capacity gain is, however, limited to the point at which all the available channels are reused in every cell (one-cell reuse).

18.3.3 Space Division Multiple Access

Figure 18.6 shows a similar situation to Figure 18.1. A second combiner has been added, with inputs from the same antenna elements. The second combiner weights are chosen to eliminate the signal from mobile 1 and retain the signal from mobile 2. In this situation the system would be simultaneously communicating with two mobiles in the same cell on the same frequency/time/code channel. This is *space division multiple access* (SDMA), which offers the potential for greatly increasing system capacity in future mobile systems, even beyond the capacity of SFIR.

Clearly, there will be times when the multiple beams produced by the base station overlap, making it impossible to separate the mobiles completely Figure 18.7. Perhaps even more importantly, the scattering nature of the propagation channel will cause the signals received from the mobile to be broadened in arrival angle, making them overlap even if the mobiles have some angular separation. Thus the capacity of an SDMA system is limited by the capabilities of the adaptive array and by the characteristics of the channel.

Implementation of SDMA requires major changes to the base station and is difficult to implement with systems for which SDMA was not originally foreseen.

18.3.4 Multiple-Input Multiple-Output Systems

The most recent, and arguably most powerful, application of adaptive antennas is the class of systems known as MIMO systems, which use space-time coding (STC) to realise substantial performance and capacity gains.

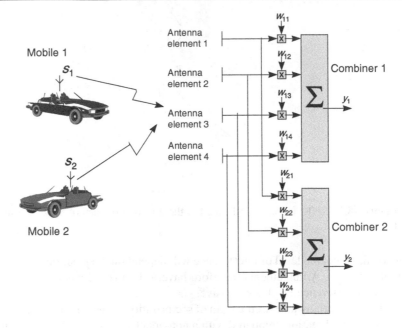

Figure 18.6: Dual combining for two-channel SDMA

In such systems, multiple antennas must be available at both transmitter and receiver. MIMO systems can then use STC to exploit the rich spatial multipath scattering present in many radio channels. MIMO systems may be seen as the ultimate extension of adaptive antenna technology to include diversity and SDMA as special cases, but being more powerful than either.

A simple example is given in Figure 18.8. At the transmitter, an input bit stream is divided into three parallel bit streams, each of which occurs at one-third the rate of the original. Each of these is transmitted via an independent beamformer, which produces three separate radiation patterns as a combination of the three transmit array elements. The radiation patterns are chosen

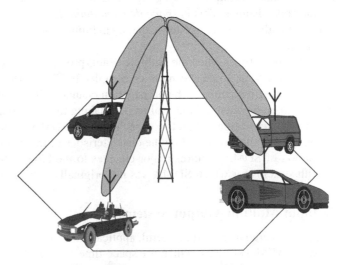

Figure 18.7: Space division multiple access

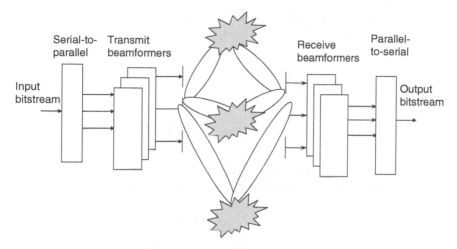

Figure 18.8: Basic MIMO example

to produce a maximum gain in the directions of three scatterers. Re-radiation from each of these scatterers is received by three separate beams formed at the receiver by three further beamforming networks. After demodulation, the three bit streams can be recombined to produce the original input bit stream. Provided the scatterers are sufficiently separated in angle to be resolved by the beams formed, the three bit streams can all occupy the same spectrum without interfering and the channel capacity has been increased by a factor of 3. Clearly, the capacity gain from such a system is limited by the amount of scattering in the multipath environment and by the number of antenna elements available at both transmitters. Nevertheless, the potential to increase channel capacity by a potentially unlimited degree offers great potential for high bit-rate systems. MIMO systems are examined in depth in Section 18.5.

18.4 OPTIMUM COMBINING

18.4.1 Formulation

The calculation of adaptive combiner weights in Section 18.3.1 was specific to the case of two mobiles and no additive noise. This section formulates the general problem of choosing the best weights in the presence of multiple interferers and additive noise.

The formulation is simplified using vector notation. In the situation previously illustrated, (Figure 18.4), the channels for the desired and interfering mobile can be arranged to form vectors $\mathbf{u}_1$ and $\mathbf{u}_2$, defined as follows:

$$\mathbf{u}_1 = \begin{bmatrix} \alpha_{11} \\ \alpha_{12} \end{bmatrix}, \mathbf{u}_2 = \begin{bmatrix} \alpha_{21} \\ \alpha_{22} \end{bmatrix} \tag{18.6}$$

Similarly, the additive noise components for branch 1 and branch 2, denoted n_1 and n_2, can be vectorised as

$$\mathbf{n} = \begin{bmatrix} n_1 \\ n_2 \end{bmatrix} \tag{18.7}$$

The received signals x_1 and x_2 can then be represented by the simple vector equation

$$\mathbf{x} = \begin{bmatrix} x_1 \\ x_2 \end{bmatrix} = s_1 \mathbf{u}_1 + s_2 \mathbf{u}_2 + \mathbf{n} \qquad (18.8)$$

The weights are then written

$$\mathbf{w} = \begin{bmatrix} w_1 \\ w_2 \end{bmatrix}.$$

The action of multiplying the received signals by the weights and summing is then given simply by the vector dot product between $\mathbf{x}$ and $\mathbf{w}$, yielding the scalar output y:

$$y = \mathbf{x}^T \bullet \mathbf{w} \qquad (18.9)$$

where T represents the transpose of the vector. This is represented by the vectorised system model shown in Figure 18.9.

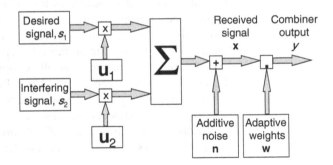

Figure 18.9: Vectorised system model

The representation can easily be generalised to include any number of branches and any number of signal sources, by simply varying the number of elements in each of the vectors and by adding more terms of the form $s\mathbf{u}$ to the expression for $\mathbf{x}$. In the general case of a desired signal s_d, N branches and L interfering signals, the result is

$$\mathbf{x} = \begin{bmatrix} x_1 \\ x_2 \\ \vdots \\ x_N \end{bmatrix} = s_d \mathbf{u}_d + \sum_{k=1}^{L} s_k \mathbf{u}_k + \mathbf{n} \qquad (18.10)$$

The problem now to be solved is how to determine the weights $\mathbf{w}$ so that the output y is as close as possible to the desired signal s_d. This is an example of a linear optimum filter problem, whose solution is the Wiener filter [Haykin, 96]. The Wiener solution requires knowledge of the *correlation matrix*, of the input signal $\mathbf{x}$, defined as

$$\mathbf{R}_{xx} = E \mathbf{x} \mathbf{x}^H \qquad (18.11)$$

where H is the transposed complex conjugate, or *Hermitian*. The Wiener solution is then

$$\mathbf{w}_{opt} = \mathbf{R}_{xx}^{-1}\mathbf{u}_d \qquad (18.12)$$

The correlation matrix can be found from the previous expression for the input signal as

$$\mathbf{R}_{xx} = \mathbf{u}_d\mathbf{u}_d^H + \sum_{j-1}^{L} \mathbf{u}_k\mathbf{u}_k^H + \sigma^2\mathbf{I} \qquad (18.13)$$

where $\mathbf{I}$ is the identity matrix

$$\begin{bmatrix} 1 & 0 & \cdots & 0 \\ 0 & 1 & \cdots & 0 \\ \cdots & \cdots & \ddots & \vdots \\ 0 & 0 & \cdots & 1 \end{bmatrix}$$

and the following assumptions have been made:

- All signals are uncorrelated with each other, i.e. $E[S_m S_n^*] = 0, m \neq n$.
- All propagation channels are uncorrelated with each other, i.e. $E[u_{mn}u_{jk}^*] = 0, m \neq j$ and $n \neq k$.
- All signals have unit variance, i.e. $E[|S_m|^2] = 1$.
- The noise variance in each branch is σ^2, i.e. $E[|n_m|^2] = \sigma^2$.

A combiner which either implements the Wiener solution directly or some close approximation to it is known as an *optimum combiner*. It may be approximated using iterative algorithms such as LMS or RLS (Chapter 16). The optimum combiner is closely related to the maximum ratio combiner (MRC) in a diversity system, except that the optimum combiner essentially maximises the SINR, whereas the MRC only optimises the SNR. Indeed, it can be shown that the optimum combiner reduces to the MRC in the special case where no interference is present.

Clearly the optimum weights depend directly on the propagation channels. In order to examine the characteristics of the optimum combiner, the simplest possible case of a propagation channel for an adaptive antenna system is first considered.

18.4.2 Steering Vector for Uniform Linear Array

In the case where a base station is operated in a free space environment, with no scatterers present, and where the wave arriving at the array can be represented by simple plane waves with constant amplitude, then the propagation channel depends only on the direction of the wave arrival. In such a case the propagation channel is given by a *steering vector* **a** (sometimes called the *array manifold*) whose components are the relative phases and amplitudes of a single wave at each of the antenna elements.

The simplest case is for a uniform linear array (ULA) of elements with a spacing of d, as shown in Figure 18.10. A plane wave arriving at an angle θ has a different phase at each of the elements, but has the same amplitude. If one of the elements is chosen as a reference

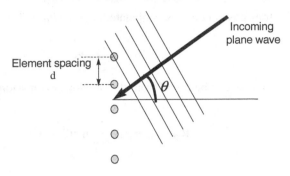

Figure 18.10: Uniform linear array

element (zero phase), then the other elements have phases given by simple trigonometry (Figure 18.11). The result is a steering vector given as

$$\mathbf{a}(\theta) = [a_1, a_2, \ldots, a_m, \ldots, a_N]^T \tag{18.14}$$

where the mth element is

$$a_m = \exp\left(-j\frac{2\pi}{\lambda}md\sin\theta\right) = \exp(-jkmd\sin\theta) \tag{18.15}$$

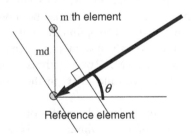

Figure 18.11: Calculation of phase progression for uniform linear array

18.4.3 Steering Vector for Arbitrary Element Positions

The steering vector concept can easily be extended to include arbitrary element positions and arrival angles. Figure 18.12 shows a wave arriving with a direction specified by the unit vector $\hat{\mathbf{r}}$, and the mth antenna element whose position is specified by the position vector $\mathbf{d}_m$. If the origin of coordinates is taken as the reference point, then the phase of the wave at the mth element is given by

$$a_m = \exp(-jk\hat{\mathbf{r}} \bullet \mathbf{d}_m) \tag{18.16}$$

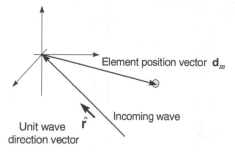

Figure 18.12: Steering vector for arbitrary element positions and arrival angles

18.4.4 Optimum Combiner in a Free Space Environment

The steering vector calculated for the ULA can now be used to investigate the behaviour of the Wiener solution for any such array. The directions of the desired and interfering mobiles are calculated, and the steering vectors in each case are computed. The correlation matrix is then found, and the Wiener weights are calculated from $\mathbf{w}_{opt} = \mathbf{R}_{xx}^{-1}\mathbf{u}_d$. Once the Wiener weights are calculated, the array response in any direction is found by calculating the magnitude of the output y when the channel is set to the steering vector for that direction, i.e.

$$y = \mathbf{x} \bullet \mathbf{w}_{opt}^{\mathrm{T}} \text{ with } \mathbf{x} = \mathbf{a}(\theta) \qquad (18.17)$$

Some examples are given in Figure 18.13. All show a four element $\lambda/2$ spaced ULA, except part (J); here the spacing is 2.8 λ. The following points are apparent from these examples:

- With no interferers, the combiner forms a lobe towards the desired signal, maximising the power received from it (Part (a)).
- With no noise present and with up to three interferers, the optimum combiner completely removes the interferers by placing nulls towards them. In this case, however, the power received from the desired signal is not necessarily maximised. (Parts (b), (c) and (d)).
- When more than three interferers are present, the combiner is unable to completely null the interferers and has to compromise in order to maximise the SINR. This is because the combiner only has four weights available to optimise. When the total number of signals exceeds the number of *degrees of freedom*, compromises must be made (Parts (e) and (f)).
- When significant noise power is present, the interferers are not completely nulled. This is because the noise has no particular direction associated with it, but its level is still minimised by the optimum combiner (Parts (g) and (h)).

- When noise is present, the combiner's ability to separate two mobiles depends on the angular separation between them, as the width of the main lobe produced by the array is limited by the array size (Part (i)).
- When the element spacing is significantly greater than $\lambda/2$, *grating lobes* are produced. These are lobes having significant amplitude away from the direction of any desired signal. The reason for such lobes is that the large phase difference produced by waves arriving at widely spaced antennas leads to an ambiguity in the direction of lobes. This is a spatial version of the phenomenon of *undersampling* which arises from time sampling of a

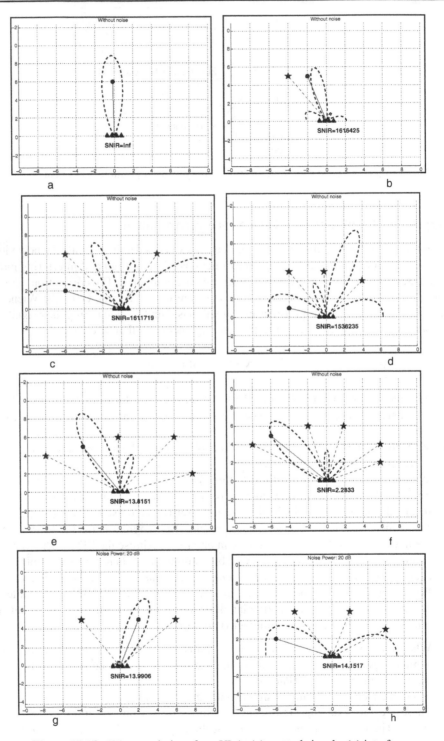

Figure 18.13: Wiener solutions for a ULA: (•) wanted signals, (∗) interferers

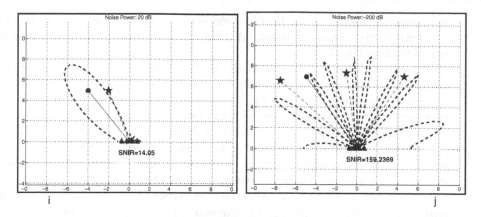

Figure 18.13: (*Continued*)

waveform at less than twice the highest frequency (Nyquist rate). Grating lobes can cause unwanted power to be radiated towards other co-channel cells, reducing the efficiency of SFIR systems (Part (j)).

In summary, an N-element adaptive array with optimum combining can completely suppress up to $(N-1)$ interferers in the absence of noise. Hence the maximum number of mobiles which could be supported in an SDMA system without interference is N. In the presence of noise the SINR is degraded.

18.4.5 Optimum Combiner in a Fading Environment

When the fading channel is considered, the properties of an antenna array with an optimum combiner are rather different to the non-fading case.

Firstly, the fast-fading component introduces a random phase and amplitude to each element, which perturbs the steering vector. In the case of Rayleigh fading, the phase can take any value, so the direction of arrival of the waves may be impossible to determine from short-duration observations of the received signals. Similarly, the concept of an array radiation pattern relies on plane waves which are incident on the array elements with constant amplitude, so in the fading environment it may not be useful to consider the array as creating lobes and nulls towards desired and interfering sources.

When the fast fading is highly correlated between the elements, it may be considered as a single scalar which multiplies the steering vector, affecting all elements equally. It may then be possible to recover the steering vector. On the contrary, no diversity gain can be obtained, as diversity relies on uncorrelated fading. The correlation between elements decreases with element spacing, so there is a clear conflict between the avoidance of grating lobes and the need for diversity gain.

In the absence of the concept of radiation patterns, the significant measure of 'distance' between two sources is actually the *correlation* of their propagation channels. Two mobiles may be effectively separated in SDMA provided they are essentially uncorrelated. In a classical Rayleigh channel, the necessary distance between two mobiles is then of the order only half a wavelength.

When the N elements have completely uncorrelated fading and no interferers are present, the optimum combiner performance is identical to N-branch maximum ratio diversity. When an interferer is introduced, the diversity order is reduced by 1 because one degree of freedom is lost in nulling the interferer. In general, it is possible to obtain $(N - L)$ branch diversity at the same time as nulling L interferers.

The number of *significant* interferers which require nulling at any instant is actually reduced by the presence of uncorrelated fast fading, as not all of the fast-fading signals between the interferers will be above the noise level at any instant. This improves the interference performance compared with all of the signals having a constant level.

It is clear that the presence of a fading channel can have important advantages for the performance of adaptive antennas. The key disadvantage, however, is that the transmit performance of the combiner is likely to be worse due to the reduced correlation in time, frequency or space between the receive and transmit channels.

In the presence of wideband fading on each branch, the combiner must be modified to produce acceptable performance. The simplest way to do this is to treat the significantly delayed multipath components as interferers and null them using the adaptive combiner, but this uses up degrees of freedom and reduces the interference tolerance of the system. A higher performance approach is to treat resolvable multipath components as containing potentially useful energy. The function of the array combiner then becomes merged with that of an equaliser. This can be performed using either a linear structure, where each antenna channel contains an adaptive FIR filter [Balaban, 91], or using a non-linear approach such as MLSE for optimal results [Bottomley, 95].

18.4.6 Implementation of Adaptive Antennas

In practice it is not possible to directly implement the optimum combiner for the following reasons:

- Exact knowledge of the channels is required to form the correlation matrix. These channels can be estimated only in the presence of noise and interference.
- The channel may be different between uplink and downlink because these may be separated in time, frequency or space.
- The channel may change rapidly in time, so only a limited amount of data is available for estimation.

These issues may be partially dealt with as follows:

- Weights must be recalculated with a frequency of around 10 times the maximum Doppler frequency of the channel.
- Instead of implementing the optimum combiner on the reverse link, the fast fading may be averaged in time to produce estimates of the angles-of-arrival of the signal sources. As these angles change more slowly, more reliable results may be obtained at the expense of sub-optimal performance.

18.4.7 Adaptive Antenna Channel Parameters

In order to model the performance of adaptive antennas in practical environments in any detail, it is necessary to have a clear understanding of the channel properties which may affect their performance. This characterisation merges elements of the narrowband and wideband channels studied in previous chapters, and adds the concept of space and correlation.

The correlation properties of a large array of antenna elements can be calculated from knowledge of the distribution of the angles of arrival of waves at the array. This has already been described in Chapter 15 within the context of diversity systems. With arrays the idea of correlations between a single pair of antenna elements is generalised to allow computation of the complete correlation matrix for N elements, given the angle of arrival p.d.f. $p(\theta)$. This distribution interacts with the steering vector of the array, yielding

$$\mathbf{R_{xx}} = \int_{\theta=0}^{2\pi} \mathbf{a}(\theta)\mathbf{a}^H(\theta)p(\theta)\mathrm{d}\theta \qquad (18.18)$$

Additionally, there is a close relationship between the angle-of-arrival distribution and the time of arrival, as can be seen by reconsidering the equal-delay ellipses shown in Figure 11.2. For short delays, the scatterers are located within a very narrow ellipse, close to the direct path between the base station and mobile. As the delay increases, the range of angles encountered increases, so there is a general tendency for angle spread to increase with delay. As a consequence of (18.18), this yields decreasing correlation between elements for the paths with longest delay. If scattering is assumed to occur with uniform probability at all locations, this relationship can easily be calculated; Figure 18.17 showed the way that $p(\theta)$ changes as a function of the excess delay, tending towards a uniform distribution as the excess delay increases.

Although theoretical calculations like this can yield some useful general conclusions about the likely characteristics of the environment, the true situation must be tested via measurements. Figure 18.14 shows a measured *scattering map*. Such maps represent the variation of power (represented by shading in Figures 18.14–18.16) with delay and angle and are thus

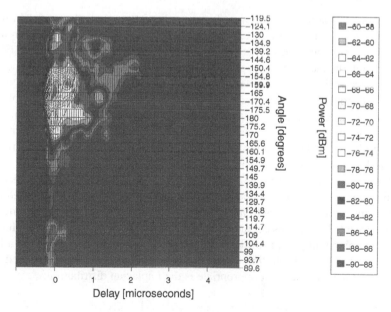

Figure 18.14: Measured scattering map in Central London–low spread example (Reproduced by permission of Nortel plc.)

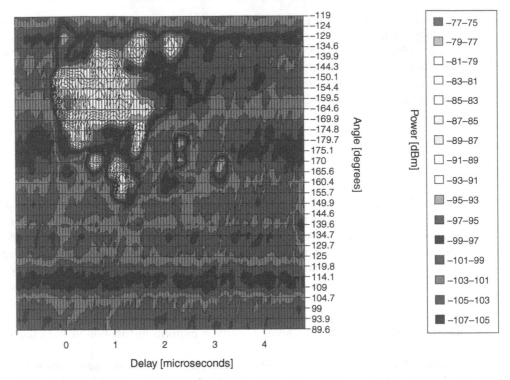

Figure 18.15: Measured scattering map in Central London – high spread example (Reproduced by permission of Nortel plc.)

analogous to the power delay profile, generalised to include the angle. In this example, the measurements were made in Central London [Ward, 96] at 1.8 GHz from a high macrocell environment. The peak of the power comes from a region close to −165°, which corresponded in the measurement very closely to the geometrical direction to the mobile. There is angular spreading over at least 20° and delay spread of around 0.5 μs, but very little power is observed from other directions. This contrasts considerably with the example in Figure 18.15, where much greater scattering in both angle and delay is evident. In a third example, Figure 18.16 shows a case measured in a more open or suburban area, where the angle spread is small but the delay spread extends to at least 5 μs. Such extreme cases must be considered when designing adaptive antenna systems.

One particular point to note from these examples is that individual peaks of scattering are themselves spread in both angle and delay. This is partly due to the limited resolution of the measuring equipment, in both angle and delay, but is also indicative of some spreading in the environment. This creates degradation in the performance of adaptive beamforming, especially when forming nulls, as it is difficult to make nulls sufficiently broad to null all of the energy in these regions [Thompson, 96]. Numerous distributions have been proposed to represent this effect, but perhaps the most appropriate is the Laplacian distribution:

$$p(\theta) = \frac{c}{\sqrt{2}\sigma}\exp\left[-\frac{|\theta|\sqrt{2}}{\sigma}\right] \qquad (18.19)$$

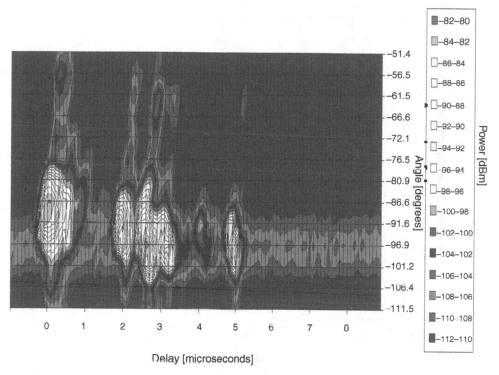

Figure 18.16: Low angle spread: Harlow area (Reproduced by permission of Nortel plc.)

where σ represents the angular spread relative to the direction of the scatterer, taken as $\theta = 0$ (see Figure 18.17). In rural environments σ has been measured at around 1^{0} [Pedersen, 97]; the value increases to many tens of degrees in indoor environments [Spencer, 97]. A detailed review of spatial channel models for antenna arrays is given in [Ertel, 98].

18.5 MULTIPLE-INPUT MULTIPLE-OUTPUT SYSTEMS

The application of beamforming principles to produce an MIMO system is described in this section. Such systems are being applied practically to advanced wireless LAN systems, such as the IEEE 802.11n standard and to WiMax systems such as IEEE802.16e. For deeper theoretical details, the reader is referred to texts such as [Oestges, 07] and [Gesbert, 03].

18.5.1 MIMO Signal Model

A general MIMO system is shown in Figure 19.21, with N transmit antennas and M receive antennas, receiving M copies of the N transmitted signals via the $M \times N$ distinct paths in the channel, represented by the $M \times N$ channel matrix $\mathbf{H}$. The received signal vector $\mathbf{r}$ is then given by

$$\mathbf{r} = \mathbf{Hs} + \mathbf{n} \tag{18.20}$$

where $\mathbf{s}$ is the $N \times 1$ transmitted signal vector and $\mathbf{n}$ represents a $M \times 1$ additive noise vector, produced by the channel and receiver front end. Each element of the $\mathbf{H}$ matrix, denoted as h_{ij}, represents the transfer function between the j-th transmit and i-th receive antenna element. In

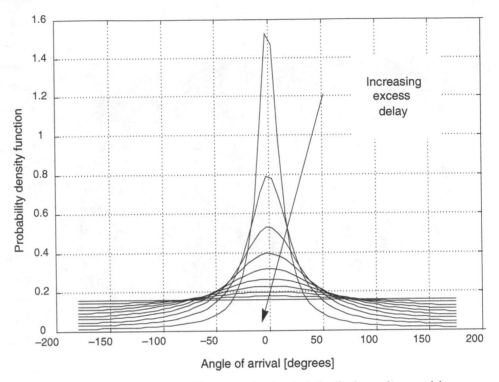

Figure 18.17: Relationship between angle-of-arrival distribution and excess delay

the following analysis the channel is assumed narrowband, but this assumption can be relaxed by making the h_{ij} frequency-dependent.

The transmitted signals at each antenna may depend on the whole set of bits to be transmitted, or they may represent a single independent bit, depending on the scheme employed. The transmitters may consist of conventional modulators/coders followed by appropriate beamforming, or the coder, modulator and beamformer may all be seen as a single process. In any case, the coder and the decoder must be designed to take account of the channel properties to allow recovery of the transmit bit stream with maximum fidelity and maximum channel capacity. The elements of the channel matrix **H** will be subject to all of the usual time- and space-varying channel fading and dispersion characteristics, so the optimum transmit and receive parameters will vary with time and location.

To maximise the channel capacity, it is necessary that the transmit and receive weights transform the routes from the transmit bit stream to the receive bit stream into as many independent paths as possible, over each of which a bit stream can be simultaneously transmitted without interference. In order to investigate this, it is useful to apply a *singular value decomposition* of the matrix **H**. Via such a decomposition, there always exist unitary[1] matrices **U** ($M \times M$) and **V** ($N \times N$) such that **H** can be written in the form:

$$\mathbf{H} = \mathbf{U}\mathbf{V}^{\mathrm{H}} \qquad\qquad (18.21)$$

---

[1]A unitary matrix **V** is one whose conjugate transpose, or Hermitian, denoted $\mathbf{V}^{\mathrm{H}}$, is also its inverse, so that $\mathbf{V}\,\mathbf{V}^{\mathrm{H}} = \mathbf{I}$

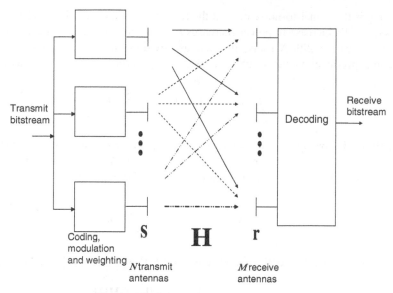

Figure 18.18: MIMO system architecture

where is a diagonal matrix, whose non-zero elements $\sqrt{\lambda_i}$ are the *singular values* of **H**, while λ_i are the eigenvalues of **H**. There are t distinct, real positive eigenvalues, where

$$t = \min(M, N) \tag{18.22}$$

If the **U** and **V** matrices were used as receive and transmit beamformer matrices, respectively, then (18.20) can be rewritten as:

$$\tilde{\mathbf{r}} = \ddot{\mathbf{s}} + \ddot{\mathbf{n}} \tag{18.23}$$

by writing $\tilde{\mathbf{r}} = \mathbf{U}^H \mathbf{r}$, $\tilde{\mathbf{s}} = \mathbf{V}^H \mathbf{s}$ and $\tilde{\mathbf{n}} = \mathbf{U}^H \mathbf{n}$. As $\varLambda$ is diagonal, (18.23)   represents t separate spatial channels of the form

$$r = \sqrt{\lambda_i \tilde{s}_i} + \tilde{n}_i \tag{18.24}$$

Each of these channels can be used separately and simultaneously for signalling of a separate bitstream, resulting in a capacity gain. This is similar in some ways to a CDMA system, where codes are chosen to be orthogonal to allow multiple signals to be transmitted in the same bandwidth and separated by decorrelation at the receiver. The MIMO system exploits spatial decorrelations, rather than decorrelations induced by codes, and does not require that the occupied bandwidth is increased beyond that required by the signalling rate for each channel.

18.5.2 MIMO Channel Capacity

For a memoryless one-dimensional system with single antennas at transmitter and receiver, the channel capacity C_{11} in bits/s/Hz is given by

$$C_{11} = \log_2(1 + \gamma|h^2|) \tag{18.25}$$

where γ is the signal-to-noise ratio at the receive antenna and h is the channel gain. For high SNRs, a 3 dB increase in power increases the channel capacity by 1 bit/s/Hz. For a transmit diversity system with N transmit antennas, dividing the power equally amongst all of the antennas produces a channel capacity (strictly, the *mutual information*) as follows:

$$C_{N,1} = \log_2\left(1 + \frac{\gamma}{N}\sum_{i=1}^{N}|h_i|^2\right) \tag{18.26}$$

The channel capacity is clearly increased, although it grows relatively slowly only with N due to the logarithmic term. The receive diversity case with M receive antennas is a little better, as the power no longer needs to be divided amongst the channels:

$$C_{M,1} = \log_2\left(1 + \gamma\sum_{i=1}^{M}|h_i|^2\right) \tag{18.27}$$

In the most general case, the capacity for any MIMO system is [Telatar, 95]

$$C_{N\times M} = \log_2(\det|\mathbf{I}_M + \mathbf{HQH}^*|) \tag{18.28}$$

where $\mathbf{Q} = E[\mathbf{s}\,\mathbf{s}^H]$ is the covariance matrix of the transmitted signal $\mathbf{s}$, $\mathbf{I}_M$ is the $M \times M$ identity matrix, $\mathbf{H}^H$ is the Hermitian of the channel matrix $\mathbf{H}$, and $\det|\cdot|$ is the determinant. The best channel capacity is then achieved if the power is distributed amongst the spatial channels proportionate to the size of the corresponding eigenvalues, so that the good channels receive more power. Each channel is filled with power so that the sum of the transmit power in a channel and the inverse of the channel gain is at a constant level D while keeping the total transmit power constant:

$$\frac{1}{\lambda_1} + P_1 = \frac{1}{\lambda_2} + P_2 = \cdots = \frac{1}{\lambda_1} + P_t = D \tag{18.29}$$

The capacity is then given by

$$C_{N\times M, WF} = \sum_{i=1}^{t}\log_2(\lambda_i D) \tag{18.30}$$

This 'water-filling' solution [Telatar, 95] requires knowledge of the channel at the transmitter, or a sufficiently slowly changing channel to permit feedback of the channel state from receiver to transmitter. If this information is not available the best solution is to divide the power equally amongst the antennas, $\mathbf{Q} = (\gamma/N)\mathbf{I}_N$ and the channel capacity is given by [Foschini, 98]

$$C_{N\times M, EP} = \log_2\left(\det\left|\mathbf{I}_M + \frac{\gamma}{N}\mathbf{HH}^*\right|\right) \tag{18.31}$$

Using the previous singular value decomposition we can rewrite (18.31) in terms of the channel eigenvalues as

$$C_{N\times M, EP} = \sum_{i=1}^{t}\log_2\left(1 + \frac{\lambda_i}{N}\right) \tag{18.32}$$

It is clear from this that MIMO channel capacity grows with the size and number of the non-zero eigenvalues. Provided these eigenvalues are significant in size, the result in (18.32) now suggests that the capacity grows linearly with $t = \min(M, N)$ rather than logarithmically and every 3 dB increase in SNR produces an additional t [bits/s/Hz].

The capacity can be examined numerically for any channel configuration using the preceding equations; for example, the individual channel matrix elements can be assumed uncorrelated complex Gaussian random variables with zero mean and unit variance in order to simulate uncorrelated Rayleigh channels between all pairs of elements. We can then examine both the mean capacity, to give a sense of the overall throughput available from the system and the statistics of capacity outage, to determine how often the channel may be unable to support a desired minimum throughput. As an alternative to simulation, several approximate expressions are available. An accurate example is the expression given in [Rapajic, 00] for the mean channel capacity, as follows:

$$C_{N \times M} \approx \min(N, M) \times \left[\log_2(w\Gamma) + \left(\frac{1 - y}{y} \right) \log_2 \left(\frac{1}{1 - v} \right) - \frac{v}{y} \right] \qquad (18.33)$$

where $y = M/N, M \le N; y = N/M, N < M$ and

$$w = \frac{1}{2} \left[\left(1 + y + \frac{1}{\Gamma} \right) + \sqrt{\left(1 + y + \frac{1}{\Gamma} \right)^2 - 4y} \right]$$

$$v = \frac{1}{2} \left[\left(1 + y + \frac{1}{\Gamma} \right) - \sqrt{\left(1 + y + \frac{1}{\Gamma} \right)^2 - 4y} \right] \qquad (18.34)$$

Here Γ is the mean signal to noise ratio. This expression is shown in Figure 18.19 and Figure 18.20, clearly illustrating the large capacity gains potentially available when compared with the single-antenna case.

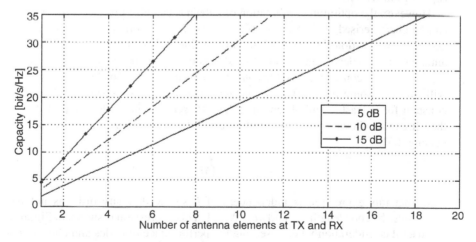

Figure 18.19: Mean MIMO channel capacity versus the equal numbers of antennas at both ends of the link for signal-to-noise ratios $\Gamma = 5, 10, 15$ dB

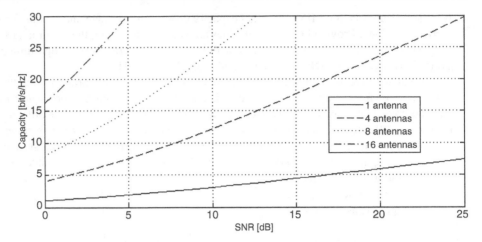

Figure 18.20: Mean MIMO channel capacity with equal numbers of antennas at both ends of the link versus SNR for varying numbers of antennas

These remarkable results suggests that the capacity of wireless systems, when operated using a MIMO approach, can be increased simply by adding more antennas, analogous to simply adding extra cables between the two terminals for wired systems.

18.5.3 Trade-Off Between Diversity and Capacity for MIMO

It is clear that MIMO systems can provide a capacity gain arising from spatial multiplexing. Following (18.32), the capacity can be expressed as

$$C = r \log SNR + k \tag{18.35}$$

where r is the capacity gain, which following [Foschini, 98], has a maximum value of $r_{max} = \min(M, N)$.

However, the optimum combination of antennas can also produce a diversity gain, which can be characterised, as in Section 16.7, by the *diversity order d* if the bit error probability decreases proportional to $1/SNR^d$ for high values of SNR. If all paths between pairs of antennas are independent, then the diversity order reaches its maximum value at $d_{max} = M \times N$, consistent with the considerations in Chapter 16. Any particular STC scheme will produce some trade-off between these two extremes. Assuming that the channel is constant for at least l symbols and that the STC scheme has a block-length $l \geq M + N - 1$, then it can be shown that the highest possible diversity order at a given capacity gain r is [Zheng, 03]

$$d_{max}(r) = (M - r)(N - r) \tag{18.36}$$

Note that this expression is independent of l, so the diversity order is not increased by increasing the code length beyond $M + N - 1$. This expression is shown in Figure 18.21. This is a useful consideration when assessing the performance of codes and choosing appropriate system parameters and trade-offs. For example, it makes clear that adding one antenna at both ends of a link increases the capacity gain by 1 at any diversity order.

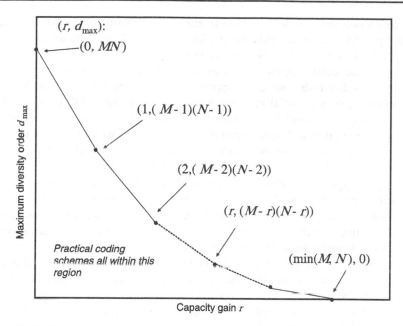

Figure 18.21: Fundamental trade-off between diversity order and capacity gain for MIMO systems

18.5.4 Particular STC Schemes

Although the singular value decomposition described in Section 18.5.1 produces beamformers which can successfully act as an ideal MIMO system, the calculations assumed that the channel matrix is completely known at both the transmitter and receiver. This is not usually practical, and anyway the necessary matrix manipulations may be complex and hard to calculate with sufficient precision. This section therefore examines some practical coding schemes. Many more have been proposed than can be covered here.

A particularly simple space-time code for transmission from two antennas was introduced by [Alamouti, 98] and later generalised to an arbitrary number of antennas [Tarokh, 99]. A version of the two-antenna code has been widely applied for 3G systems [Derryberry, 02]. A stream of symbols to be transmitted is split into pairs $[s_1, s_2]$. In the first time instant, the two symbols are simply transmitted simultaneously from the two antennas. In the second instant, the first antenna transmits $-s_2^*$ and the second transmits s_1^*. Thus two symbols have been sent in two time instants and the rate of transmission is identical to a standard system without this coding. At the receiver, which has a single antenna, let the received signals in the two time instants be r_1 and r_2. It can then be shown, provided the channel can be assumed constant over two time instants, that maximum likelihood estimates of the transmitted symbols are given by

$$\begin{bmatrix} \hat{s}_2 \\ \hat{s}_1 \end{bmatrix} = \begin{bmatrix} h_2^* r_2 - h_1 r_1^* \\ h_1^* r_2 + h_2 r_1^* \end{bmatrix} \tag{18.37}$$

where h_1 and h_2 are the receiver's estimates of the channel from the first and second antennas, respectively. These estimates can be shown to have the same performance as maximal ratio combining diversity. This code is an example of a wide family of space-time block codes.

Although only this simple case provides transmission with no bandwidth expansion, all such codes provide a high diversity order without the need for complex processing and with a relatively small number of receive antennas.

When the central consideration is maximising the channel capacity rather than its reliability, other codes are more appropriate. One well-known example is V-BLAST[2], a prime example of a spatial multiplexing scheme [Foschini, 96]; [Wolniansky, 98]. Spatial multiplexing is a special class of space-time block code in which independent streams of data are transmitted over different antennas to increase the data rate. The particular code defines the sequence and timing of the transmissions. In the case of V-BLAST, the transmitter simply splits the symbols to be transmitted into N independent streams and transmits them simultaneously through the N transmit antennas, with no coding across the antennas. At the receiver, beamforming weights are first applied for the strongest eigenvalue of the channel matrix and the receiver detects the bit stream which the detector estimates as a result. The receiver then calculates the interference which these signals would have caused to the received signal and subtracts it from the total. The residual signal which results is then used to calculate a new set of beamforming weights and to detect the next-strongest set of symbols. This process proceeds until all N symbols are detected. The architecture is therefore similar to a decision-feedback equaliser (see Chapter 18) but operating in space rather than time. V-BLAST achieves a high data rate, but the independent treatment of each of the bitstreams comes at the expense of a loss of diversity.

There are several other types of space-time code, including space-time trellis codes and space-time turbo codes, which provide high diversity order and high bit rate simultaneously, but often with a very large degree of complexity [Gesbert, 03]. Finding codes with high performance but acceptable complexity is a key current research topic.

18.5.5 MIMO Channel Modelling

The large gains potentially available from MIMO systems may not be provided in practice if the channel properties do not satisfy the assumptions of uncorrelated, independent, equal power scattering MIMO channel characterisation is then essential to determine the properties of the matrix of transfer functions.

As illustrated by the considerations of the previous section, the fundamental consideration is the eigenvalue distribution of the channel matrix. If all of the $N \times M$ paths from transmitter to receiver exhibit decorrelated fading and are of equal mean gain, then the channel matrix is of full rank and there exist the full set of $t = \min(M, N)$ eigenvalues of equal magnitude, thereby maximising the channel capacity even without information at the transmitter on the channel state.

Unlike beamsteering systems, then, MIMO systems perform best in rich multipath environments, where scattering components leave the transmitter over a wide range of angles, are scattered widely in space, mix together and are received at a similarly wide angular range at the receiver. If paths from a transmit antenna to two receive elements undergo scattering from almost the same scatterers, it is likely they will be correlated, producing a linear dependence between two rows of the scattering matrix. One of the eigenvalues then drops to zero and the effective number of independent channels reduces by one.

---

[2]V-BLAST stands for Vertical-Bell Labs Layered Space-Time Architecture

Other factors such as antenna impedance matching, array size and configuration, element pattern and polarisation properties, multipath propagation characteristics and mutual coupling can all affect MIMO system performance [Jensen, 04].

According to [Gesbert, 03], key modelling parameters are required to be measured, such as path loss, shadowing, Doppler spread profile, delay spread profile, Ricean k-factor distribution, joint antenna correlations at transmit and receive ends and the channel matrix singular value distribution.

For *narrowband* channels, the channel matrix $\mathbf{H}$ can be modelled as the sum of a coherent and incoherent component, as follows:

$$\mathbf{H} = \mathbf{H}_{\text{CO}} + \mathbf{H}_{\text{INCO}} \tag{18.38}$$

For MIMO systems, the higher the Ricean k-factor, the more dominant the $\mathbf{H}_{\text{CO}}$ component is, and as it is time-invariant, its effect is to drive up antenna correlation – hence singular value spread is increased [Gesbert, 03] and the MIMO capacity decreases. A like-for-like comparison is valid only if the mean power is the same, however, and it may be preferable in macrocell situations to have a strong coherent LOS component in order to increase the mean SNR, hence compensating for the loss in capacity in the link budget. For indoor environments, and to a lesser extent microcells, the multipath is so rich that the k-factor is almost always low, giving rise to very good MIMO performance.

As with diversity systems, a wide angle spread around the antenna array produces a large decorrelation of spatial channels and hence good diversity performance. With MIMO, however, one may be concerned with a situation in which very difficult angle spreads are present around the transmitter and receiver. One example is an elevated macrocell base station with a relatively narrow angle spread, transmitting to a mobile which is surrounded by scatterers and hence experiences an angle spread close to 360°. In such situations, even given a low correlation between adjacent elements at one end of the link, the overall MIMO performance may not necessarily be good, due to an effect known as a pinhole channel. Here the signal paths all pass through a narrow region or pinhole somewhere between the transmitter and receiver, producing a low channel matrix rank despite low correlation between antennas at either the transmitter or receiver.

In order to create a complete channel model for MIMO based on the underlying propagation effects, the channel has to be represented as a sum of multipath contributions with different delays and amplitudes, leaving the transmitter at a variety of angles of departure, Ω and arriving at the receiver from a variety of directions of arrival, Ψ. The channel is then described by a double-directional impulse response for each transmit-receive element pair (i, j) as

$$h_c(i,j) = \sum_{i=1}^{n_{mp}} \alpha_l e^{j\varphi_l} \delta(\tau - \tau_l)\delta(\Omega - \Omega_l)\delta(\Psi - \Psi_l) \tag{18.39}$$

where n_{mp} is the number of multipaths, α_l are the amplitudes and φ_l are the phases. Entries in the full channel matrix $\mathbf{H} = [h(i,j)]$ can be determined by further weighting the multipath contributions by the gains of the transmit and receive antennas in the Ω and Ψ directions:

$$h(i,j) = \sum_{i=1}^{n_{mp}} \alpha_l e^{j\varphi_l} \delta(\tau - \tau_l)\delta(\Omega - \Omega_l)\delta(\Psi - \Psi_l) g_T(\Omega_l) g_R(\Psi_l) \tag{18.40}$$

It is then necessary to characterise the statistical parameters of the multipaths according to the physical environment in order to determine the overall system performance. Many measurements have been conducted in a variety of environments, and several models have been suggested as standard reference environments for comparing coding schemes and antenna configurations. Such models are complicated by the finding from measurements that parameters such as delay spread, angular spread and shadowing are all correlated with each other, and that these correlations can have a significant impact on system outage capacity. For more detail on channel modelling for MIMO systems, see [Jensen, 04] and, [Molisch, 04].

18.5.6 MIMO Channel Models for Specific Systems

Some of the models which are in use for specifying the performance of some important practical systems are described below. A good assessment of the issues involved in selecting MIMO channel models is available in [Correia, 06], whereas other environmental issues relating to the implementation issues for MIMO are described in [Bliss, 02].

Wireless LAN Channel Model

This model was created for comparing the performance of MIMO techniques for the IEEE 802.11n wireless LAN ('Wi–Fi') standard [IEEE, 04]. The model is intended to represent short-range indoor environments such as residential homes and small offices. The key features of the model are summarised as follows:

- Several power-delay profiles with RMS delay spread varying from 15 to 150 ns.
- Profiles are assumed to follow clusters with exponential delays as described in Section 13.10. The number of clusters is assumed to vary between 2 and 6 according to the environment.
- The first tap follows the Rice distribution in the case of the presence of a line-of-sight, with a Rice k-factor of up to 6. All other taps are assumed to be independently Rayleigh fading.
- The power angular spectrum of each cluster is assumed to follow a Laplacian distribution as in Eq. (18.19) with an angular spread of around 30°. The mean angle of arrival and departure for each cluster is assumed to be uniformly distributed over all horizontal angles, with all arrivals in the horizontal plane.
- The specific RMS delay spread and angular spread of a particular cluster is assumed to be positively correlated, so that clusters with a low delay spread also have a small angle spread.
- A Doppler spread for each tap of around 3 Hz at 2.4 GHz carrier frequency and 6 Hz at 5.25 GHz, even when the terminals are stationary, arising from the motion of people around the building. In one instance of the model, an additional Doppler component is introduced to present a vehicle moving past the environment and creating scattering corresponding to $40 \, \mathrm{km \, h^{-1}}$.

3GPP Spatial Channel Model

This model was developed for use in comparing proposed MIMO extensions to the 3G mobile standards [3GPP, 03] for outdoor environments at around 2 GHz carrier frequency and system bandwidth of 5 MHz. Key features are as follows:

- The model is based on physical considerations of scatterers which are different for three environment types: urban macrocell, suburban macrocell and urban microcell. The models parameters are different for each environment.

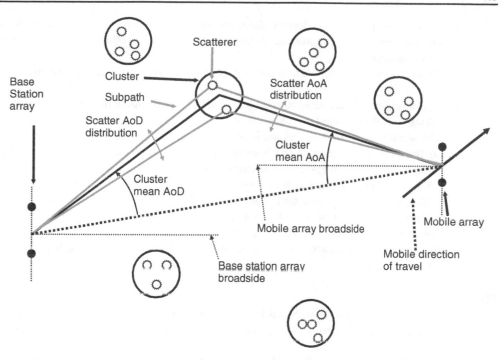

Figure 18.22: Scatterer geometry for 3GPP Spatial Channel Model. Note that in the model each cluster actually contains 20 scatterers

- Channel dispersion is modelled via six channel taps, each of which arises from a subtap produced by a cluster of 20 scatterers which are close enough to produce essentially the same delay but with different angles of arrival and departure (AoA and AoD). See Figure 18.22 for details of the geometry which this implies.
- The mean AoA or AoD is calculated geometrically for the assumed location of the cluster, while the AoA and AoD of the individual clusters is randomly generated from a Gaussian distribution whose variance is a parameter of the model depending on the environment.
- Each of the scatterers in a cluster has fixed amplitude but random phase, and the sum produces either Rayleigh or Rice fading.
- An overall system simulation is conducted as a series of 'drops' where the locations of clusters are randomly placed. During the duration of each drop the large-scale channel parameters, such as angle spread, mean DoA and shadowing are held constant. The mobile station positions are randomly selected at the start of each drop.
- Additional optional features allow modelling of polarisation, distant scattering clusters, line-of-sight situations for the microcell case and a modified angular distribution to emulate propagation in an urban street canyon.

COST 273 MIMO Channel Model

This model [Correia, 06] has been developed to be applicable to a very broad range of future broadband systems and environments. It has broad similarity to the 3GPP model, but provides enhancements in a number of respects:

- It provides parameters, derived from extensive measurements, for a large number of environments including macrocellular, microcellular, picocellular, peer-to-peer and fixed wireless access cases.
- Scatterers are redistributed according to a Poisson distribution (i.e. uniform in location at a given rate per area).
- The angular spread, delay spread and shadowing are modelled as lognormal random variables, all correlated with each other.
- The scatterers within a cluster are individually used to determine both the AoA and AoD, so the parameters seen at the transmitter and receiver are closely linked, unlike the independent AoA and AoD distributions in the 3GPP model.
- Each radio environment is made up of a number of local propagation environments, with approximately constant parameters. These parameters are in turn random realisations of a set of global parameters defining the propagation environment. Thus the model can reproduce the characteristics of the fine structure encountered in real propagation environments with continuity of random mobile motion between the local environments.

Fixed Links: 802.16a

Other models have been developed for the broadband fixed wireless access standard (fixed WiMax) IEEE 802.16a, using a three-tap power delay profile and fixed correlations amongst antenna elements at the mobile and base station [IEEE, 03]. Such models have been further developed to provide more physical insight and the ability to scale the models to a wider range of distances and environments [Oestges, 03].

18.5.7 Impact of Antennas on MIMO Performance

The MIMO channel transfer matrix **H** depends not only on the propagation environment, but also on key parameters of the antenna arrays at the transmitter and receiver. These parameters are: array configuration, mutual coupling, radiation pattern and polarisation, which in turn affect correlation and mean received power (see [Jensen, 04]). These are the same basic issues as for diversity reception, but the complexity is increased by the large number of pairs of elements which must be considered and the need to ensure that the arrays are sufficiently compact for practical operation.

Therefore, to determine which is the optimal array configuration for all cases can become difficult, as this strongly depends on the propagation characteristics of the environment. After conducting an extensive study where different array types were investigated for both the mobile and the base station in an outdoor environment [Martin, 01], the results suggest that average capacity is relatively insensitive to array configuration. Although these results suggest that it is difficult to establish a strong dependence of MIMO capacity on array configuration, an alternative *intrinsic capacity* has been formulated recently for specific antenna apertures and a specific channel [Wallace, 02]. This opens interesting research opportunities to find optimal array configurations which could maximise MIMO system performance. This search may be assisted by considering *angle diversity*, *polarisation diversity* and *mutual coupling*.

Angle diversity is produced when antennas have distinct radiation patterns but may have closely spaced phase centres. If, by design, the elements are chosen to have highly orthogonal patterns to create low correlation, then capacity gains are possible. A rather interesting topic for active research is to find out newer antenna topologies which can produce high orthogonality between the radiation patterns.

Antenna polarisation is another factor which strongly influences MIMO capacity bounds. The work produced by [Andrews, 01] and [Svantesson, 04] suggests that in a strong multipath environment, the combined polarisation and angle diversity offered by three orthogonally oriented electric and magnetic Hertzian dipoles at a point can lead to six uncorrelated signals. However, from a practical standpoint, constructing a multi-polarised antenna which can achieve this bound could be problematic, and further research to construct new practical antennas which can exploit polarisation is envisaged.

Finally, *antenna mutual coupling* has a strong effect on MIMO systems and has been studied extensively, for which two main conclusions reported in [Jensen, 04] are drawn. First, because of the induced angle diversity combined with improved power collection capability of coupled antennas, the capacity of two coupled dipoles can be higher than that of uncoupled dipoles, on the basis that a proper termination exists and that the spacing between the dipoles is small, hence producing high coupling. Second, for a fixed-length array, the high coupling between antenna elements within the same physical space leads to an upper bound on capacity performance.

18.6 ADAPTIVE ANTENNAS IN A PRACTICAL SYSTEM

It is expected that the adaptive antenna schemes described in this chapter will form a major enhancement to both fixed and mobile systems in practical use in the coming years. As an example which utilises a wide range of the features described here, Figure 18.23 shows the techniques which are available as part of the IEEE 802.16e mobile WiMax standard [Piggin, 06]. Two broad approaches are available: *Adaptive antenna system* (AAS) which is essentially conventional beamforming as described in Section 18.3, providing coverage extension and

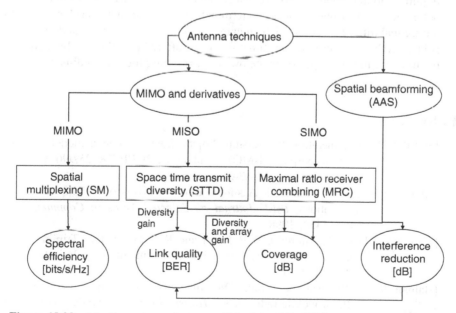

Figure 18.23: Adaptive antenna features within the mobile WiMax system, IEEE 802.16e

interference reduction via spatial filtering, which does not necessitate multiple antennas at the mobile station. Alternatively, diversity schemes, which can be regarded as simplified MIMO schemes are available for quality and coverage improvement, using both conventional receiver combining via MRC or using the Alamouti transmit diversity scheme described in Section 18.5.4. Full MIMO schemes are also available, including the provision of channel information fed back to the transmitter, providing a high degree of spatial multiplexing (SM) and hence capacity gain. Although this is likely to be most powerful when multiple antennas are available at the mobile, a *collaborative* MIMO scheme is also defined in which multiple mobiles use appropriate coding in the uplink to allow them to transmit simultaneously. This increases system capacity rather than the capacity of individual mobile-base links.

The best choice of schemes from amongst these is dependent on the capabilities of mobiles, the services to be offered and the environment. For example, in rural areas the main drive is likely to be to extend cell range, without particular limitations on system capacity, so AAS may be used. In urban areas, where cells are anyway small and user densities high, the greatest need may be for a high capacity at the expense of link budget, so the full MIMO/SM scheme may be applied. These need map well to the channel characteristics, as the rich scattering required for efficient SM is most likely to be present in the urban case. These schemes may even be chosen dynamically, by allocating different schemes to different user links according to the subcarrier/ time interval which the user occupies within WiMax's OFDMA structure.

18.7 CONCLUSION

The application of adaptive antennas to mobile systems has significant advantages in terms of coverage, capacity and quality. In order to realise these advantages, however, it is necessary to have a good understanding of the propagation channel and to use this understanding to design systems which have performance benefits outweighing the extra costs involved. Doing so requires a joint optimisation of RF, hardware and signal processing technology, to realise the potential benefits a reasonably sized terminal with multiple antenna elements at low cost and more capabilities to accommodate integrated multimedia and broadband services. Such effort is justified by the enormous spectral efficiencies offered by MIMO, which arguably provides the ultimate in system performance by fully exploiting the spatial dimension.

REFERENCES

[3GPP, 03] 3rd Generation Partnership Project, Spatial channel model for multiple input multiple output (MIMO) simulations, 3GPP TR 25.996 V6.1.0, September 2003. Available: http://www.3gpp.org/

[Alamouti, 98] S. M. Alamouti, A simple transmit diversity technique for wireless communications, *IEEE Journal on Selected Areas in Communications*, 16 (8), 1451–1458, 1998.

[Andrews, 01] R. A. Andrews, P. P. Mitra and R. de Carvalho, Tripling the capacity of wireless communications using electromagnetic polarisation, *Nature*, 409, 316–318, 2001.

[Balban, 91] P. Balaban and J. Salz, Dual diversity combining and equalisation in digital cellular mobile radio, *IEEE Transactions on Vehicular Technology*, 40, 342–354, 1991.

[Bliss, 02] D. W. Bliss, K. W. Forsythe, A. O. Hero, A. F. Yegulalp, *IEEE Transactions on Signal Processing*, 50 (9), 2128–2142, 2002.

[Bottomley, 95] G. E. Bottomley and K. Jamal, Adaptive arrays and MLSE equalisation, in *Proceedings on IEEE Vehicular Technology Society Conference*, 50–54, 1995.

[Correia, 06] L. M. Correia (ed.), Mobile Broadband Multimedia Networks: Techniques, Models and Tools for 4G, Elsevier, 2006, ISBN 0-12-369422-1.

[Derryberry, 02] Transmit Diversity in 3G CDMA Systems, *IEEE Communications Magazine*, 69–75, April 2002.

[Ertel, 98] R. B. Ertel, P. Cardieri, K. W. Sowerby, T. S. Rappaport and J. H. Reed, Overview of spatial channel models for antenna array communication systems, *IEEE Personal Communications*, 10–22, 1998.

[Foschini, 96] G. J. Foschini, "Layered Space-Time Architecture for wireless communication in a fading environment when using multiple antennas", *Bell Laboratories Technical Journal*, 1 (2), 41–591, 1996.

[Foschini, 98] G. J. Foschini and M. J. Grans, On limits of wireless communications in a fading environment when using multiple antennas, *IEEE Wireless Personal Communications*, 6 (3), 311–335, 1998.

[Gesbert, 03] D. Gesbert, M. Shafi, D. Shiu, P. J. Smith and A. Naguib, From theory to practice: An overview of MIMO space-time coded wireless systems, *IEEE Journal on Selected Areas in Communications*, 21 (3), 281–302, 2003.

[IEEE, 03] IEEE 802.16 Broadband Wireless Access Working Group, Channel models for fixed wireless applications, http://ieee802.org/16, 2003.

[IEEE, 04] IEEE P802.11 Wireless LANs, TGn Channel Models, IEEE 802.11-03/940r4, http://4grouper.ieee.org/groups/802/11/, 2004.

[Haykin, 96] S. Haykin, *Adaptive Filter Theory*, 3rd edn., Prentice Hall, Englewood Cliffs NJ, ISBN 0-13-322760-X, p.194 onwards, 1996.

[Jensen, 04] M. A. Jensen and J. W. Wallace, A review of antennas and propagation for MIMO wireless communications, *IEEE Transactions on Antennas and Propagation*, 52 (11), 2810–2824, 2004.

[Martin, 01] C. C. Martin, J. H. Winters and N. R. Sollenberger, MIMO radio channel measurements: Performance comparison of antenna configurations, in *Proceedings on IEEE 54th Vehicular Technology Conference*, Vol.2, Atlantic City, NJ, Oct 7–11, pp. 1225–1229, 2001.

[Molisch, 04] A. F. Molisch, A Generic Model for MIMO Wireless Propagation Channels in Macro- and Microcells, *IEEE Transactions on Signal Processing*, 52 (1), 61–71, 2004.

[Monzingo, 80] R. A. Monzingo and T. W. Miller, *Introduction to Adaptive Arrays*, Wiley-Interscience, New York, 1980.

[Oestges, 03] C. Oestges, V. Erceg, and A. Paulraj, "A physical scattering model for MIMO macrocellular broadband wireless channels," *IEEE Journal on Selected Areas in Communications*, 21 (5), 721–729, 2003.

[Oestges, 07] C. Oestges and B. Clerckx, MIMO Wireless communications: from real-world propagation to space-time code design, Academic Press, New York, Elsevier, ISBN 0123725356, 2007.

[Pedersen, 97] K. I. Pedersen, P. E. Mogensen and B. H. Fleury, Power azimuth spectrum in outdoor environments, *Electronics Letters*, 33 (19), 1583–1584, 1997.

[Piggin, 06] P. Piggin, Emerging Mobile WiMax Antenna Technologies, Communications Engineer, *Institution of Engineering and Technology* (IET), 4 (5), 29–33, 2006.

[Rapajic, 00] P. B. Rapajic and D. Popescu, Information capacity of a random signature multiple-input multiple-output channel, *IEEE Transactions on Communication*, 48 (3), 1245–1248, 2000.

[Spencer, 97] Q. Spencer, M. Rice, B. Jeffs, and M. Jensen, A statistical model for angle of arrival in indoor multipath propagation, in *Proceedings on IEEE Vehicular Technology Conference*, 3, 1415–1419, 1997.

[Svantesson, 04] T. Svantesson, M. A. Jensen and J. W. Wallace, Analysis of electromagnetic field polarisations in multi-antenna systems, *IEEE Transactions on Wireless Communications*, 3, 316–318, 2004.

[Tarokh, 98] V. Tarokh, N. Seshadri, and A. R. Calderbank, Space-time codes for high data rate wireless communication: Performance criterion and code construction, *IEEE Transactions on Information Theory*, 44 (2), 744–765, 1998.

[Tarokh, 99] V. Tarokh, H. Jafarkhani, and A. R. Calderbank, Space-time block codes from orthogonal designs, *IEEE Transactions on Information Theory*, 45 1456–1467, 1999.

[Telatar, 99] I. E. Telatar, Capacity of Multi-antenna Gaussian Channels, Technical Memorandum, Bell Laboratories, Lucent Technologies, October 1995. Published in *European Transactions on Telecommunications*, 10 (6), 585–595, 1999.

[Thompson, 96] J. S. Thompson, P. M. Grant, and B. Mulgrew, Smart antenna arrays for CDMA systems, *IEEE Personal Communications*, 3 (5), 16–25, 1996.

[Wallace, 02] J. W. Wallace and M. A. Jensen, Intrinsic capacity on the MIMO wireless channel, in *Proceedings on IEEE 56th Vehicular Technology Conference*, 2, 701–705, 2002.

[Ward, 96] C. Ward, M. Smith, A. Jeffries, D. Adams and J. Hudson, Characterising the radio propagation channel for smart antenna systems, *IEE Electronics and Communication Engineering Journal*, 8 (4), 191–202, 1996.

[Wolniansky, 98] P. W. Wolniansky, G. J. Foschini, G. D. Golden, R. A. Valenzuela, V-BLAST: An architecture for realizing very high data rates over the rich-scattering wireless channel, *Proceedings of ISSSE-98*, Pisa, Italy, 1998.

[Zheng, 03] L. Zheng and D. N. C. Tse, Diversity and Multiplexing: A fundamental tradeoff in multiple-antenna channels, *IEEE Transactions on Information Theory*, 49 (5), 1073–1096, 2003.

PROBLEMS

18.1 Verify that the weights which were found in (18.5) can also be found from the Wiener solution.

18.2 A new mobile radio system is being designed. One requirement is that it should support the use of SDMA. What duplex schemes and access schemes would be most appropriate?

18.3 Prove the MIMO capacity Eq. (18.28).

18.4 Suggest physical situations, typical of picocells, microcells and macrocells, in which pinhole channels could impair the capacity of MIMO systems.

18.5 Prove that Eq. (18.19) provides an estimate of the transmitted symbols.

18.6 Replot the MIMO trade-off graph Figure 18.21 for the (2.1) case and add the performance of the Alamouti code described in Section 18.5.4.

19

Channel Measurements
for Mobile Systems

*'To measure is to know. . .If you can not measure it, you can
not improve it'.*
Lord Kelvin

19.1 INTRODUCTION

In planning a mobile communication network or developing mobile equipment, it is essential
to characterise the radio channel to gain insight into the dominant propagation mechanisms
which will define the performance experienced by users. This characterisation allows the
designer to ensure that the channel behaviour is well known prior to the system deployment,
to validate the propagation models used in the design process and to ensure that the equipment
used provides robust performance against the full range of fading conditions likely to be
encountered. Furthermore, after a system has been deployed, measurements allow field
engineers to validate crucial design parameters which show how the system is performing
and how it may be optimised.

This chapter provides a description of the motivations for conducting measurements and a
practical overview of how to design a measurement campaign to ensure that the results
obtained are reliable. It does not attempt to provide details of how measurement equipment
such as wideband channel sounders is designed, for which texts such as [Parsons, 00] should
be consulted.

19.2 APPLICATIONS FOR CHANNEL MEASUREMENTS

19.2.1 Tuning Empirical Path Loss Models

This is usually the starting point for the rollout of any new network, whether in a new
frequency band or in a new environment. Detailed measurements are made of path loss for
sites in a range of environments, and models of the propagation path loss are fitted to them
for later application within planning tools. For example, Figure 19.1 shows typical
measured values of path loss as individual data points, after removing fast fading. The
curve in the same figure represents an empirical power-law model of the form described in
Chapter 8, fitted to minimise the mean and the standard deviation of the error between the
model and the measurements. The mean error can always be reduced to zero by choosing
the model offset appropriately, so the usual way of assessing the goodness of such a model
is via the standard deviation of the differences between the measurements and predictions.
In the case illustrated, the error standard deviation is 6.7 dB, which is typical of a

Antennas and Propagation for Wireless Communication Systems Second Edition   Simon R. Saunders and
Alejandro Aragón-Zavala
© 2007 John Wiley & Sons, Ltd

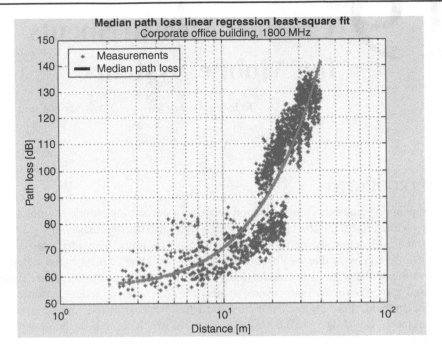

Figure 19.1: Typical path loss model fitted to measurements

well-fitted model. Accurate measurements are therefore crucial in achieving an accurate path loss model. See Section 19.3 for analysis of the impact of measurement errors.

19.2.2 Creating Synthetic Channel Models

When developing and characterising base or mobile station equipment, it is often necessary to subject the equipment to the full range of signal variations which will be encountered in practical operation. However, it is often desirable to do this in a repeatable fashion, so that different equipment can directly be compared without introducing measurement uncertainties. In such situations a *synthetic channel model* may be used, implemented in hardware and/or software, which produces entirely realistic channel variations in a laboratory environment. An example of such a model was the baseband fast-fading generator illustrated in Figure 10.36. For a more complete description of various channel simulators, both wideband and narrowband, see [Parsons, 00]. Although the channel is synthetic in such systems, the starting point is always extensive field trials to accurately determine the variation of the statistical parameters of the channel with the environment.

Measurements for this purpose will typically involve the detailed capture of both wideband and narrowband fading statistics, carefully indexed to the locations and environments in which they are recorded. This facilitates detailed subsequent analysis to determine the dependence of channel parameters on the environments in which they are experienced. These parameters are stored and loaded into the channel simulator to replicate given environments. The GSM channel parameters in Figure 11.9 are an example covering several environment types.

19.2.3 Existing Coverage

Ongoing measurements are conducted by network operators to determine the performance of their network. These may be compared to previous measurements, revealing the impact of network changes intended to optimise performance or the impact of changing user traffic patterns. They may also be used to benchmark network performance against alternative operators or network technologies. Such measurements may simply be of signal strengths, but more usually they involve measuring a large number of parameters simultaneously in order to determine the experience of a real customer and to relate this to the network parameters, both RF and logical. For example, it is common to place repetitive voice calls using mobile phones in an engineering mode and to determine the proportions of such calls which are dropped or subject to blocking. It is increasingly common to also measure the data performance which may be experienced by users in the form of measurements of throughput, latency and other *quality of service* indicators. Simultaneously, the measurement system must log the network state and signalling traffic to determine the network messages, such as handover commands and mobile measurement reports, which will help to diagnose the causes of any problems and suggest actions to be taken to optimise performance in the future.

19.2.4 Design Survey

When designing a new portion of a mobile network for any cell type (macro, micro or picocell), it is common to conduct site-specific surveys to ensure that any design assumptions made are valid. For example, while propagation models may be used to determine an appropriate general location for a new macrocell, a survey may be conducted to ensure that a specific proposed base station location provides coverage in particular locations, which is difficult to do with complete confidence with any propagation model. While a good macrocell prediction model may achieve an 8 dB standard deviation of error, well-conducted and calibrated measurements should be repeatable with a standard deviation as low as around 3 dB.

Similarly, in an indoor environment, it is often important to conduct surveys from some or all of the proposed antenna locations in order to validate initial assumptions made concerning the properties of the building materials or the detailed building construction geometry. It is also important to check the coverage of the existing systems, particularly for licence-exempt bands such as those used by wireless LAN systems, for which no central record of potential interferers exists.

19.3 IMPACT OF MEASUREMENT INACCURACIES

The impact of measurement inaccuracies depends on the precise application for the measurements, although good insight can be gained by analysing the impact of errors on the fitting or *tuning* of a path loss model. The measurements have to be made with a carefully calibrated system to allow the measured signal strength values to be related accurately to path loss values. The impact of errors depends critically on the size of the error compared with the path loss exponent. If we model path loss with a simple slope-intercept model of the form

$$L = A + B \log r \tag{19.1}$$

and use subscripts t to indicate true parameters and m to indicate the result of measurement, then it is simple to show that the ratio of the numbers of sites required for an erroneous model versus the true result is

$$\frac{N_t}{N_m} = 10^{2\left(\frac{L_t - A_t}{B_t} - \frac{L_m - A_m}{B_m}\right)}$$

(19.2)

Thus, a simple offset error of 1 dB will increase the number of sites required by $10^{2/40}$ or 12% given a path loss exponent of 4, assuming the system is limited by coverage, rather than by interference. The cost implication of such errors is very considerable.

Additionally, the measurement routes and the site locations have to be chosen carefully to ensure that the model is applicable to all environment types in which the network is to be operated. Around 50 sites are typically required in order to produce reasonably acceptable models for a range of clutter types sufficient to represent a national network roll-out. This produces sufficient data to tune the model with reasonable confidence using, say, two-thirds of the sites, while testing the resulting model using the remainder of the sites as an independent check. Using the same data to test the model as to tune it is not valid.

The model tuning process itself should allow the model parameters to be varied in order to minimise the errors. It is typical to use statistical optimisation techniques such as regression analysis to minimise the mean-square errors. This fundamentally minimises the error standard deviation, but it may be useful to simultaneously consider more explicit measures of detailed accuracy such as the correlation between the predictions and measurements at particular locations; these *hit-rate metrics* [Owadally, 01] have been found to be a very sensitive and revealing indicator of modelling goodness. They avoid the situation that a very simple model, producing a smooth variation of path loss with distance, produces a better standard deviation than a more detailed model which accounts for a wide range of propagation effects. The simple model will perform very poorly when used for interference analysis.

Once the model has been tuned, its accuracy must be taken into account when designing the network. The variability of measurements around the model has to be accounted for via a *fade margin*, which reflects the fact that the model only predicts the loss at 50% of locations at a given distance and clutter class. This fade margin is composed partly of the *shadowing margin*, calculated as discussed in Chapter 9, combined with other effects introduced by the measurement and modelling process. These effects include the fact that the measurement route encompasses only a subset of prediction locations, the inaccuracies in the positioning of the measurement system and also any other non-systematic rounding and averaging errors. Typically, these effects can be considered statistically independent from the shadowing issues, so they can be combined with the true location variability to yield a total variability. As an example, a total computation may determine the total variability σ_T as a function of the measurement uncertainties σ_M and the variabilities in the mobile equipment receiver sensitivity σ_E as follows:

$$\sigma_T = \sqrt{\sigma_L^2 + \sigma_M^2 + \sigma_E^2}$$

(19.3)

This value of σ_T is then used to calculate the fade margin via Eq. (9.7). The coverage probability is no longer purely a location variability, but is an overall *confidence* of achieving acceptable coverage.

19.4 SIGNAL SAMPLING ISSUES

Although the requirements of the measurement system will depend in detail on the specific application for the measurements, there are some generic issues related to the way in which mobile signals must be sampled which are common to all applications.

In modern receiver systems, signals are recorded by digital sampling, producing a series of discrete samples rather than a continuous signal record. The available sample rate is typically limited by the speed of the associated analogue-to-digital converters, the available storage space and (in the case of scanning multiple channels) the retuning rate of the receiver. It is then important to determine a sample rate which represents the signal sufficiently accurately for the application in hand.

For path loss modelling, the interest is mainly in determining the local mean of the signal, removing the fast fading component while providing a high-confidence estimate of the underlying power associated with the overall path loss and shadowing processes. This implies that all of the samples gathered must be taken within a time period over which the mobile receiver is well within the shadowing correlation distance defined in Section 9.6.1, otherwise, the local mean will not represent the shadowing variations adequately, which loses the detail of the system coverage which the test is aimed at revealing. There must also be enough samples so that the receiver noise floor does not excessively affect the estimate. On the contrary, it is important that the samples taken are not so closely gathered that they have a high probability of being in a fading null, or peak, which will produce a significant over- or under-estimation of local mean. In this application, only the signal amplitude is of direct relevance, although sampling of both in-phase and quadrature components simultaneously will effectively produce twice as many independent samples of noise and of fast fading within the same distance.

If, however, the measurement work is intended to establish the statistics of the fast fading, a considerably higher sample rate is needed. For the first-order statistics alone, it would be sufficient to gather samples which represented all of the signal levels encountered at their relative frequency, while determination of second-order statistics requires a near-continuous signal record in order to accurately deduce the rates of change of the signal with distance.

The following sections examine several aspects of this problem: Section 19.4.1 examines methods of combining multiple samples to estimate the local mean; Section 19.4.2 examines the number of samples required for confidence in various applications of measurement data. Later, Section 19.7.4 discusses the necessary size and shape of the area used for averaging.

19.4.1 Estimators of the Local Mean

So far we have loosely referred to 'averaging' of the signal strength samples gathered over a short range in order to produce an estimate of the local mean. This section describes several alternative ways in which this averaging process can be performed. In all cases, the aim is to determine the mean value of the signal voltage, $\bar{r} = E[r]$ from the samples r. Often the decibel version of these values will be used, denoted $R = 20 \log r$.

Median

The median is simple to compute by sorting the values in order and determining the value of the middle sample in the list or taking the average of the middle two samples if an even number of samples exists. As well as being simple to compute, it has the great benefit that it is independent of any monotonic transformation of the signal strengths. For example, the same value is obtained whether the median is found from decibel values or from linear signal strengths, avoiding the need to perform complicated arithmetic operations on all of the samples. This also avoids the need for any non-linearities in the receiver gain characteristic to be calibrated out, provided that the final median value can be related repeatably to the actual input signal strength.

However, the difference between the median and the true mean is dependent on the fading distribution, which may not in itself be known. In the absence of fading, the mean and the median are the same. In the case of the Rayleigh distribution, the ratio between the mean and the median is (from Section 10.8)

$$\frac{\text{Mean}(r)}{\text{Median}(r)} = \frac{\sigma\sqrt{\frac{\pi}{2}}}{\sigma\sqrt{\ln(4)}} \approx 1.06 \qquad (19.4)$$

That implies a 0.54 dB error, which is unlikely to be significant. More importantly, however, this estimator for the mean has a high variance. Most of the samples at the tails of the distribution have no impact on the outcome, while a small amount of noise in the values close to the median will have a large impact on the estimated value. The median is thus rarely preferred as an estimator in modern measurement work.

Decibel Mean

It is common for measurement receivers to report signal strengths in decibels. This is often because the receiver uses an amplifier with a logarithmic response in order to increase the available dynamic range. In such cases, it is then natural to estimate the local mean by taking the average of the decibel samples. This situation was analysed by [Parsons, 00], correcting earlier analysis by [Lee, 85].

If the samples are taken from a receiver with a logarithmic (decibel) characteristic, then the sample average $\bar{R}$ of N independent logarithmic samples is defined by

$$\bar{R} = \frac{1}{N}\sum_{i=1}^{N} R_i \qquad (19.5)$$

Assuming Rayleigh fading of the underlying signal voltage, then the standard deviation of $\bar{R}$ is [Wong, 99]

$$\sigma_{\bar{R}} = \frac{\pi}{\ln 10}\sqrt{\frac{50}{3N}} \approx \frac{5.57}{\sqrt{N}}\,\text{dB} \qquad (19.6)$$

It is desirable to understand the number of samples required to ensure with high probability that the sample average so derived lies within a given small error from the true value. If it is assumed that the sample average $\bar{R}$ computed from N samples is itself a normal random variable with mean given by the true value R_{true}, then the probability P_{good} that $\bar{R}$ lies less than K decibels from the true value can be calculated using the Q function defined in Appendix B as follows:

$$P_{good} = \Pr(R_{true} - K \leq \bar{R} \leq R_{true} + K) = 1 - 2 \times Q\left(\frac{K}{\sigma_{\bar{R}}}\right) \qquad (19.7)$$

Thus the number of samples required to achieve a given confidence in the sample mean is given by

$$N = \left\lceil \left[\frac{5.57}{K} Q^{-1} \left(\frac{1 - P_{good}}{2} \right) \right]^2 \right\rceil \tag{19.8}$$

where $\lceil t \rceil$ is the next highest integer to t. For $K = 1$ dB and $P_{good} = 90\%$, this produces $N = 84$ samples and $N = 120$ samples for $P_{good} = 95\%$.

Linear Voltage Mean

If a linear power is available, or if it is feasible to convert the decibel values to powers, then we can directly calculate the sample-mean of the voltage values as follows:

$$\bar{r} = \frac{1}{N} \sum_{i=1}^{N} r_i \tag{19.9}$$

The accuracy of this estimator can be estimated in a similar fashion to the previous section, this time replacing the value of 5.57 in Eq. (19.8) with the decibel value of the variance of r using Eq. (10.22),

$$\sigma_r|_{dB} = -10 \log \left(\frac{4 - \pi}{2} \right) \approx 3.67 dB \tag{19.10}$$

For $K = 1$ dB and $P_{good} = 90\%$, this produces $N = 37$ samples and $N = 52$ samples for $P_{good} = 95\%$.

Linear Power Mean

If a receiver which responds to linear power is available, or it is feasible to convert the decibel values to powers, then we can directly calculate the sample-mean of the power values as follows:

$$\bar{r} = \sqrt{\frac{1}{N} \sum_{i=1}^{N} r_i^2} \tag{19.11}$$

The variance of the power quantity is 3.0 dB, so for $K = 1$ dB and $P_{good} = 90\%$, Eq. (19.8) now produces $N = 25$ samples and $N = 35$ samples for $P_{good} = 95\%$.

Optimal Estimation

If the linear power values are not available and the uncertainty associated with the decibel mean is too high, then the optimum estimator proposed by [Wong, 99] is a good alternative. It is defined by

$$E_{op} = 10 \left[\log T - \frac{H_N - 1}{\ln 10} \right] \tag{19.12}$$

where $T = \sum_{i=1}^{N} 10^{X_i/10}$ and $H_N = 1 + \frac{1}{2} + \frac{1}{3} + \ldots + \frac{1}{N}$. The standard deviation of this estimator is calculated as

$$\sigma E_{op} = \frac{10}{\ln 10} \sqrt{\frac{\pi^2}{6} - \sum_{i=1}^{N-1} \frac{1}{i^2}} \qquad (19.13)$$

Comparison of Methods

The estimators described in the preceding sections are compared as shown in Figure 19.2, which shows how the error on either side of the mean for each estimator varies with the number of samples used for 90% confidence. Table 19.1 shows the number of samples required to provide errors within ± 1 dB for 90 and 95% confidence. It should be recalled that all of these values assume Rayleigh fading statistics and it is assumed that all samples experience statistically independent fading.

19.4.2 Sampling Rate

Independent Sampling

It is now necessary to consider the rate at which samples must be collected in order to produce the accuracies calculated in the preceding section. The analysis so far assumed that

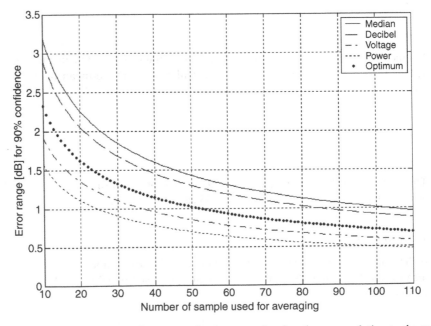

Figure 19.2: Comparative performance of estimators, showing the error relative to the mean for 90% confidence

Table 19.1: Number of samples required to provide ± 1 dB error for various estimators of the local mean

Estimator	Number of samples for 90% confidence	Number of samples for 95% confidence
Median	102	145
Decibel mean	84	120
Linear voltage mean	37	52
Linear power mean	25	28
Optimal estimator	52	73

all samples were statistically independent. In order for this to be the case, the autocorrelation function of adjacent samples must be sufficiently small. This is exactly the criterion which was examined in Chapter 15 for two diversity branches to yield a high gain. If we assume a classical autocorrelation function, with uniform angle of arrival of multipaths, then the autocorrelation function is (from Eq. (10.48))

$$\rho(d) = J_0\left(\frac{2\pi d}{\lambda}\right) \tag{19.14}$$

If we take the maximum permissible value of the autocorrelation function as 0.7, we find that $d \approx 0.18\lambda$. A stronger criterion is to ensure that samples are space apart far enough to coincide with the first zero of the classical autocorrelation function at $d = 0.38\lambda$. This also allows a margin of safety for channels where the range of angular arrivals is limited or where there are dominant multipaths, resulting in non-Rayleigh statistics. As a result, the number of samples used in the local mean estimator needs to be spread over a sufficient distance to provide the necessary independence. Table 19.2 shows the minimum electrical distance for each of the estimator types. In some cases, the electrical distance may be a large fraction of the shadowing serial correlation defined in Chapter 9, which is typically some tens of meters for outdoor environments. In such cases, which are mostly likely to occur at lower frequencies, selection of a good estimator is important to avoid averaging the detail of the shadowing information. For indoor environments, where the shadowing correlation

Table 19.2: Minimum electrical distance over which samples should be collected and averaged for various estimators of the local mean

Estimator	Averaging length for 90% confidence	Averaging length for 95% confidence
Median	39 λ	55 λ
Decibel mean	32 λ	46 λ
Linear voltage mean	14 λ	20 λ
Linear power mean	10 λ	11 λ
Optimal estimator	20 λ	28 λ

distance is of the order of a few metres, it may be impossible to meet this criterion and a compromise between averaging fast fading and not disturbing shadowing must be reached [Fiacco, 00]. One approach is to average all the samples in a small area rather than along a line [Valenzuela, 97].

The spatial sampling rate also sets the minimum required sampling frequency f_s for a given collection mobile speed v_{max},

$$f_s \geq \frac{v_{max}}{0.38\lambda} \tag{19.15}$$

Note that it is never harmful to collect samples at a higher rate than this minimum; although some of the samples will not be statistically independent, all of the information is still present provided all the samples collected within the distance specified in Table 19.2 are included in the estimation process. The extra samples will serve to provide additional averaging of noise in the measurement receiver.

Nyquist Sampling

Another approach to determine appropriate sample rates for channel sampling is to consider the *first Nyquist criterion*. This was defined in detail in Chapter 17, but can be more simply stated for the purposes of this chapter as follows: for a band-limited signal waveform to be reproduced accurately, each cycle of the input signal must be sampled at least twice. In the context of a narrowband fading signal, this implies that at least two samples per wavelength are required. Under such conditions, an interpolation procedure can be applied to calculate the values of the original, continuous signal with any degree of accuracy, as follows.

If the original signal is $r(t)$, then denote the signal sample occurring at $t = iT$ as r_i, sampled at a frequency $f_s = 1/T$. Then the original signal can be expressed in terms of the samples as

$$r(t) = \sum_{i=-\infty}^{\infty} r_i \times T \times h(t - iT) \tag{19.16}$$

where

$$h(t) = \frac{\sin(\pi f_s t)}{\pi t} \tag{19.17}$$

This approach allows the original signal to be reconstructed essentially perfectly, with as high a sampling rate as desired. This provides a completely alternative approach to the estimators described in Section 19.4.1. Provided the signal samples can be stored and processed subsequent to the measurements, the signal statistics can be examined in detail as if the signal were continuous. Additionally, if complex samples are gathered, the reconstructed signal represents the original in terms of its high-order statistics and is thus suitable for determining channel properties such as Doppler spectra and autocorrelation functions in addition to the basic local mean. Note that the sampling frequency needs to be at least enough to gather two samples per wavelength at the fastest collection speed for a narrowband signal. If the signal is modulated, then the sample rate needs additionally to be at least twice the maximum signal bandwidth.

19.5 MEASUREMENT SYSTEMS

19.5.1 Narrowband Channel Sounding

To characterise the narrowband behaviour of the mobile radio propagation channel, including all of the behaviour described in Chapter 10, continuous wave (CW) measurements are often performed by transmitting an unmodulated single tone carrier. Note that path loss is essentially a narrowband effect so that CW measurements are adequate for most path loss models, although the UWB system path loss described in Section 13.8 is a possible exception.

A fixed base station transmitter is often employed for such measurements, and the mobile receiver is located either in a car, in a train, carried by a pedestrian user or even on a mobile robot platform. Occasionally, the receiver remains fixed and the base station is moved to perform the narrowband measurements, which is particularly useful if it is desired to compare signal paths between multiple base stations and a mobile location [Fiacco, 00].

A generic narrowband channel sounding system is presented in Figure 19.3. Notice that depending on the environment in which the measurements are to be performed (indoor or outdoor), the navigation and positioning system changes. These issues are discussed in detail in Sections 19.7 and 19.8.

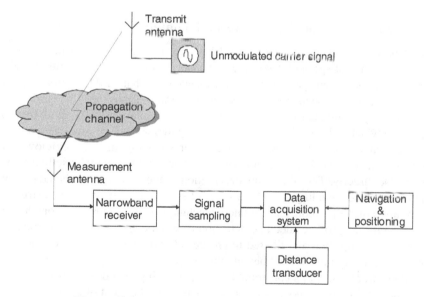

Figure 19.3: Generic narrowband channel sounding system

The data acquisition system is normally composed of a laptop computer which has specialised data acquisition software installed, and includes all the interfacing drivers to communicate with the receiver and navigation and positioning system. The distance transducer can be used to trigger the sampling mechanism at equally spaced intervals, although it is often easier to sample at constant time intervals and use the distance transducer to select the correct length for averaging in subsequent processing.

Signal envelope or phase measurements can be performed with the above mentioned system, and the design of the receiver determines whether this amplitude or phase is measured. If phase is of interest, then this phase can be measured relative to a fixed reference by demodulating the signal in two channels, i.e. in-phase and quadrature. In some applications, the absolute phase might be required and then it is required that the local oscillators of transmitter and receiver are phase-locked [Parsons, 00], and hence only phase variations due to propagation characteristics are recorded.

When selecting measurement receivers, *dynamic range* is an essential parameter to take into account, since this will determine both the maximum signal strength that can be detected as well as the receiver noise floor. For example, for measurement campaigns in which large distances are to be covered, a large dynamic range is desired, especially if macrocell coverage range is determined. For in-building measurements, where distances are not as large as for outdoor environments, the maximum signal strength that can be recorded is often of interest, as walk tests are performed very close to the transmitting antennas. In all cases, the full extent of signal fast fading has to be considered within the required dynamic range.

For a detailed analysis of narrowband channel sounding systems as well as receiver design, the interested reader will need to consult the recommended literature [Gonzalez, 96], [Hickman, 97], [Parsons, 00].

19.5.2 Wideband Channel Measurement Techniques

Wideband channel parameters such as delay spread, delay profile, average delay and coherence bandwidth are of special interest when channel characterisation is desired, and are especially important for system performance, as explained in Chapter 11. Various techniques may be employed, some of which use the principle of transmitting many narrowband signals, either sequentially or simultaneously. However, due to the limitations of the above mentioned methods, genuine wideband sounding techniques are required, as will be explained in this section.

Several approaches to wideband channel sounding may be employed. For example, *periodic pulse sounding* transmits pulses of short duration, which excite the channel impulse response directly. This allows the observation of the time-varying behaviour of the channel. Periodic pulse sounding provides a series of snapshots of the multipath structure. The major limitation of the periodic pulse sounding technique is its requirement for a high peak-to-mean power ratio to provide adequate detection of weak echoes. The pulses have to be carefully generated, amplified and filtered to ensure full excitation across the bandwidth of interest without causing interference beyond this range.

As power is the major constraint for pulse sounding methods, *pulse compression* is a viable alternative. In this method, white noise is applied to the input of a linear system. If the output $y(t)$ is cross-correlated with a delayed replica of the input, then the resulting cross-correlation coefficient is proportional to the impulse response of the system, denoted as $h(\tau)$, evaluated at the delay time, as follows:

$$E\{n(t)n^*(t-\tau)\} = R_n(\tau) = N_0\delta(\tau) \tag{19.18}$$

$R_n(\tau)$ represents the autocorrelation function of the noise and N_0 is the noise power spectral density. As stated above, the signal output is given by the convolution relationship,

$$y(t) = \int h(\psi)n(t-\psi)d\psi \tag{19.19}$$

Hence, the cross-correlation of the output and the delayed input is given by

$$E\{y(t)n^*(t-\tau)\} = N_0 h(\tau) \qquad (19.20)$$

which is just a scaled version of the desired channel impulse response. This proves that the impulse response of a linear system can be determined by using white noise and some method of correlation processing. This white noise is often generated in the laboratory by using deterministic waveforms with a noise-like character, having excellent auto-correlation properties.

More usually in modern systems, a correlating sounder is employed, which transmits a pseudo-random bit pattern using an appropriate modulation scheme. The sequence is chosen to have good auto-correlation properties, so that the convolution of the signal with itself is as close to a single impulse as possible. The received signal is correlated with a replica of the transmitted signal to produce an estimate of the channel. The duration of the signal must be short compared with the coherence time of the channel, whilst being long enough to provide adequate processing gain and hence signal-to-noise ratio in the correlation process. This overall approach is essentially identical to the channel estimation process which is performed using a training sequence in the equalisers described in Chapter 17.

19.5.3 Other Measurements

Along with narrowband and wideband measurements, other specialist measurements may be required. In particular, it is increasingly common to conduct wideband measurements between multiple antennas at both the transmit and the receive ends. Processing of the resulting signals enables channel parameters in both space and time to be determined, including scatterer locations, angle-of-arrival and angle-of-departure measurements, eigenvalue spreads and the ways in which all of these parameters vary with channel delay. This allows characterisation of the performance of the MIMO systems described in Chapter 18 as well as more conventional systems. The resulting measurement systems can be very complicated and expensive, and great care and careful planning is required to ensure that the resulting data is meaningful. See, for example, [Steinbauer, 00].

19.6 EQUIPMENT CALIBRATION AND VALIDATION

19.6.1 General

As illustrated in Section 19.3, the calibration of the measurement system is essential if the resulting data is to be of any value. The calibration process aims to determine accurate calibration offsets and performance characteristics for a complete measurement system. Active equipments, such as transmitters and receivers, require individual calibration at minimum intervals defined by their manufacturers. These calibrations must usually be completed by an accredited test house. Channel measurement work, however, subjects equipment to extreme environmental conditions such as heat, cold, vibration, dust and humidity. As a result, it is often necessary to calibrate at far more frequent intervals, and to more stringent requirements, than the manufacturer's minimum standards.

It is often helpful not to validate each element of an RF measurement system individually but instead to validate the system as a whole. For example, in a test vehicle, channel sounders, cables, navigation and positioning systems, splitters and antennas are all interconnected. Disconnecting them repeatedly for calibration purposes will induce uncertainties, add errors and degrade the system performance over time. Measuring a whole system at once ensures that measurement errors associated with measuring power or loss are only encountered once, reducing the overall uncertainty compared with individual measurements of all the system components, which can accumulate to produce a large overall error.

It is important to make an explicit assessment of the accuracy of the overall measurement system. When adding individual error contributions, the approach will depend on the nature of the errors. Worst-case errors can be determined by simply adding individual error ranges, but, if errors are independent, it is usually more appropriate to determine a *standard error* by adding the root-sum-square of the component errors, just as in Eq. (19.3). A measurement system cannot reliably detect differences between quantities less than the standard error.

19.6.2 Transmitters

For transmitters, equipment validation needs to be performed, especially to make sure that

- the transmitter's output power is within the specified range in the presence of the whole range of potential impedance load conditions, temperatures and supply voltages;
- when various output frequencies are selected, the transmitter maintains a peak power within acceptable at the expected frequency with minimum total harmonic distortion, according to its specifications;
- power supply stabilisation and conditioning is important if the transmitter is to be operated from a generator, which may produce transients, noise and voltage variations which may adversely affect the performance;
- the transmitter is stable in both frequency and output power if operated over large periods of time – comparable with those expected in the measurement campaign. It is best to perform an explicit power stability measurement using a spectrum analyser in time domain mode with a long sweep time (>3 h). The spectrum analyser connected to the transmitter via an appropriately rated RF attenuator, and the power output is recorded as the transmitter warms up and stabilises. Multiple traces may be used configured with minimum, RMS and maximum detectors in operation to help in detecting short-lived spikes and dropouts which could introduce large uncertainties into propagation measurements.

19.6.3 Receivers

For receivers, periodic validation needs to be performed to detect whether the receiver is operating within design specifications. Ideally, the calibration measurement should be conducted with the receiver located *in situ* within the test vehicle, using the whole measurement system, apart from the antenna but including the data logging system to minimise calibration errors.

A signal of known power from a well-calibrated signal generator can be injected at the frequency of interest, and samples of the recorded signal strength by the scanner will need to be within these injected values. The frequency range is varied, as well as the transmitter power, to cover the entire dynamic range of the radio scanner. The transmitter employed should be a carefully calibrated laboratory signal generator, and the received signal strength can be compared with a calibrated spectrum analyser as the known reference.

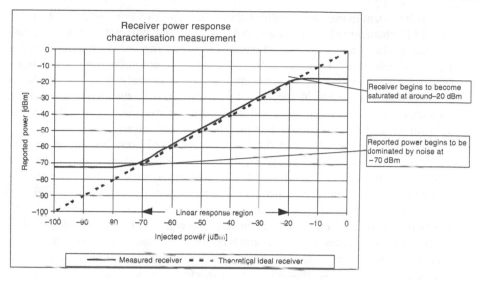

Figure 19.4: Example power response characterisation measurement for a test receiver. This receiver provides a linear dynamic range of around 50 dB and the reported signal is 0.7 dB higher than the injected signal in the linear region

Figure 19.4 shows a typical receiver calibration response. This establishes that the receiver is linear over the region −20 to −70 dBm, giving a 50 dB maximum dynamic range. Often receivers will include a switched attenuator to avoid overloading for strong signals. The switching process often takes some time and introduces transients, so it is best to plan the measurement campaign so that the attenuator switches only rarely or is disabled for the test.

If channels are to be scanned over a wide frequency range, then it is important to ensure that the receiver is calculated at least at the extremes and centre of the measurement range to determine whether any significant differences exist.

The fading statistics of the received signal must be taken into account such that the linear dynamic range of the receiver can cover the full range of fading statistics; for example, at least 20 dB should be allowed for Rayleigh fading effects.

19.6.4 Passive Elements

Passive elements, such as cables, attenuators, antennas and splitters form an integral part of a measurement system. These elements, although reasonably stable, do suffer degradation with time and therefore periodic validation of their main RF properties is required. Passive elements are frequently the cause of measurement failures in field systems. RF connectors must be tightened to the correct torque, as recommended by the manufacturer. If they are too loose they will produce intermittent errors which are hard to diagnose and introduce additional loss into the system. If they are too tight they will deform, causing a change in the presented electrical impedance and inducing unwanted reflections within the transmission line. Mechanical vibration in the mobile then effectively varies the transmission line impedance dynamically, causing severe and unpredictable variation in the system calibration. Most RF connectors are only rated to several hundred mating cycles, so cables and connectors used for field measurement work should be replaced at regular intervals.

It is important to use cables with an appropriate specification: for example, cables with 50 and 75 Ω characteristic impedance look identical but will not operate correctly. Cables degrade swiftly when used in field measurement systems, producing additional losses and unwanted reflections. This may occur, for example, due to corrosion or due to cable being bent to too small a radius. A measurement with a network analyser can check both the loss and the impedance of a cable. A specialised time domain *reflectometer* is also a useful tool for characterising and locating any cable faults. This operates by sending short pulses along a cable and measuring the time and amplitude of the reflections.

Antennas are elements which should not be neglected from this validation process as for very accurate propagation measurements, a few decibels in antenna gain or decimal places in VSWR can be critical. Therefore, ideally, a radiation pattern characterisation at least in azimuth and elevation at the operating frequency would be desired, as well as a VSWR measurement over its operating frequency range to evaluate how much power will not be radiated from the antenna. This is essential for antennas that have been designed specifically for the measurement campaign plan and is optional if the antennas are purchased from known and recognised antenna manufacturers. Ideally, the antenna radiation pattern would be characterised *in situ*, but typically it is sufficient to use a standard test route with known characteristics to establish an appropriate value for the mean effective gain (MEG). The key point is to regularly monitor the antenna characteristics over time and ensure that the same setup is used through a given measurement campaign.

19.7 OUTDOOR MEASUREMENTS

19.7.1 General

This section considers the practical considerations needed to conduct a reliable outdoor measurement, typically conducted via a drive test from a vehicle containing appropriate equipment. Figure 19.5 shows a typical example of the results from a test of a single base station site.

19.7.2 Measurement Campaign Plan

A *measurement campaign plan* is a document which establishes the requirements and planning needed for the propagation testing to be performed. This plan encompasses essential issues such as motivation, general measurement requirements, base station locations, testing routes, testing guidelines, workplan, and required equipment checklist. Some of these issues will be discussed in more detail in this chapter.

The importance of having a measurement campaign planned in advance relies on more than being a simple checklist document. The aims and goals of what is to be achieved after the testing are often overlooked and should not be minimised. Time is sometimes critical, especially when the measurements are performed in public areas where permission is given only for a limited time frame. Therefore, careful planning is essential if successful results are to be obtained.

19.7.3 Navigation

It is important to accurately record position when conducting outdoor measurements. Figure 19.6 shows a typical configuration for a measurement vehicle. A Global Positioning

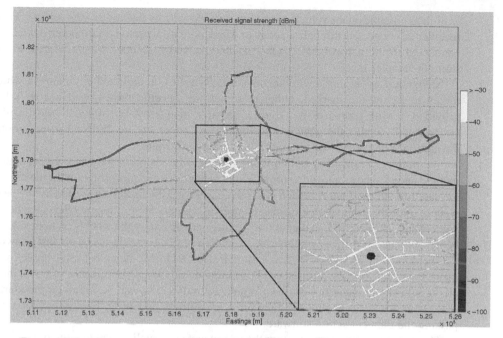

Figure 19.5: Measurement route example for outdoor measurement campaign. The shading indicates signal strength measured from a mobile receiver normalised to 0 dBi antenna gain. The transmitter location is marked with a ●

System (GPS) antenna and receiver are used to determine the fixed position of the vehicle in two dimensions. The vehicle's height above ground level can also be extracted from the GPS receiver if a sufficient number of satellites are visible, although it is usually more accurate to use a terrain database to look up heights from position. The GPS antenna should be mounted

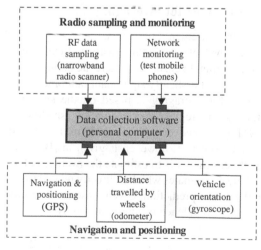

Figure 19.6: Navigation and positioning system used for outdoor channel measurements

as high as possible on the vehicle to minimise shadowing by the body of the vehicle. This GPS receiver is connected through a serial or USB port to a personal computer, which is also connected to the narrowband channel sounder or radio scanner which collects the signal strength samples.

Navigation in outdoor environments normally relies on digitised city maps in which the routes to follow are marked, although sometimes specific routes need not be followed, provided a wide spread of routes over the area-of-interest is sufficient. Most collection systems will display the vehicle's position in real-time on the digitised map, simplifying navigation. Figure 19.5 shows a typical outdoor drive route.

On routes surrounded by tall buildings, the satellite visibility may be poor, so alternative methods need to be employed to provide adequate positioning information for the duration of the satellite outage. One approach is to use *odometry*. The distance travelled by the vehicle's wheels since the last known location is sensed using optical or other encoders attached to the vehicle's axles. The comparison of the distances advanced by the wheels on both sides of the vehicle can be used directly to give an estimation of position relative to the last known point via the odometry equations [Klarer, 88], [Crowley, 91]. Odometry is fairly simple to implement, but used alone the position errors accumulate rapidly as the distance of travel increases due to wheel slip and other causes. Commercial odometers can be found on the market for reasonably low prices, although they often are sold as part of an integral solution encompassing the data collection software.

The distance measurements from the odometers are usually complemented by a direction measurement using a gyroscope, which is subject to entirely different errors, and sometimes also by a magnetic compass, carefully calibrated to minimise the impact of the metalwork of the vehicle. The combination of the various measurements results in *dead reckoning*, which is a method for estimating a new position from a previous known position using available information regarding direction, speed, time and distance of travel. The details of the dead reckoning procedure depend on the relative errors of the sources, so that, for example, a decision has to be made as to how best to combine the direction estimates when the odometer and gyroscope estimates differ significantly. One approach is described in [Borenstein, 96].

19.7.4 Size and Shape of Area for Averaging

When planning a measurement, the extent and shape of the area driven needs to be considered. The removal of fast fading vastly reduces the available number of path loss samples, and it is important to ensure that there are enough samples, appropriately distributed in location, to provide a statistically significant outcome. As was stated in Section 19.3, one reasonable area size for averaging is based on the shadowing correlation distance since, by definition, this is the region over which the shadowing does not vary significantly. [Gudmundson, 91] gave between 44 and 122 m for this distance in particular tests at 900 MHz. Another perspective, however, is given in [Bernardin, 00]. Here it is argued that, if the goal of a path loss model is to estimate the average cell radius, then there is a minimum resolution of local mean estimates which is directly related to that radius, in order to give a level of confidence in the radius prediction. Given reasonable assumptions, the error in the average cell radius r_{av} for 95% confidence is approximated by

$$\delta r_{av} = \frac{\Delta r_{av}}{r_{av}} \approx \frac{3.821\sigma_L + 4.619}{N} \qquad (19.21)$$

where N is the number of independent samples of the local mean. For location variabilities σ_L between 6 and 12 dB, keeping the independent samples $N \geq 3000$ produces less than 3% error in the cell radius. Assuming that samples in the first 50% of the cell radius contribute little to the cell radius and that tests are conducted out to a distance of 12% beyond the cell radius (corresponding to an extra 25% of the cell area), this implies that each sample should be the result of averaging an area no smaller than $r_{av}/40$. Areas larger than this produce unacceptable quantisation noise into the estimation process and cannot be overcome by a larger volume of test data, while smaller areas limit the available number of samples. Note that the authors assert that this value is independent of the terrain or clutter variations and of the resolution of either the path loss predictions or the underlying terrain or clutter databases.

Regarding the shape of the area averaged, this is often taken to be a line along the test route, which is simple to achieve since samples are received in that order and can be grouped by distance travelled and then assigned to the appropriate bin. However, this approach leads to additional sampling inaccuracies, since two elements of the test route may pass through the same averaging bin, producing two values where a single one should be obtained and artificially increasing the apparent location variability and estimation error. It is recommended instead that all samples measured within a square area of the chosen bin width are averaged together into a single value.

These findings are summarised together in Figure 19.7.

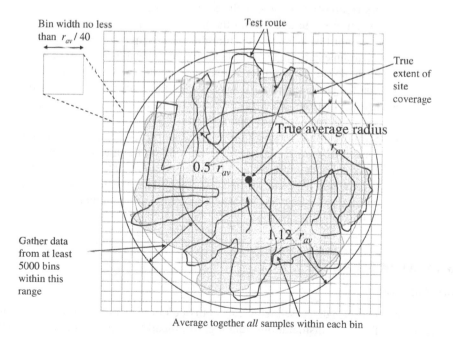

Figure 19.7: Guidelines for drive testing to accurately estimate average cell radius or for tuning a path loss model

19.7.5 Outdoor Testing Guidelines

For outdoor measurements, the following recommendations are made when designing a measurement campaign plan.

- Select a base station antenna location which is representative of the environment to be characterised. Compromises often have to be made due to the availability of access, but it is particularly important to avoid any local clutter which can shadow the antenna and affect the coverage for a very large portion of the coverage area.
- The antenna height should be carefully measured from the local ground level to the radiation centre of the antenna, either by triangulation (a carefully measured distance from the base of the antenna combined with a measurement of the angle from the ground to the antenna) or directly via a tape measure if this is practical.
- The mobile antenna should be chosen to provide a good omnidirectional pattern. This is not only a function of the antenna design but also of the location of the antenna on the vehicle rooftop. Even for an electrically large ground plane, the antenna needs to be very close to the centre of the roof to avoid significant azimuth pattern distortion.
- Test routes should be chosen to include roads in both radial and circumferential directions around the base station to ensure the result are not biased by, for example, street waveguiding effects. Routes should extend to well beyond the distance at which the minimum signal threshold is first encountered to allow for shadowing. See the drive test route in Figure 19.5, which includes some extended radial routes into the rural environment outside of the site, and the detailed test route in the suburban area where the base site is located.
- Use the maximum transmit power possible, subject to licensing and equipment restrictions. This will help to maximise the use of the available receiver dynamic range and extend the useful measurement distance.
- Care must be taken that the transmitter is adequately characterised for any adverse environmental conditions that may be present at the transmit site. If a transmitter is to be deployed in a specific environment (e.g. at $-1°C$ in 20 kph winds) it is important that the performance of the unit has been characterised under such conditions.
- Ensure that the drive speed does not exceed that used to compute the required sampling rates based on the methods in Section 19.4.
- Stop the test from time to time and save the associated data files to minimise the impact that any file corruption may have on the measurements.
- Ensure that the time of day for the test is appropriate due to the effect of traffic etc.
- Ensure that all details of the test (time, date, weather, equipment used etc.) are carefully recorded and stored with the measurement data so that the data can have value long after the test.

19.8 INDOOR MEASUREMENTS

19.8.1 General

Many of the issues discussed with respect to outdoor measurements also apply indoors, but there are further considerations relating specifically to the methods of navigation and choice of measurement locations. An example indoor measurement result is shown in Figure 19.8.

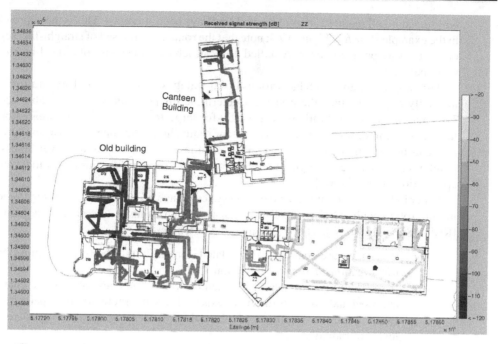

Figure 19.8: Typical in-building measurement route and signal strength. The transmitter location is marked with a ●

19.8.2 Navigation

When performing channel measurements indoors, unlike for outdoor scenarios, GPS data is not available, and it is therefore necessary to look for alternative methods for calculating this position.

The most common approach to navigation indoors is called *way-point navigation*. This method has been widely used in many commercial in-building collection tools developed by several radio data collection software and hardware manufacturers.

The approach consists of having a digital representation of parts of the building in consideration, which may be split by floors or regions. This representation is typically a bitmap image showing the floor layout, but some collection tools are capable of navigating using CAD formats which provide vector representations of walls and other building features. The user interacts with the data collection software using a 'pen' input device on the touch-screen of the collection computer. This input selects their position at a given point in time on the floor layout displayed on the collection system's graphical user interface. The user's entry of their start position begins recording of regular samples from the RF receiver hardware and data collection continues as the user walks in a straight line to a new position within the building, and then stops. The user then indicates this end-point to the collection software on the floor layout using the touch-screen. The collection software then uniformly assigns position to each discrete sample collected over the period between the start time and end time of the walk segment the user has just completed, assuming the user has walked at a constant speed. This process is repeated segment by segment, as the user moves around the building, until all required areas of the building have been surveyed. This is the approach used in the example shown in Figure 19.8; note that the route is composed of straight-line segments due to the way-point approach. A skilled user can achieve accuracies of order 1 m from way-point navigation.

This way-point approach has various disadvantages. The use of high-resolution maps is normally required, since the user must determine his position with reasonable accuracy. This may not be often available, especially for large facilities with complicated building layout. Also, in cases where a measurement route should be repeated, way-point navigation can provide neither the repeatability nor the accuracy required. Nevertheless, way-point navigation is very simple to implement and is sufficiently accurate for most common applications [Aragón, 03].

Other indoor positioning and navigation systems have been employed and suggested using mobile robots with a 'fifth wheel' mechanism, which samples at equally spaced intervals [Radi, 98], [Aragón, 99]. This configuration is often desired.

- When measurements are to be isolated from the human body, for example the MEG investigations described in Section 15.2.7.
- If repeatability is required, for example when time variations in the channel are analysed and different runs over exactly the same route are to be performed.
- In hazardous environments for the field engineer, for example exposure to high levels of radiation.
- If high accuracy in position is necessary at the expense of longer time to conduct the measurements, with requirements for absolute position updates and limitations on the surface on which the robot can navigate.
- Whenever autonomous navigation is needed, providing an accurate and inexpensive method of navigation for indoors.

Figure 19.9 shows one of the mobile robot platforms employed for these special measurement campaigns. One follows a route marked-out on the floor with a high degree of accuracy, while the other is pushed by the user and uses odometry and gyroscopes to determine its position.

19.8.3 Selection of Walk Routes

In order to properly characterise a building, the walk routes should be carefully selected. The following guidelines may help.

- The walk route should include both line-of-sight and non-line-of-sight situations.
- Data should be collected over paths which include penetration through multiple walls of various construction types and also through floors above and below the test antenna location.
- The walk should cover the area of interest fairly evenly, walking all the way across rooms where practical, rather than only close to the edges.
- The walk should be in straight lines given the waypoint navigation and multiple routes should be conducted in large open areas, typically with around 10–20 m between routes in a grid or zigzag pattern.
- Walk through as many doorways as possible, that is if a room has two different entrances, walk through each door at least once.
- Walk smoothly at a constant pace to minimise the navigation error.

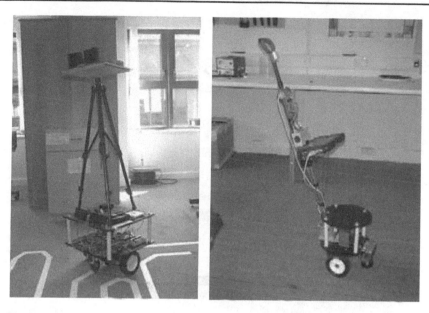

Figure 19.9: Examples of mobile robots employed for indoor channel measurements

- If it is desired to directly compare multiple instances of the same route, then consider placing markers on the floor to act as accurate reminder of the waypoints used.
- Survey antennas should be low-gain omni directional antennas to illuminate the whole environment, unless a specific directional antenna characteristic is of interest.
- An outdoor measurement route should usually be conducted to determine signal leakage from the building.
- Survey for different types of environment (e.g. cluttered, open, densely populated etc.)

Some recommended walk routes are shown in Figure 19.10. In (a), a side view of the building is presented, which indicates the floor penetration requirement. In (b), a top view of the building is shown, where the areas of LOS, NLOS and propagation through walls can be observed.

19.8.4 Equipment

Essential equipment for an indoor measurement includes the following. All equipment should be marked and recorded so that the measurements are fully traceable and repeatable.

- Transmit equipment
  - Tripod for supporting the antenna, preferably made of nylon or wood to avoid disturbing the propagation environment being measured.
  - Signal source, carefully calibrated.
  - Full-charged main and spare battery packs.
  - Coaxial jumper cables.
  - Transmit antennas.

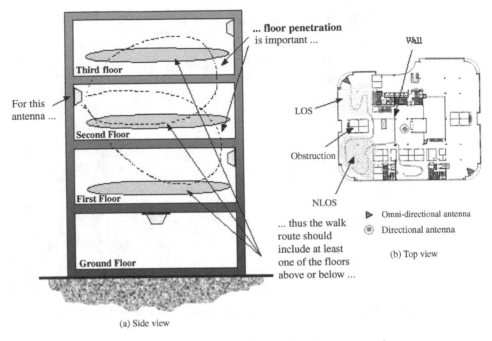

Figure 19.10: Recommended internal walk route scenarios

- Receive equipment located in a rucksack containing
  - full-charged main and spare battery packs;
  - measurement receiver;
  - antenna and mounting arrangements (e.g. attached to rucksack). Antenna height should be clear of head height if propagation measurements are to be performed. Although this will not be the case for practical user equipment, it is essential to gain repeatable measurements and isolate channel effects from those due to the collection personnel. If a ground plane antenna, such as a monopole or helix, issued, great care needs to be taken to ensure the ground plane is of an appropriate size not to distort the radiation pattern. It is usually better to use an antenna which does not require a ground plane, such as a sleeve dipole;
  - portable computer, conveniently in a tablet/palmtop form which can be operated with one hand. The computer should be running the appropriate collection software and should include the electronic format of the building's floor plan (and external areas where necessary).

- Accessories including
  - power meter to confirm transmit power;
  - digital camera to record the locations of the transmit antenna;
  - temporary security barriers if equipment is to be left unattended in a public place for any length of time;
  - paper copy of building floor plans;

- o clipboard;
- o measuring tape or electronic laser measure;
- o tape measure and compass.

19.8.5 Documentation

It is essential to document the survey in a form which will allow the measurements to be used at a much later date, perhaps years later, with all important parameters and conditions recorded. A survey completion form is typically designed for the measurement campaign to facilitate this. It should typically include

- site details, contact names and access arrangements;
- the type of survey being conducted;
- details of which floors are to be surveyed and the corresponding data filenames;
- the purpose of the survey (propagation model calibration, existing, coverage, design survey etc.);
- the building description ;
- the on-site dates and time of surveys along with which floor they were on and the names of the associated data files and photographs should be completed;
- a complete list of the serial numbers of the measurement equipment should be completed;
- the form should be signed and dated by the engineer;
- the survey completion form should be scanned and stored with the project data.

19.9 CONCLUSION

The use of measurements to gain insight into real-world propagation effects is an essential part of propagation analysis for mobile systems. While the techniques presented in this chapter may often seem fairly simplistic, paying close attention to them can save very expensive mistakes when conducting a major propagation campaign. This also adds value to the measurement process by making the resulting data a long-term resource for ongoing investigation and hopefully ever-better propagation models and higher-performance mobile networks.

REFERENCES

[Aragón, 99] A. Aragón-Zavala and S. R. Saunders, Autonomous positioning system for indoor propagation measurements, *Proceedings of IEEE Vehicular Technology Conference*, Amsterdam, the Netherlands, 1999.

[Aragón, 03] A. Aragón-Zavala, In-building cellular radio system design and optimisation using measurements, Ph.D. thesis, University of Surrey, Centre for Communication Systems Research, UK, 2003.

[Bernardin, 00] P. Bernardin and K. Manoj, The post-processing resolution required for accurate RF coverage validation and prediction, *IEEE Transactions on Vehicular Technology*, 49 (5), 1516–1521, 2000.

[Borenstein, 96] J. Borenstein and L. Feng, Gyrodometry: a new method for combining data from gyros and odometry in mobile robots, *Proceedings of the IEEE International Conference on Robotics and Automation*, Minneapolis, USA, 1996, pp. 423–428.

[Crowley, 91] J. R. Crowley and P. Reignier, Asynchronous control of rotation and translation for a robot vehicle, *Robotics and Autonomous systems*, **10**, 243–251, 1991.

[Fiacco, 00] M. Fiacco, S. Stavrou and S.R. Saunders, Measurement and modelling of shadowing cross-correlation at 2 GHz and 5 GHz in indoor environments, *Proceedings of Millenium Conference in Antennas and Propagation*, Davos, Switzerland, 2000.

[Gonzalez, 96] G. Gonzalez, Microwave *Transistor Amplifiers: Analysis and Design*, Prentice Hall, 2nd edition, 1996, ISBN 0-1325-4335-4.

[Gudmundson, 91] M. Gudmundson, Correlation model for shadow fading in mobile radio systems, *Electronics Letters*, 27, 2145–2146, 1991.

[Hickman, 97] I. Hickman, *Practical Radio Frequency Handbook*, 2nd edn, Newnes, pp. 160–176, ISBN 0-7506-3447-2, 1997.

[Klerer, 88] P. R. Klarer, Simple 2D navigation for wheeled vehicles, Sandia Report SAND88-0540, Sandia National Laboratories, Albuquerque, New Mexico, USA, 1988.

[Lee, 85] W. C. Y. Lee, Estimate of local average power of a mobile radio signal, *IEEE Transactions on Vehicular Technology*, 34 (1), 22–27, 1985.

[Owadally, 01] A. S. Owadally, E. Montiel and S. R. Saunders, A comparison of the accuracy of propagation models using hit rate analysis, *IEEE Vehicular Technology Conference (Fall)*, Atlantic City, 4 (54), 1979–1983, 2001.

[Parsons, 00] J. D. Parsons, *The Mobile Radio Propagation Channel*, 2nd edn, John Wiley and Sons, Ltd, New York, 2000, pp. 226–232, ISBN 0-471-98857-X.

[Radi, 98] H. Radi, M. fFiacco, M. Parks and S. R. Saunders, Simultaneous indoor propagation measurements at 17 and 60 GHz for wireless local area networks, *Proceedings of IEEE Vehicular Technology Conference*, Ottawa, Canada, 1998.

[Steinbauer, 00] M. Steinbauer, D. Hampicke, G. Sommerkorn, A. Schneider, A. F. Molisch, R. Thoma and E. Bonek, Array measurement of the double-directional mobile radio channel, *Proceedings of IEEE Vehicular Technology Conference*, VTC 2000 Spring, Vol. 3, Tokyo, Japan, May 2000, pp. 1656–1662.

[Valenzuela, 97] R. Valenzuela, O. Landron and D. L. Jacobs, Estimating local mean signal strength of indoor multipath propagation, *IEEE Transactions on Vehicular Technology*, 46 (1), 203–212, 1997.

[Wong, 99] D. Wong and D. C. Cox, Estimating local mean power level in a Rayleigh fading environment, *IEEE Transactions on Vehicular Technology*, 48 (3), 956–959, 1999.

PROBLEMS

19.1 Prove Eq. (19.2), given Eq. (19.1). If the optimum number of sites for a system is 100, how many more would be required if the measurement system indicated a path loss exponent of 4.2 instead of a true value of 4.0 and was subject to a 1.5 dB offset? How many more sites would be needed if the measurement system increased the system standard deviation of path loss uncertainty from 8 to 12 dB for 95% cell-edge availability?

19.2 Outdoor propagation measurements are to be taken using a vehicle equipped with a radio scanner, a GPS navigation system and a portable computer to collect signal strength samples at 900 MHz. Calculate the maximum vehicle speed using a 1 dB uncertainty with 90% confidence estimation of the local mean assuming a 300 Hz sampling receiver for decibel averaging and optimum estimation.

19.3 In an indoor environment the shadowing autocorrelation distance may be as small as around 3 m. How could you collect samples to give a high-confidence estimate of the local mean without averaging out the shadowing?

19.4 Design a measurement campaign plan for propagation measurements if you want to estimate the coverage of a potential indoor site for WLAN 802.11g, in a corporate four-storey building. You are also interested in assessing the potential interferers which may arise from neighbouring WLAN sites and other possible sources of interferers. Assume a layout for the building; for example could be the building in which you work.

19.5 Establish a set of measurement guidelines for testing PAN (personal area networks) devices, such as Bluetooth, in outdoor and indoor environments. You are interested in determining data rate, coverage area and carrier-to-interference ratios.

20

Future Developments in the Wireless Communication Channel

'The best way to predict the future is to invent it'.
Alan Kay

'The future, according to some scientists, will be exactly like the past, only far more expensive'.
John Sladek

'Radio has no future'.
Lord Kelvin

20.1 INTRODUCTION

This brief chapter provides a speculative and highly personal view of the likely developments in future understanding and treatment of the wireless communication channel. Although the fundamental properties of radiowaves which arise from Maxwell's equations have been well understood for over a century, the application of these properties to practical wireless systems is an evolving field. The points raised in this chapter should indicate some of the key future challenges. I have restricted this discussion to the issues relating to the wireless channel specifically; for my views and those of others on future developments on the technology and use of wireless communications more generally, see [Webb, 07].

20.2 HIGH-RESOLUTION DATA

The demand for high-resolution data for use as input to propagation predictions has increased substantially in recent years. This has coincided with the expansion in the use of small macrocells and microcells, which require a greater level of detail for useful propagation predictions. The data available includes terrain, building height data, vector building data, land usage categories and meteorological information to increasingly fine resolution and high accuracy. Nevertheless, the costs of such data are still prohibitively high for many applications when compared with making measurements. These costs are expected to fall over time as the demand increases further and the technology for acquiring such data becomes cheaper.

Antennas and Propagation for Wireless Communication Systems Second Edition   Simon R. Saunders and Alejandro Aragón-Zavala

20.3 ANALYTICAL FORMULATIONS

There is an ongoing need for the solution of basic electromagnetic problems in order to permit propagation models to be based on sound physical principles. These are essential in order to gain value from the high-resolution input data described above. Particular needs are in the consideration of three-dimensional problems involving diffraction and rough scattering from multiple obstacles. Such formulations will permit rapid evaluation compared to brute-force numerical techniques and allow predictions to make the best possible use of any available data.

20.4 PHYSICAL-STATISTICAL CHANNEL MODELLING

Section 14.6 described a physical-statistical propagation prediction methodology within the context of mobile satellite systems and Section 12.4.4 described another example of this modelling approach. These methods are also attractive in terrestrial applications, particularly, when full deterministic input data may be too expensive for a given application or when the channel is randomly time-varying due to the motion of vehicles, people or other scatterers. It is expected that physical-statistical models will increasingly be seen as an appropriate compromise between the accuracy and applicability of deterministic physical models and the coarse but rapid results produced by empirical models.

20.5 MULTIDIMENSIONAL CHANNEL MODELS

Mainly, this book examined the variations of the channel in space, time and polarisation as separate topics. However, future systems will increasingly require knowledge of the *joint* behaviour of channels with respect to these variables in order to optimise their performance, particularly to support large numbers of users with high data rates. Clear examples are the joint angle-of-arrival/time-of-arrival scattering maps and MIMO channel models in Chapter 17. This will require the creation of new models which account for the correlations between various channel parameters in detail, and which permit the design, characterisation and verification of advanced processing systems such as space-time beamforming or multi-user detectors.

20.6 REAL-TIME CHANNEL PREDICTIONS

In the past, system designers made predictions of propagation parameters and system performance as part of the initial design process for a new system. However, this approach requires that considerable fade margins are included, leading to system operation which is far from optimum at any particular point in time. Significant performance gains are possible if the wireless system is permitted to evolve its parameters (e.g. power level, modulation and coding rates, antenna patterns, channel reuse) over time to meet changing constraints resulting from the behaviour of the users and the propagation channel. This can best be done by allowing the system to maintain an evolving model of the propagation characteristics with which it can test the likely best choice of parameters. Such models are likely to take a very different form from those currently used in non-real-time situations, as they will necessarily need to run much faster and also to integrate both theoretical predictions and measurements. The measurements will include measurement reports from users, whose locations will be tracked continuously and used to integrate the data with model-based predictions, updated frequently to include the most up-to-date information available. These models will act as a key component of a closed-loop control system which adjust the radio parameters of an entire network, optimising power, channel and antenna directionality minute-by-minute to make the best use of system resources in delivering capacity and coverage to the changing service needs, locations and devices of users.

20.7 INTELLIGENT ANTENNAS

In order to meet the changing needs of various wireless systems, antenna structures developed from the generic families of antennas described in Chapter 4 are being developed on an ongoing basis. A more radical shift, however, is emerging from the possibilities of merging the functionality of digital signal processing with multiple antenna elements. This creates *intelligent antennas*, whose characteristics are varied over time to optimise the antenna characteristics with respect to specific system goals, such as coverage, capacity or quality. Intelligent antennas include the diversity and adaptive antennas concepts discussed in Chapters 15 and 19 as special cases, but they can also merge the concepts of equalisers to overcome wideband effects, adaptive matching to improve power delivery and interference cancellation to enhance system capacity. Initially, these concepts are being applied primarily at base stations, where the extra processing and relatively large numbers of antenna elements are relatively easily accommodated, but developments in compact multi-element antenna structures and available signal processing power will enable intelligent antennas to be a standard feature of future mobile terminals as well. MIMO and space-time coding systems are current examples of this technology, but the full potential of such systems in increasing user densities and channel data rates beyond those possible with today's technology will take many years to realise.

20.8 DISTRIBUTED AND AD-HOC CELL ARCHITECTURES

Today's macrocells, microcells and picocells together constitute a complex hierarchical architecture to deliver capacity and coverage in an appropriate form to a variety of local environments. However, these ultimately represent a narrow set of the possible cellular architectures available and therefore constrain the available performance. In all cases, they restrict the signal processing and other system resources to a single base station location, and they assume a simple tree-architecture where connections between users always take place via a base station, even when the best route would be the direct path between pairs or groups of users. Additionally, the backhaul from the base stations is usually conducted via fixed links, whether wired or wireless, to major switch locations. Meanwhile, the optimum path for a given user interaction may follow a route which is unavailable.

Future systems will overcome these constraints in a variety of ways. Cells will become distributed, enabling the system resources (channels, codes, power, spatial dimensions) to move between widely-separated antenna locations according to the need. This will create an intelligent version of the distributed antenna systems described in Chapter 13, but applied to outdoor situations as well as indoors. Transmission will take place from multiple locations in the radio vicinity of a user to provide macro diversity and beamforming, often using space-time coding in addition. The antenna locations will often be interconnected wirelessly, via routes which will vary according to conditions and demand. Such systems are already starting to emerge in the form of 'mesh' networks, although these are currently fairly narrowly applied to Wi-Fi systems. Additionally, more use will be made of the radio paths between mobile users, extending the coverage area available from fixed locations and saving system resources when such paths provide low path loss or good isolation to minimise interference.

Beyond MIMO technologies for individual links, only such flexible architectures with large numbers of small but intelligent cells can increase the capacity and coverage of wireless systems to avoid limitations on the potential growth of services and applications, given the finite available bandwidth and ongoing regulatory constraints on the use of spectrum.

20.9 CONCLUSION

The fast-moving developments in the field of wireless communications make it inevitable that this book will not reflect the state of the art by the time it is read. It is hoped, however, that by focusing on fundamental physical mechanisms and a wide range of wireless system types, the book has been able to give an indication of methodologies for analysing and understanding antennas and propagation which will be useful for many years to come. For more information on the latest developments in the areas listed above and in the rest of the book, the reader is invited and encouraged to visit the following web site:
http://www.simonsaunders.com/

REFERENCES

[Webb, 07] W. Webb (ed), *Future Wireless Communications: Trends and Technologies*, John Wiley & Sons, Ltd, Chichester, January 2007, ISBN 0-470-03312-6.

Appendix A: Statistics, Probability and Stochastic Processes

A.1 INTRODUCTION

The major results used within the body of this book in the field of statistics are summarised here. For more detail, see texts such as [Spiegel, 75] and, [Peebles, 93].

A.2 SINGLE RANDOM VARIABLES

A continuous random variable (r.v.) x takes values over some range, with relative frequencies specified by its *probability density function* (p.d.f.), written as $p(x)$. The probability that x lies between two constant a and b is written as $\Pr(a < x < b)$ and is calculated as the area under the p.d.f. between limits a and b. Formally

$$\Pr(a < x < b) = \int_{x=a}^{b} p(x)\mathrm{d}x \tag{A.1}$$

and, by definition of probability, all p.d.f.s must satisfy

$$\int_{x=-\infty}^{\infty} p(x)\mathrm{d}x = 1 \tag{A.2}$$

Thus the probability that x is less than a is

$$\Pr(-\infty < x < a) = \int_{x=-\infty}^{a} p(x)\mathrm{d}x = P(a) \tag{A.3}$$

where $P(x)$ is the cumulative distribution function (c.d.f.) of x. Thus

$$\Pr(a < x < b) = P(b) - P(a) \tag{A.4}$$

and inverting (A.3) shows

$$p(x) = \frac{\mathrm{d}}{\mathrm{d}x} P(x) \tag{A.5}$$

Antennas and Propagation for Wireless Communication Systems Second Edition   Simon R. Saunders and Alejandro Aragón-Zavala
© 2007 John Wiley & Sons, Ltd

The *expectation* of some function f of an r.v. x is

$$E[f(x)] = \int_{-\infty}^{\infty} f(x)p(x)\,dx \tag{A.6}$$

Thus the *mean* μ of an r.v. x is given by

$$\mu = E[x] = \int_{-\infty}^{\infty} xp(x)dx \tag{A.7}$$

The *variance* or *second central moment* of x is σ^2 ; it is given by

$$\sigma^2 = E[(x - E[x])^2] = E[x^2] - (E[x])^2 = \int_{-\infty}^{\infty} xp(x)dx \tag{A.8}$$

The *standard deviation* is the positive square root of the variance, i.e. σ. The *median* of an r.v. is the value m for which x is equally likely to take values greater than m as less than m, i.e.

$$P(m) = \frac{1}{2} \tag{A.9}$$

A.3   MULTIPLE RANDOM VARIABLES

Given two r.v.s, x and y, their *joint* c.d.f. is defined by

$$\Pr(x < a, y < b) = P(a, b) \tag{A.10}$$

and their joint p.d.f. is

$$p(a, b) = \frac{\partial^2}{\partial x\,\partial y} P(x, y) \tag{A.11}$$

Hence

$$P(a, b) = \int_{x=-\infty}^{a} \int_{y=-\infty}^{b} p(x, y)dy\,dx \tag{A.12}$$

The two r.v.s are said to be *independent* if

$$\Pr(x < a, y < b) = \Pr(x < a) \times \Pr(y < b) \tag{A.13}$$

from which it follows that

$$P(x, y) = P(x) \times P(y) \tag{A.14}$$

and

$$p(x, y) = p(x) \times p(y) \tag{A.15}$$

A.4 GAUSSIAN DISTRIBUTION AND CENTRAL LIMIT THEOREM

The *Gaussian* or *normal* p.d.f. is given by

$$p(x) = \frac{1}{\sigma\sqrt{2\pi}} \exp\left[-\frac{(x-\mu)^2}{2\sigma^2}\right] \tag{A.16}$$

The corresponding normal c.d.f. can be expressed in several equivalent forms:

$$\begin{aligned}
\Pr(x \le a) &= \frac{1}{2} + \frac{1}{2}\mathrm{erf}\left(\frac{a-\mu}{\sigma\sqrt{2}}\right) \\
&= 1 - \frac{1}{2}\mathrm{erfc}\left(\frac{a-\mu}{\sigma\sqrt{2}}\right) \\
&= 1 - Q\left(\frac{a-\mu}{\sigma}\right)
\end{aligned} \tag{A.17}$$

The erf, erfc and Q functions are defined in Appendix B.

Consider the sum y of N independent random variables x_i

$$y = x_1 + x_2 + \cdots + x_N \tag{A.18}$$

The *central limit theorem* states broadly that the distribution of y always approaches a Gaussian distribution as $N \to \infty$, provided that no one of the x_i dominates, whatever the distribution of the x_i.

A.5 RANDOM PROCESSES

A time-variant process $x(t)$ (which may be complex) is a random process when the value of x at any fixed time $t = t_1$ is a random variable with the properties listed in Section A.2. As t_1 can take any value, the mean and other properties listed in Section A.2 may themselves be functions of time. The process $x(t)$ is said to be *stationary* if its statistical properties do not change with time. That is, for a stationary process,

$$p(x; t_1) = p(x; t_1 + \tau) \tag{A.19}$$

where $p(x; t_1)$ is the p.d.f. of x at $t = t_1$ and τ is any arbitrary time shift.

The autocorrelation function of $x(t)$ is defined by

$$R_{xx}(t_1, t_2) = E[x(t_1)x^*(t_2)] \tag{A.20}$$

The process $x(t)$ is *wide-sense stationary* if the autocorrelation function is a function of only the time shift τ; that is, it may be written as

$$R_{xx}(\tau) = E[x(t)x^*(t+\tau)] \tag{A.21}$$

If $x(t)$ is wide-sense stationary, then its power spectrum is given by the Fourier transform of its autocorrelation function

$$S(f) = \mathsf{F}[R_{xx}(\tau)] = \int_{\tau=-\infty}^{\infty} R_{xx}(\tau)e^{-j2\pi f\tau}\,d\tau \tag{A.22}$$

Given two random processes, $x(t)$ and $y(t)$, the *cross-correlation function* is defined by

$$R_{xy}(t_1, t_2) = E[x(t_1)y^*(t_2)] \tag{A.23}$$

This is frequently normalised by the variances of x and y to yield the normalised cross-correlation

$$\rho_{xy}(t_1, t_2) = \frac{E[x(t_1)y^*(t_2)]}{\sigma_x \sigma_y} \tag{A.24}$$

If

$$R_{xy}(t_1, t_2) = 0 \tag{A.25}$$

for all times, then $x(t)$ and $y(t)$ are *orthogonal*.

The cross-covariance function of $x(t)$ and $y(t)$ is

$$C_{xy}(t_1, t_2) = R_{xy}(t_1, t_2) - E[x(t_1)]E[y^*(t_2)] \tag{A.26}$$

If $C_{xy}(t_1, t_2) = 0$ then $x(t)$ and $y(t)$ are *uncorrelated*. Thus, independent processes are always uncorrelated, although the converse is not always true.

REFERENCES

[Peebles, 93] P. Z. Peebles, *Probability, random variables, and random signal processes*, 3rd edn., McGraw-Hill, New York, ISBN 0-07-049273-5, 1993.

[Spiegel, 75] M. R. Spiegel, *Theory and problems of statistics*, Schaum, New York, ISBN 0-07-060220-4, 1975.

Appendix B: Tables and Data

B.1 NORMAL (GAUSSIAN) DISTRIBUTION

Given a normal random variable x with mean μ and standard deviation σ, the probability density function of x is shown in Figure B.1 and given by

$$p(x) = \frac{1}{\sigma\sqrt{2\pi}} e^{-(x-\mu)^2/2\sigma^2} \tag{B.1}$$

The probability that x is greater than some value x_0, as shown in Figure B.1, is then

$$\Pr(x > x_0) = Q\left(\frac{x-\mu}{\sigma}\right) = \frac{1}{\sigma\sqrt{2\pi}} e^{e^{-(x-\mu)^2/2\sigma^2}} \tag{B.2}$$

where $Q(t)$ is the *complementary cumulative normal distribution* of a normal random variable t, with zero mean and unit standard deviation:

$$t = \frac{x-\mu}{\sigma} \tag{B.3}$$

$Q(t)$ is bounded from above and below by

$$\left(1 - \frac{1}{t^2}\right)\frac{1}{t\sqrt{2\pi}} e^{-t^2/2} < Q(t) < \frac{1}{t\sqrt{2\pi}} e^{t^2/2} \tag{B.4}$$

which are useful as approximations to $Q(t)$ for $t > 1.5$ (Figure B.2). $Q(t)$ has the following property:

$$Q(-t) = 1 - Q(t) \tag{B.5}$$

Table B.1 shows some useful values of $Q(t)$.

Antennas and Propagation for Wireless Communication Systems Second Edition   Simon R. Saunders and
Alejandro Aragón-Zavala
© 2007 John Wiley & Sons, Ltd

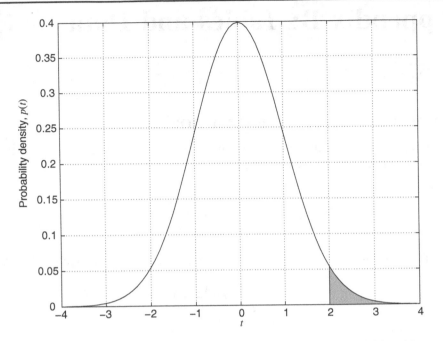

Figure B.1: Probability density function of a normally distributed random variable t with zero mean and unit standard deviation. The shaded area is $Q(2)$, the probability that t is greater than 2

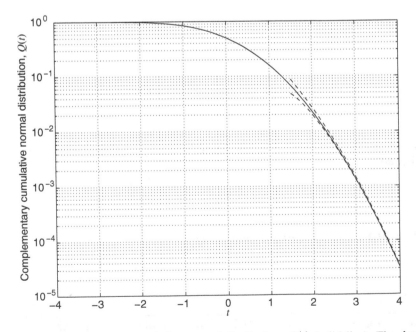

Figure B.2: Complementary cumulative normal distribution, $Q(t)$ (solid line). The dotted lines show the upper and lower bounds given by (B.4)

Table B.1: Values of the complementary cumulative normal distribution

t	$Q(t)$	$100\,Q(t)$ (%)
0.00	0.500 00	50.000 00
0.25	0.401 29	40.129 37
0.50	0.308 54	30.853 75
0.75	0.226 63	22.662 74
1.00	0.158 66	15.865 53
1.25	0.105 65	10.564 98
1.50	0.066 81	6.680 72
1.75	0.040 06	4.005 92
2.00	0.022 75	2.275 01
2.053 74	0.020 00	2.000 00
2.25	0.012 22	1.222 45
2.326 34	0.010 00	1.000 00
2.50	0.006 21	0.620 97
2.75	0.002 98	0.297 98
3.00	0.001 35	0.134 99
3.090 23	0.001 00	0.100 00
3.25	0.000 58	0.057 70
3.50	0.000 23	0.023 26
3.75	0.000 09	0.008 84
4.00	0.000 03	0.003 17

B.2 ERROR FUNCTION

The *error function* is defined by

$$\mathrm{erf}\, z = \frac{2}{\sqrt{\pi}} \int_{x=0}^{z} e^{-x^2} dx \tag{B.6}$$

and the complementary error function is

$$\mathrm{erfc}\, z = \frac{2}{\sqrt{\pi}} \int_{x=z}^{\infty} e^{-x^2} dx \tag{B.7}$$

These two functions are related by

$$\mathrm{erfc}\, z = 1 - \mathrm{erf}\, z \tag{B.8}$$

They are related to the complementary cumulative normal distribution by

$$Q(z) = \frac{1}{2}\left[1 - \mathrm{erf}\left(\frac{z}{\sqrt{2}}\right)\right] = \frac{1}{2}\mathrm{erfc}\left(\frac{z}{\sqrt{2}}\right) \tag{B.9}$$

$$\mathrm{erfc}(z) = 2Q(z\sqrt{2}) \tag{B.10}$$

$$\mathrm{erf}(z) = 1 - 2Q(z\sqrt{2}) \tag{B.11}$$

B.3 FRESNEL INTEGRALS

The definition of the Fresnel integrals used in this book is

$$
\begin{aligned}
C(u) + jS(u) &= \int_{x=0}^{u} e^{j\pi x^2/2} dx \\
&= \int_{x=0}^{u} \cos\frac{1}{2}\pi x^2 dx + j \int_{x=0}^{u} \sin\frac{1}{2}\pi x^2 dx
\end{aligned}
\tag{B.12}
$$

An asymptotic expansion for $x \gg 1$ gives

$$
\begin{aligned}
C(u) &\approx \frac{1}{2} + \frac{1}{\pi u}\sin\frac{1}{2}\pi u^2 \\
S(u) &\approx \frac{1}{2} - \frac{1}{\pi u}\cos\frac{1}{2}\pi u^2
\end{aligned}
\tag{B.13}
$$

Therefore, as $u \to \infty$, $C(u) \to \frac{1}{2}$ and $S(u) \to \frac{1}{2}$. For non-negative values of u, the Fresnel integrals can be calculated using the following approximations:

$$
\begin{aligned}
C(u) &= \frac{1}{2} + f(u)\sin\left(\frac{\pi}{2}u^2\right) - g(u)\cos\left(\frac{\pi}{2}u^2\right) \\
S(u) &= \frac{1}{2} - f(u)\cos\left(\frac{\pi}{2}u^2\right) - g(u)\sin\left(\frac{\pi}{2}u^2\right)
\end{aligned}
\tag{B.14}
$$

where

$$
\begin{aligned}
f(u) &= \frac{1 + 0.926u}{2 + 1.792u + 3.104u^2} + \varepsilon(u) \\
g(u) &= \frac{1}{2 + 4.142u + 3.492u^2 + 6.670u^3} + \varepsilon(u)
\end{aligned}
\tag{B.15}
$$

with $|\varepsilon(u)| \le 0.002$ in both cases (Abramowitz, 1970). For negative values of u, the same approximations can be used in conjunction with the following symmetry relations:

$$
C(u) = -C(-u) \qquad S(u) = -S(-u)
\tag{B.16}
$$

Equations (B.14) are plotted for $u \in [-5, 5]$ in Figure B.3, producing a figure known as the *Cornu Spiral*. The magnitude of the Fresnel integral is the distance from the origin to the Cornu spiral for a given value of u and this may be compared to the Fresnel diffraction integral in Figure 3.15.

B.4 GAMMA FUNCTION

The gamma function is used in the Nakagami-m distribution introduced in Chapter 10. It is defined by the following integral:

$$
\Gamma(z) = \int_{t=0}^{\infty} t^{z-1}e^{-t} dt
\tag{B.17}
$$

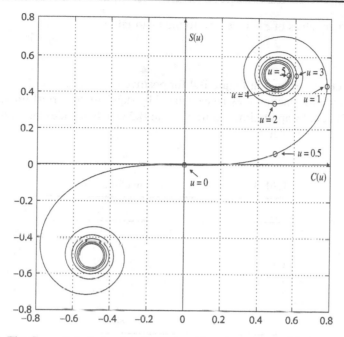

Figure B.3: The Cornu Spiral, showing the real and imaginary parts of the Fresnel integral

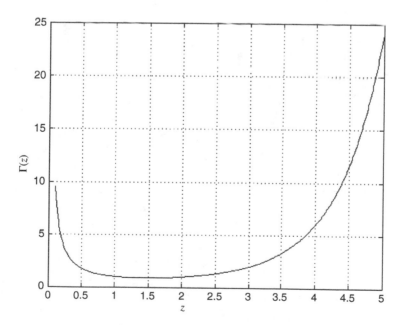

Figure B.4: The gamma function

For integer values of z it is related to the factorial by

$$\Gamma(n) = (n-1)! \tag{B.18}$$

The function is shown in Figure B.4 for positive values of z.

The gamma function may be calculated by standard numerical libraries, by reading from Figure B.3, or by approximating from Table B.2 which includes the values used in producing the graphs in Chapter 10.

Table B.2: Values of the gamma function

z	$\Gamma(z)$
0.5	1.77
1.0	1.00
1.5	0.89
2.0	1.00
3.0	2.00
5.0	24.00
10.0	362 880.00

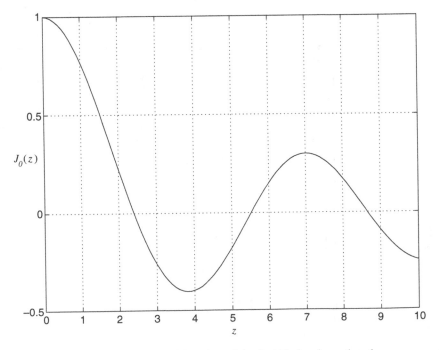

Figure B.5: The Bessel function of the first kind and zeroth order

B.5 BESSEL FUNCTION

The Bessel function of the first kind and zeroth order is defined by

$$J_0(z) = \frac{1}{\pi} \int_0^\pi \cos(z \sin \theta) d\theta = \frac{1}{\pi} \int_0^\pi \cos(z \cos \theta) d\theta \tag{B.19}$$

The function is plotted in Figure B.5.

REFERENCE

[Abramowitz, 70] M. Abramowitz and I. A. Stegun, *Handbook of Mathematical Functions*, Dover Publications, New York, 1970, ISBN 0-486-61272-4.

Abbreviations

AAS	Adaptive Antenna System
AI	Air Interface
AMPS	Advanced Mobile Phone System
AoA	Angle of Arrival
AoD	Angle of Departure
AR	Axial Ratio
ARQ	Automatic Repeat reQuest
BER	Bit Error Rate
BPSK	Binary Phase Shift Keying
BS	Base Station
BSC	Base Station Controller
BSS	Base Station Subsystem
BTS	Base Transceiver Station
cdf	Cumulative Distribution Function
CDMA	Code Division Multiple Access
CIR	Carrier to Interference Ratio
CINR	Carrier to Interference plus Noise Ratio
cm	centimetre(s)
CP	Circular Polarisation
CW	Continuous Wave (or Carrier Wave)
DAS	Distributed Antenna System
dB	decibel(s)
DECT	Digital Enhanced Cordless Telecommunications (originally Digital European Cordless Telephone)
DFE	Decision Feedback Equaliser
DS-CDMA	Direct Sequence Code Division Multiple Access
E	erlangs
FDD	Frequency Division Duplex
FDMA	Frequency Division Multiple Access
FEC	Forward Error Correction
FSL	Free Space Loss
GEO	Geostationary Earth Orbit
GHz	gigahertz
GPRS	General Packet Radio Service
GPS	Global Positioning System

Antennas and Propagation for Wireless Communication Systems Second Edition   Simon R. Saunders and Alejandro Aragón-Zavala
© 2007 John Wiley & Sons, Ltd

GSM	Global System for Mobile Communications (originally Groupe Spéciale Mobile)
GTD	Geometrical Theory of Diffraction
HLR	Home Location Register
HPBW	Half-Power Beam Width
Hz	Hertz
IEE	Institution of Electrical Engineers (now IET)
IEEE	Institute of Electrical and Electronics Engineers
IET	Institution of Engineering and Technology
IF	Intermediate Frequency
I-QHA	Intelligent Quadrifilar Helix Antenna
IS95	Interim Standard 95 (US CDMA narrowband standard, now known as cdmaOne)
ISM	Industrial, Scientific and Medical
ITU	International Telecommunications Union (an agency of the United Nations)
ITU-R	International Telecommunications Union - Radio communications Bureau
JTACS	Japanese Total Access Communication System
kHz	kilohertz
km	kilometre(s)
LAN	Local Area Network
LEO	Low Earth Orbit
LHCP	Left Hand Circular Polarisation
LMS	Least-Mean Square
LNA	Low Noise Amplifier
LOS	Line Of Sight
m	metres
MAHO	Mobile Assisted HandOver
MAN	Metropolitan Area Network
MAPL	Maximum Acceptable Path Loss
MEG	Mean Effective Gain
MEO	Medium Earth Orbit
MHz	Megahertz
MIMO	Multiple Input Multiple Output
MISO	Multiple Input Single Output
ML	Maximum Likelihood
MLSE	Maximum Likelihood Sequence Estimator
MS	Mobile Station
MSC	Mobile Switching Centre
NLOS	Non Line Of Sight
NMT	Nordic Mobile Telephone
NTT	Nippon Telegraph and Telephone Corporation
OFDM	Orthogonal Frequency Division Multiplexing
OFDMA	Orthogonal Frequency Division Multiple Access
OS	Ordnance Survey
pdf	Probability Density Function
PDP	Power Delay Profile
PE	Parabolic Equation

PEL	Plane Earth Loss
PHS	Personal Handyphone System
PIFA	Planar Inverted-F Antenna
PIM	Passive InterModulation
PSTN	Public Switched Telephone Network
Q	Quality factor
QAM	Quadrature Amplitude Modulation
QHA	Quadrifilar Helix Antenna
QPSK	Quaternary Phase Shift Keying
RF	Radio Frequency
RH	Right Hand
RHCP	Right Hand Circular Polarisation
RLS	Recursive Least Square
r.v.	random variable
RMS	Root Mean Square
SAR	Specific Absorption Rate
SDMA	Space Division Multiple Access
SFIR	Spatial Filtering for Interference Reduction
SIMO	Single Input Multiple Output
SINR	Signal to Interference plus Noise Ratio
SIR	Signal to Interference Ratio
SM	Spatial Multiplexing
SMS	Short Message Service
SNR	Signal to Noise Ratio
SRLS	Square-root Recursive Least Square
STC	Space-Time Coding
STEC	Slant Total Electron Content
STTD	Space Time Transmit Diversity
TACS	Total Access Communication System
TDD	Time Division Duplex
TDMA	Time Division Multiple Access
TEC	Total Electron Content
TVT	Time-Variant Transfer function
ULA	Uniform Linear Array
USDC	United States Digital Cellular (IS54)
USGS	United States Geographical survey
UTD	Uniform geometrical Theory of Diffraction
UWB	Ultra Wide Band
V-BLAST	Vertical Bell Labs Space Time
VLR	Visitor Location Register
VSWR	Voltage Standing Wave Ratio
VTEC	Vertical Total Electron Content
W	Watts
WCDMA	Wideband CDMA
Wi-Fi	Wireless Fidelity. Product certification for IEEE 802.11 WLAN interoperability

WiMax	Worldwide Interoperability for Microwave Access - Industry body promoting the IEEE802.16 standard for wireless MAN technology
WLAN	Wireless LAN
WSSUS	Wide-Sense Stationary Uncorrelated Scattering
XPD	Cross Polar Discrimination
XPI	Cross Polar Isolation

Index

Antennas and Propagation for Wireless Communication Systems Second Edition Simon R. Saunders and
Alejandro Aragón-Zavala
© 2007 John Wiley & Sons, Ltd